Guide to
Sustainable Development
and Environmental
Policy

Guide to
Sustainable Development
and Environmental
Policy

EDITED BY

NATALIA MIROVITSKAYA

AND WILLIAM ASCHER

DUKE UNIVERSITY PRESS   Durham and London, 2001

© 2001 Duke University Press
All rights reserved
Printed in the United States of America on acid-free paper ∞
Library of Congress Cataloging-in-Publication Data
appear on the last printed page of this book.

Contents

1 BASIC CONCEPTS OF DEVELOPMENT AND ENVIRONMENT

1.1 Development Theory: Introductory Concepts, Criteria, and Measurements

---

1.2 Natural Resources and Environmental Services

1.3 Epistemology of the Field

---

2 SUSTAINABILITY

2.1 Basic Sustainability Concepts

2.2 Measurements of Sustainability

2.3 Philosophical Approaches and Social Movements

Alternative Normativism
Alternative Structuralism
Animal Rights
Anti-Environmentalism
Back-to-Nature Movement
Bioregionalism
Core-Periphery Theory
 (Center-Periphery Theory)
Deep Ecology
Dependency Theory
Ecoanarchism
Ecodevelopment
Ecofascism
Ecofeminism
Eco-Marxism
Ecosocialism
Gaiasophy
Gandhian Concept of Development

(Swaraj Doctrine; "Hindu
 Development")
Gender and Development Approach
 (GAD)
Holistic Ecology
Kropotkin, Piotr Alekseyevich
Land Ethic
Lifeboat Ethics
Radical Environmentalism
Social Ecology
Spaceship Economy
Stewardship
Transpersonal Ecology
Women and Development (WAD)
 Approach
Women in Development (WID)
 Approach

2.4 Sustainable Development Outlooks and Models

Boserupian Thesis
Buddhist Model of Development
Development Models
Ecological Energetics Paradigm
Environmental Convergence Theory
Global 2000 Report

"Golden Billion" Development
IPAT Model
Leontieff Report
Limits to Growth Report
Outbreak-Crash Model of the Future

2.5 Sustainable Development Strategy and Agenda Statements

Agenda 21
Blueprint for Survival
Caring for the Earth
Declaration on the Human
 Environment
Development Decades
Earth Charter
Global Marshall Plan for the
 Environment

New International Economic Order
 (NIEO)
Small Is Beautiful
United Nations Declaration of a New
 International Economic Order
UN Development Decades
United Nations World Charter for
 Nature
World Charter for Nature
World Conservation Strategy (WCS)

3.5 Worldviews and Social Traps

4 INTERNATIONAL POLITICAL ECONOMY OF ENVIRONMENT AND DEVELOPMENT

4.1 Globalization

4.4 Regional Economic Cooperation and Integration

4.5 Security and Sustainable Development

6 MAJOR PROBLEMS OF ENVIRONMENTAL DEGRADATION AND DEVELOPMENT

6.1 Agricultural Decline and Pollution

6.4 Climate Change

6.5 Deforestation and Forestry

International Council for Research in
  Agroforestry (ICRAF)
International Tropical Timber
  Agreement (ITTA)

International Tropical Timber
  Organization (ITTO)
Rainforest Action Network (RAN)

6.6 Desertification

Club du Sahel
Convention to Combat Desertification
  (CCD)
Desertification
Global Mechanism

Intergovernmental Authority on
  Drought and Development (IGADD)
Plan of Action to Combat
  Desertification (PACD)

6.7 Energy Shortages and Pollution

Ecological Energetics Paradigm
Energy Valuation
Energy-GNP Decoupling
International Atomic Energy Agency
  (IAEA)
International Energy Agency (IEA)
Organization of Petroleum Exporting
  Countries (OPEC)
Soft Energy Path

6.8 Fresh Water Scarcity and Pollution

Absolute Territorial Integrity,
  Doctrine of
Canada/USA International Joint
  Commission (IJC)
Clean Water Act (U.S. Federal Water
  Pollution Control Act)
Community of Interests, Doctrine of
Convention on the Law of the Non-
  Navigational Uses of International
  Watercourses
Convention on the Protection and Use
  of Transboundary Watercourses and

International Lakes (Helsinki
  Convention)
Equitable Utilization, Doctrine of
Harmon Doctrine
Helsinki Rules on the Use of Waters of
  International Rivers
International Watercourses
Riparian Rights
Transboundary Waters
Water Environment Federation (WEF)
Water-Resource Management
Water Use Charges

6.13 Outer Space and Electromagnetic Spectrum

6.14 Ozone Depletion

6.15 Transportation Energy Consumption and Pollution

Introduction

This *Guide to Sustainable Development and Environmental Policy* is designed to serve as a road map for understanding the issues and debates in the overlapping fields of environment and development. Tremendous strides have been made in the past twenty years in elevating the issues of the environment and sustainable development in public and governmental consciousness, in developing institutions for addressing these issues, and in implementing practices and policy options. There has been an explosion of efforts to address conservation and economic growth issues, from the smallest-scale efforts to approach rural development in a sustainable, integrated way, to global agreements on resource sharing and environmental protection of the oceans, deserts, and forests. The philosophies and movements for placing humans in the broader ecosystem context have also proliferated, as have the models and schools of thought for understanding and steering development in sustainable directions.

The complexity of issues, practices, institutions, policy options, and agreements has become truly bewildering. There are literally thousands of environmental organizations and development programs, hundreds of international environmental treaties and conventions, and innumerable labels for the approaches and strategies for managing pollution, resource conservation, and socioeconomic development. This complexity makes it very difficult for policymakers, activists, students, journalists, and other citizens to make sense of the policies and debates surrounding environmental protection and sustainable development. Sometimes the difficulty comes simply in understanding the exploding terminology. The editors of this *Guide* have between them more than fifty years of experience working in various subfields of environmental and resource policy and development, and yet, before undertaking this project, would have had difficulty in accurately defining many of the terms that appear in this volume. Sometimes the difficulty is in knowing how to put all of the practices, theories, and institutions into proper perspective. Of the thousands of environmental organizations, which carry the most weight? Of all the conservation approaches, which are mainstream, which are avant garde, which are accepted practice, which are highly controversial? Of all the international environmental agreements, which have been effectively implemented, which are likely to be relegated to dusty shelves?

The *Guide* meets these challenges in two ways. First, it provides brief definitions and descriptions of over one thousand entries. These are organized within a conceptual structure that begins with clusters of entries designed to help the reader understand fundamental concepts of environmental issues and sustainable development, and then to compare broad models and approaches to achieving environmental protection, conservation, and sustainability. For example, the approaches to environmental knowledge and theory are gathered in one section, and the philosophical movements for understanding the place of humankind within nature are gathered together in another. Next, it groups entries useful for understanding the trends and forces behind the broadest environmental changes, such as population growth, insecure property rights, poverty, affluence, and globalization. This is followed by sections that allow the reader to compare the broad instruments for dealing with environmental challenges, such as the different means of measuring and valuing environmental elements and resources, or the approaches that the business sector can take to make their operations more environmentally friendly.

The final third of the *Guide* focuses on specific environmental and sustainability issues, such as air pollution, the decline of world fisheries, global climate change, agricultural decline, and desertification. The *Guide* is intended to help the reader capture the sense of each problem and the efforts to address it by reading the section that groups entries related to how the problem arises, how it can be addressed, and what institutions have emerged to deal with it. For example, one section deals with maritime problems (e.g., overfishing, oil spills, and conflicts over ocean mining rights), describes key policy responses and international agreements to address these problems (e.g., ranging from the UN Convention on the Law of the Seas to technologies that allow endangered sea turtles to escape from fishing nets), and enumerates the relevant organizations, such as the International Commission for the Regulation of Whaling and the International Maritime Organization.

For ease of locating each specific entry, each subsection lists its entries in alphabetical order, with cross-references to other sections for relevant entries that are listed elsewhere. The comprehensive alphabetical index denotes, through bold numbers, where entries can be found. Thus, a typical reader who is puzzled about a particular term ("What is an assurance bonding system, anyway?"), will find that entry's page number in the index, read that entry, and then read about related concepts (e.g., other instruments besides assurance bonding systems for reducing the incentives of producers to pollute), as well as pertinent cross-references. The reader who wants a quick entrée to the concepts and issues related to a particular type of environmental problem, such as ocean pollution or loss of biological diversity, can find the heading or subheading within the alphabetical index or the analytical outline in the contents. Within the text of each entry, other problems, approaches, institutions, and agreements that have their own entries are denoted by small caps to indicate to the reader that separate entries can be found for them.

Considerable effort went into trying to make the entries as definitive, reliable, and impartial as possible. This does not mean, however, that the entries avoid conveying how important or how controversial an approach or institution may be. There are frequent references to the fact that a particular practice or theory has been criticized, or that a particular institution has been ineffective. These evaluations reflect the prevailing view of the status of the approach or institution, yet we have avoided asserting our own views of the merits of the approach or institution described by the entry. Our own views, and those of other contributors, are reflected in commentaries on selected definitions.

The *Guide* approaches the challenge of putting these problems, approaches, ideas, organizations, and agreements into perspective by offering over two hundred short commentaries by practitioners and scholars in the fields of environmental policy and sustainable development. Whereas the entries are intended to be objective definitions, the commentaries are intended to allow the commentators to express their personal enthusiasms, skepticism, and passion with respect to any and all of the approaches described in the entries. The commentators are all respected figures in their fields, but we make no claim that the set of commentators reflects a complete and balanced summary of all possible views on these issues—that would have been impossible, and even a game effort to approach comprehensiveness would have made this *Guide* far too bulky. Even so, the commentaries do provide the reader with insights into the controversies that swirl around the contentious issues of environment and sustainable development.

In light of the vastness of the fields that touch on environment and development, the *Guide* is unavoidably selective. There is considerable coverage of concepts and problems of economic development, particularly in developing countries where so many resource and environmental outcomes depend on the strategies that governments and nongovernmental actors adopt in their efforts to bring their countries out of poverty. Issues of natural-resource conservation and depletion are covered extensively, in that they are integral facets of the environment and ecosystems, as well as playing a crucial role in the economic development of many countries. There is also extensive coverage of the pollution issues that afflict developed countries, and the relationships among the economic processes, policy instruments, and organizations that influence the levels of pollution.

The limitations of the *Guide* lay in the inevitably restricted treatment of institutions that play complicated development roles that go far beyond the implications for environment protection and resource conservation. For example, the multilateral development banks such as the World Bank Group have many impacts on economic development that cannot be covered in a work focused primarily on environmental issues. By the same token, financial problems are very important for economic development, and yet considerations of space required that we concentrate on issues and instruments that pertain to environmental and economic sustainability, such as debt-for-nature swaps and the conditions imposed by multilateral lenders. In short, the coverage of this *Guide* is not on all environmental issues and all development issues; rather, it puts most of the effort on covering the environmental issues and the intersection of environment and development.

Another limitation is that the *Guide* cannot enumerate all of the national and subnational environmental policies that prevail around the world. Country-specific policies are listed only if they have served as particularly important models—positive or negative—for policies in other countries or at the international level. Much effort has been placed in covering the international level, not only because of the increasingly global nature of environmental and developmental issues, but also because international agreements impinge on so many nations' environmental and development potentials.

Finally, the *Guide* reflects a concerted effort to avoid a predominantly U.S.-focused bias. A large proportion of the entries were initially written by Russian contributors, relying both on Russian and other European conceptions and experiences. Many of the commentators are from outside of the United States, and virtually all of them have worked in other countries. Americans' understanding of environment and sustainability are likely to be enriched by learning that many seemingly home-grown ideas and approaches have their counterparts in other parts of the world. In particular, Russian and Western European theorists and scientists have made very important contributions, sometimes preceding parallel U.S. developments, in both theory and practice. It is hoped that putting these approaches within the same *Guide* will illustrate the convergences that are sometimes hidden by the proliferation of labels.

Contributors

GUSTAVO J. ARCIA is Senior Economist in the Center for International Development, Research Triangle Institute, and Adjunct Associate Professor of Public Policy at Duke University. Most of his current work is in social sector policy, education reform, poverty measurement, and poverty reduction policies. During the past five years he has worked in education reform in Ecuador, Honduras, and Nicaragua, and poverty analysis in Nicaragua and Uzbekistan. He has also analyzed population-environment interactions in Costa Rica and Ecuador, and conducted extensive studies of farming systems in West Africa and Latin America. He is the author of several monographs and book chapters on education reform in Latin America, and on poverty, economic, and social policy in Central America.

WILLIAM ASCHER is Professor of Government and Economics, Dean of the Faculty, and Vice-President at Claremont-McKenna College. Previously he was Director of Duke University's Center for International Development Research. His research is on policymaking processes in developing countries, natural resource policymaking, Latin American and Asian political economy, and forecasting methodologies. His books on resources and development include *Scheming for the Poor: The Politics of Redistribution in Latin America* (Harvard University Press, 1984), *Natural Resource Policymaking in Developing Countries* (with Robert G. Healy, Duke University Press, 1990), *Communities and Sustainable Forestry in Developing Countries* (ics Press, 1994), *Why Governments Waste Natural Resources* (Johns Hopkins University Press, 1999), and *The Caspian Sea: A Quest for Environmental Security* (ed. with Natalia Mirovitskaya, Kluwer, 2000).

FIKRET BERKES is Professor of Natural Resources at the University of Manitoba, Winnipeg, Canada. His main area of expertise is common-property resources and community-based resource management, currently focusing on traditional ecological knowledge and co-management. He teaches in these areas, contributes to theory development, and carries out applied work both internationally and in the Canadian North. He contributes as well to the literature of human ecology and conservation, biodiversity, ecological economics, and environmental assessment. He has over one hundred scholarly publications, including the books *Sacred Ecology* (Taylor and Francis, 1999), *Linking Social and Ecological Systems* (Cambridge University Press, 1998, coedited with Carl Folke), and *Common Property Resources* (Belhaven, 1989).

GARRY D. BREWER is the Frederick K. Weyerhaeuser Professor at Yale University's Schools of Management and Forestry and Environmental Studies (1985–1991; 2001–). Before returning to Yale, he was Professor and Dean of the University of Michigan School of Natural Resources and Environment (1991–1998) and Professor of Environmental Policy and Management and Dean of University Extension at the University of California, Berkeley (1998–2001). In 1998 he was the King Carl XVI Gustaf Professor of Environmental Sciences and Management in Sweden, and was elected to the Swedish Royal Academy. He was also a Presidential appointee to the U.S. Nuclear Waste Technical Review Board (1992–1997). The author of nine books on policy analysis, simulation and modeling, and development, his most recent publications on natural resources and environment include "Science to Serve the Common Good," *Environment* 39(6) (July 1997); "Closing the Gap between Ecosystem Management and Ecosystem Research," *Policy Sciences* 31(1) (February 1998); and "The Challenges of Interdisciplinarity," *Policy Sciences* 32(4) (December 1999).

DANIEL W. BROMLEY is Anderson-Bascom Professor of Applied Economics at the University of Wisconsin-Madison, and has been editor of the journal *Land Economics* since 1974. He has published extensively on natural resource and environmental economics and on economic development. He has been a consultant to the Global Environment Facility, the World Bank, the Ford Foundation, the U.S. Agency for International Development, the Asian Development Bank, the Organization for Economic Cooperation and Development, and the New Zealand Ministry for the Environment. He has published nine books, most recently *Environment and Economy: Property Rights and Public Policy* (Blackwell, 1991); *The Social Response to Environmental Risk* (Kluwer, 1992); *Making the Commons Work: Theory, Practice, and Policy* (ICS Press, 1992); *The Handbook of Environmental Economics* (Blackwell, 1995); and *Sustaining Development: Environmental Resources in Developing Countries* (Elgar, 1999). He is now writing "Sufficient Reason: Institutions and Economic Change."

RONALD D. BRUNNER is Professor of Political Science at the University of Colorado at Boulder. A native of Colorado, he was educated at Yale University where he received his B.A. and Ph.D. degrees. He was a tenured member of the faculty at the University of Michigan, and has held research positions at Harvard University and in the U.S. Congress. He is interested primarily in the integration of theory and practice in the policy sciences including problems of epistemology in policy inquiry, the symbolic instruments of politics and policy, and context-sensitive methods. His publications cover a wide variety of practical problems, including energy, welfare, space policy, global climate change policy, and natural resources policy, including "Harvesting Experience: A Reappraisal of the U.S. Climate Change Program," *Policy Sciences* 32(2) (June 1999) and "A Practice-Based Approach to Ecosystem Management," *Conservation Biology* 11(1) (February 1997). On the subject of energy policy, he wrote *Community Energy Options: Getting Started in Ann Arbor* (University of Michigan Press, 1982).

ELSA CHANG is an Associate in the World Resources Institute's (WRI) Institutions and Governance Program. Over the past five years, Ms. Chang, a native of Guatemala, has led and managed WRI's collaboration on strengthening national and regional institutions' framework for environmental management and policy planning in Central America. Currently, she manages WRI's strategic policy research program aimed to catalyze actions to implement the Mesoamerican Biological Corridor in the Central American region. Before joining WRI, Ms. Chang conducted a research project on Forest Policies and Administration in Peru and Bolivia, was Executive Assistant and Technical Adviser for the Central American Commission on Environment and Development (CCAD) and Guatemala's National Commission on the Environment (CONAMA). Formerly a Mesoamerican archaeologist, she worked for various archaeological projects throughout Guatemala. Ms. Chang holds an M.A. in International Development Policy from the Sanford Institute of Public Policy at Duke University.

TIM W. CLARK is Professor (adjunct) in the School of Forestry and Environmental Studies and Fellow in the Institution for Social and Policy Studies, Yale University, and board president of the Northern Rockies Conservation Cooperative in Jackson, Wyoming. He received his Ph.D. from the University of Wisconsin-Madison in 1973. His interests include conservation biology, organization theory and management, natural resources policy, and the policy sciences. He has written over 300 papers, including a recent one addressing interdisciplinary problem solving in the Greater Yellowstone Ecosystem. His books include *Mammals in Wyoming* (coauthor, 1987), *Conservation Biology of the Black-Footed Ferret* (1989), *Tales of the Grizzly* (coauthor, 1993), *Greater Yellowstone's Future: Prospects for Ecosystem Science, Management, and Policy* (1994), *Endangered Species Recovery: Finding the Lessons, Improving the Process* (1994), *Averting Extinction: Reconstructing Endangered Species Recovery* (1997), and *Carnivores in Ecosystems: The Yellowstone Experience* (coedited). He has served as a member of the Policy Review Committee on Research and Resource Management in the U.S. National Park Service. He is also a member of three species survival commissions of the IUCN-World Conservation Union.

HELEN CORBETT is a writer and researcher who spent eighteen years working on the seal islands of the Bering Sea. Her research of the northern Aleut culture, resource economies, and Bering Sea island systems has taken her to the Pribilof Islands and Alaska (1981–1993) and the Commander Islands and Kamchatka (1994–present). She has published numerous articles and research papers about her Bering Sea work and coproduced four films with Susanne Swibold on the Pribilof Islands. She is a founder and codirector of the Amiq Institute and a research associate at the Arctic Institute of North America, University of Calgary. She holds a B.A. (University of Alberta) and Bachelor of Journalism (Carleton University).

GERALD A. EMISON is Senior Adviser to the Chief Financial Officer of the U.S. Environmental Protection Agency (EPA). Prior to this he managed EPA's Pacific Northwest regional office and directed EPA's air-quality planning program during the development of the Clean Air Act of 1970. He holds an undergraduate degree in civil engineering and graduate degrees in city planning, political science, and engineering management. He received a doctorate in City and Regional Planning from the University of North Carolina, Chapel Hill, and is a diplomate of the American Academy of Environmental Engineers and a member of the American Institute of Certified Planners.

ROBERT G. HEALY (Ph.D., Economics, University of California, Los Angeles) is Professor of Environmental Policy in the Nicholas School of the Environment, Duke University. Before coming to Duke in 1986, he was on the research staff of the Conservation Foundation/World Wildlife Fund, Resources for the Future, and the Urban Institute, all in Washington, D.C. He has written several books on land-use policy, U.S. public lands, and agricultural and forest-land use, including *Competition for Land in the American South: Agriculture, Human Settlement, and the Environment* (Conservation Foundation, 1985) and (with William Ascher) *Natural Resource Policymaking in Developing Countries* (Duke University Press, 1990). At Duke, he has been director of the Center for International Studies and the Center for North American Studies. His current research concerns the relation of protected area management to local economic development, with particular emphasis on tourism and sustainable craft production.

RANDALL A. KRAMER is Professor of Resource and Environmental Economics in the Nicholas School of the Environment and the Department of Economics at Duke University. Before coming to Duke in 1988, he taught at Virginia Polytechnic Institute and State University, and was a visiting fellow at the Economic Growth Center, Yale University, and visiting economist, Directorate General of Forest Protection and Nature Conservation, Indonesian Ministry of Forestry. He has consulted with the World Bank, U.S. Agency for International Development, the U.S. Department of the Interior, the Organization of American States, and the Asian Development Bank. He received his B.A. in Economics from the University of North Carolina, his Masters of Economics from North Carolina State University, and Ph.D. in Agricultural Economics at the University of California, Davis. His research has focused on ecosystem valuation, tropical forest management, environmental impacts of agriculture, economics of biodiversity, and natural resource management in developing countries. He is currently studying the relationship between human population change and the use of coastal resources in Indonesia. He is the author of over sixty papers and is coeditor of a recent book, *Last Stand: Protected Areas and the Defense of Tropical Biodiversity* (Oxford University Press).

NATALIA S. MIROVITSKAYA has been on the faculty at Duke University since 1995. She has specialized on the interplay among world economic trends, international relations, and natural resources management. She also leads a Duke faculty seminar on the Black and Caspian Seas and codirected an Advanced Research Workshop on the Caspian Sea in 1999. Prior to joining Duke, she was a Senior Fellow at the Institute of World Economy and International Relations, Russian Academy of Sciences. She has published extensively on international environmental cooperation, gender issues and resource management, and FSU environmental issues. She published *International Aspects of the World Food Problem* (International Relations, 1983) and *Biological Resources of the World Oceans: International Aspects* (Nauka, 1991). With William Ascher she has co-edited *The Caspian Sea: A Quest for Environmental Security* (Kluwer, 2000).

NORMAN MYERS is an independent scientist with an interdisciplinary background based on systems ecology and resource economics. He makes frequent forays into the arena of public policy, as in his recent book *Perverse Subsidies: Tax Dollars Undercutting Our Economies and Environments Alike.* He has lectured at over one hundred American universities and undertakes projects with organizations such as the U.S.

National Academy of Sciences, the World Bank, OECD, and the European Commission. His current interests lie with over- and misconsumption in affluent nations, and with the 800 million new consumers in developing and transition nations. His other recent books include *Gaia: An Atlas of Planet Management, Ultimate Security: The Environmental Basis of Political Stability,* and *Scarcity or Abundance?: A Debate on the Environment* (with Julian Simon).

RENAT A. PERELET has been Research Leader, Laboratory of Systems Ecology, Institute for Systems Analysis, Russian Academy of Sciences, since 1987. In 1996–1998 he was also Senior Expert at the Harvard Institute of the International Development, Moscow Office. He is the author of over sixty papers and books on environmental management. He has served as vice-president of the Russian Branch of the International Society for Ecological Economics and as Full Member (Academician) of the International Academy of Informatics. Dr. Perelet has consulted with the Parliament of the Russian Federation on the issues of sustainable development strategy for Russia, bills on Lake Baikal, environmental security for Russia, and a follow-up to UNCED. He is actively involved in the activities of Russian nongovernmental organizations, serving as a board member of the Russian Association for United Nations, the Ecological Union, and the Russian environmental weekly *Zeleny mir.* He is a member of the Advisory Committee of the Association for University Environmental Education in Central and Eastern Europe (Hungary).

STEVE RAYNER is Professor of Environment and Public Affairs and Professor of Sociology at Columbia University. He is also Chief Social Scientist at the International Institute for Climate Prediction. Previously he was a Chief Scientist and leader of the Global Change Group at the Pacific Northwest National Laboratory, where he was the coeditor (with Elizabeth Malone) of the four-volume series *Human Choice and Climate Change.* He has received several awards for his work on risk management and environmental policy analysis. He has served on various national and international bodies concerned with the human dimensions of global change, including the Intergovernmental Panel on Climate Change. He has testified before Congress on climate change research policy and prepared reports to Congress on climate policy and implementation.

DENNIS A. RONDINELLI is Glaxo Distinguished International Professor of Management, and Director of the Center for Global Business Research, Kenan Institute of Private Enterprise, University of North Carolina at Chapel Hill. His research (in Asia, Central Europe, Latin America, and Africa) is on corporate environmental management, global business strategy, economic reform policy, privatization, urban and regional economic development, business opportunities in emerging market economies, employment-generation policies in developing countries, international trade and investment, and program and project management for economic development. He has published sixteen books and more than two hundred monographs, book chapters, and articles in scholarly and professional journals. He has served as an advisor, consultant, or expert to the U.S. State Department's Agency for International Development, the World Bank, the Asian Development Bank, the Canadian International Development Agency, the International Labor Office, the United Nations Development Programme, and to private corporations.

DANIEL SAREWITZ is Managing Director, and Senior Research Scholar, of Columbia University's Center for Science, Policy, and Outcomes. His work focuses on understanding the connections between scientific research and social benefit, and on developing policies to strengthen such connections. He is the author of *Frontiers of Illusion: Science, Technology, and the Politics of Progress* and coeditor of *Prediction: Science, Decision Making, and the Future of Nature.* Previously he was Director of the Geological Society of America's Institute for Environmental Education. From 1989 to 1993, he worked on Capitol Hill, first as a Congressional Science Fellow, then as science consultant to the House of Representatives Committee on Science, Space, and Technology, where he was involved in issues ranging from science education to federal research policy. As a researcher in geology, his publications focused on processes of mountain building and basin formation along active plate boundaries, with field studies in Argentina, the Philippines, and Tajikistan.

SUDHIR SHETTY is Principal Economist in the Office of the Chief Economist, Africa Region, of the World Bank. He joined the World Bank in 1987, and his previous assignments have been in the Eastern and Southern Africa Departments, as well as in the East Asia and Pacific Region. He has prepared economic reports on environmental problems and policies in Thailand and the Philippines, and was a member of the core team that prepared the 1992 *World Development Report* entitled "Development and the Environment." Before joining the World Bank, he taught public policy and economics at Duke University. He has a Ph.D. in economics from Cornell University.

MARVIN S. SOROOS is Professor of Political Science and Public Administration at North Carolina State University. He has published widely in the fields of international public policy and global environmental politics and policy, including *The Global Predicament: Ecological Perspectives on World Order, The Environment in the Global Arena, Beyond Sovereignty: The Challenge of Global Politics,* and *The Endangered Atmosphere: Preserving a Global Commons.* His articles have appeared in the *Journal of Peace Research, International Studies Quarterly, International Organization, Environmental Review, Policy Studies Journal, International Political Science Journal, Environment,* and *Human Ecology.* He has been president of the Southern Regional Section of the International Studies Association and chair of the association's Environmental Studies Section. He has chaired the Environmental Studies Section of the International Studies Association and been active in the International Human Dimensions of Global Change Program.

DAVID SPENCE is Professor of Management Science and Information Systems at the University of Texas at Austin. He served as a policy analyst at the North Carolina Alternative Energy Corporation and the North Carolina Department of Administration, and as an attorney in private practice. His research has focused on environmental law, energy, and business law, with a particular emphasis on regulation and procedural reform. He received his law degree from the University of North Carolina at Chapel Hill and his doctorate from Duke University.

TODDI A. STEELMAN is Assistant Professor at North Carolina State University's Department of Forestry. Her research focuses on the role of public and community involvement in enrivonmental governance. Recent research projects and publications include work on forests, watersheds, open space protection and climate change. Recent publications include "Innovation in Local Land Protection and Governance: The Case of Great Outdoors Colorado," "The Public Comment Process: What Does the Public Contribute to National Forest Management?," "Common Property, Collective Action and Community Opposition," and "Public Involvement Methods in Natural Resources Policy Making: Advantages, Disadvantages and Tradeoffs."

Susan E. Subak is Senior Researcher with the Natural Resources Defense Council's climate program. Her primary interests are strategies for limiting greenhouse gas emissions from land use and agriculture, and regional climate impact analysis. She is the author of more than fifty professional and technical articles. She has worked as a researcher and lecturer at the School of Environmental Sciences at the University of East Anglia in England, where she is a senior fellow, and is a research associate with the Stockholm Environment Institute and Harvard University. She has served as a principal investigator on major projects sponsored by the European Commission and the United Kingdom's Department of Environment, and has worked as a consultant for a number of national and international organizations. She is a graduate of Reed College and MIT, and holds a doctorate from the University of East Anglia.

Susanne Swibold is a visual artist, photographer, filmmaker, and ecologist who is known internationally for her visual research techniques and her conservation efforts in the Bering Sea. Her research concerns northern native people (Aleuts), resource economies, and island systems; her fieldwork has focused on the Pribilof Islands and Alaska (1981–1993), and the Commander Islands and Kamchatka (1994-present). She is founder and codirector of the Amiq Institute and a research associate at the Arctic Institute of North America, University of Calgary. Her four films on Aleut culture and the environment of the Pribilof Islands received several awards and have been televised in many countries. Her photography is published widely in books and periodicals.

Ian Welsh lectures in Sociology in the School of Social Sciences, Cardiff University. He has a long-standing interest in environmental sociology, spanning nuclear, climate change, and road building issues. His books include *Environment and Society in Eastern Europe* (edited with Andrew Tickle) and *Mobilising Modernity: The Nuclear Moment*. He recently published "Risk, 'Race' and Global Environmental Regulation," in Avtar Brah et al., eds., *Global Futures; Migration, Environment and Globalization*. His research interests also include the social aspects of xenotransplantation—the use of animal tissue and organs in humans.

Note on Entry Contributors

Definition entries by contributors other than the general editors are designated by the following initials:

G. J. A.	Gustavo A. Arcia	S. R.	Steve Rayner
F. B.	Fikret Berkes	D. A. R.	Dennis A. Rondinelli
R. D. B.	Ronald D. Brunner	M. S. S.	Marvin S. Soroos
R. G. H.	Robert G. Healy	D. S.	David Spence
R. A. K.	Randall A. Kramer	T. S.	Toddi Steelman
R. P.	Renat Perelet	I. W.	Ian Welsh

Basic Concepts of Development
and Environment

1.1 Development: Introductory Concepts,
Criteria, and Measurements

1.1.1 CAPITAL

Absorptive Capacity, Economic: the ability of the economy to use capital productively. Rapid natural-resource exploitation that brings large volumes of capital into an economy often results in disappointing economic growth because low absorptive capacity limits the returns on investment. *R. P.*

Capital: 1) physical assets produced by the economy, and financial assets, available as inputs for further production. This definition of capital is the classic definition of the factor of production that combines with land and labor as inputs in the production process. The value of physical capital can be calculated from the present value of the goods and services that the capital creates over time;

2) more broadly, all inputs available for economic activity, including human capital, technological capital, and natural-resource capital.

COMMENTARIES ON CAPITAL

The proliferation of types of "capital"—cultural capital, environmental capital, human capital, natural capital, social capital, etc.—that have been added to the initial category of financial capital reflects the growing recognition of the complexity of economic growth. It was not always so; in the classic formulation of "land, labor, and capital," "capital" was much more narrowly defined. Yet in the decades following World War II, financial capital was such a predominant focus of development, even to the point of assuming a fixed "capital-output" ratio, which implied that output would increase in lockstep with increases in financial capital (e.g., the Harrod-Domar model), that the other factors of growth were neglected. This was not difficult to understand: financial capital shortages focused attention on how foreign assistance could "close the savings gap" between available financial capital and the level needed to achieve specified targets of growth. However, as understandings of the complex challenges of sustainable economic development matured, it became clear that infusions of financial capital alone were often insufficient for either economic growth or development more broadly defined. Yet in the rhetoric of development economics, the term "capital" still had tremendous sway. Hence the rhetorical strategy of extending the term to virtually all of the other "factors of production." *William Ascher*

Capital of any kind in the industrial countries is to be exploited for production and consumerism, no matter what the social or environmental costs. (Advertising is the handmaiden for the enforced distribution of goods.) This does not acknowledge the reality of the nonmaterialistic, that is, nature is something we *capitalize* on. We use it to gain monetary wealth, to make profits. Even our "heritage sites" become spoiled by tourism "markets" where land and cultures are turned into profits. Life-supporting systems may in theory be protected but they are fragments that ultimately are threatened, compromised by pollution and technology. Until we stop denying the American-style capitalism that is materialistic on all counts—human, ecological, national heritage, natural wealth—the mold on the orange will eat into every aspect of life on earth. We are committing suicide in the name of capital accounting. *Susanne Swibold*

Capital Accumulation (in ENVIRONMENTAL ACCOUNTING): an environmentally adjusted concept of capital formation that accounts for depletion and degradation of NATURAL CAPITAL. See also SECTION 5.1.1.

Capital-Resource Substitution: a measurement of the degree of substitutability between humanmade capital and natural resources in production. For example, if the elasticity of substitution is 1.0 (perfect elasticity) between tractors and hectares of arable land in agriculture, a farmer with two tractors and one hundred hectares should theoretically be able to produce as much agriculture output as one with one tractor and two hundred hectares of land. *R. P.*

Critical Natural Capital: a concept introduced by the proponents of the STRONG SUSTAINABILITY approach that some resources, such as the ozone layer, the carbon cycle, or BIODIVERSITY have primary (e.g., ecological) values as well as secondary (e.g., human-use-based, or market) values. Given that humanmade capital often cannot completely or directly substitute the ecological capital, this school argues that a minimum goal of SUSTAINABLE DEVELOPMENT is to protect critical natural capital.

See David W. Pearce, *Blueprint 3: Measuring Sustainable Development* (London: Earthscan, 1994).

Cultural Capital: 1) the factors that provide human societies with the means and adaptations to deal with the natural environment and to actively modify it. It includes the way people view the world and the universe (cosmology); environmental philosophy and ethics, including religion; traditional ecological knowledge; and social and political institutions. See *also* SOCIAL CAPITAL.

See Fikret Berkes and Carl Folke, "A Systems Perspective on the Interrelations between Natural, Human-made and Cultural Capital," *Ecological Economics* 5 (1992): 1–8. *Fikret Berkes*

2) cultural beliefs and practices held by particular groups within society, as influenced by history, education, habitual practices, income levels, etc. For example, Max Weber argued that cultural capital of specific groups provides them with orientations to succeed or fail at development (e.g., attributing economic success to the "Protestant ethic"). Differences in cultural capital can divide societies along class and ethnic lines.

See Max Weber, *The Protestant Ethic and the Spirit of Capitalism* (New York: Charles Scribner's Sons, 1958); Pierre Bourdieu, *Distinction: A Social Critique of the Judgement of Taste* (Cambridge, MA: Harvard University Press, 1984).

Ecological Capital: a notion used in ENVIRONMENTAL ACCOUNTING to define the contribution of the environment to economic activity. See also NATURAL CAPITAL/NATURAL ASSETS. See, generally, SECTION 5.1.1.

Human Capital: capital that comprises all individuals' capacities for work (skills, knowledge, health, strength, motivation) as well as the networks and organizations through which they are mobilized. Human capital therefore encompasses both individual resources and SOCIAL CAPITAL.

National Heritage (National Estate, Natural Patrimony): a designation given by governments to the components of the cultural and natural environment that are of national value and need to be preserved for the benefit of the nation. Such components possess aesthetic, historical, scientific, social, cultural, ecological, or other values. They may include rare species, parks and reserves, museums and sites of archaeological interest, or nature sites of special merit. Some components, such as Lake Baikal and the Hermitage in Russia, belong to the global world heritage and are to be preserved in accordance with the WORLD HERITAGE CONVENTION.

COMMENTARY ON NATIONAL HERITAGE

An important, subjective concept that easily gets blurred by prevailing socioeconomic pressures, the concept encourages national pride and nationalistic sentiment; therefore public policymakers often use it in political discourse or campaigns to gain public support. In some developing countries, governments and groups consider indigenous populations as part of the national heritage only insofar as national heritage is a resource to attract foreign tourism, thereby transforming cultural preservation into an economic strategy. *Elsa Chang*

Natural Capital/Natural Assets: 1) the stock of life-supporting systems, BIODIVERSITY, renewable and nonrenewable resources. Natural capital excludes humanmade capital. The contribution of the concept is in clarifying that environmental resources are as-

3

sets in the economy insofar as they contribute to economic productivity and welfare.
R. P.

2) defined narrowly, the noncommercialized part of the natural environment that provides environmental services not captured in the standard GROSS NATIONAL PRODUCT (GNP) account. In contrast, return on the commercialized environment, such as agricultural lands or oil reserves, are reflected in GNP statistics and therefore counted as part of a country's economic production and wealth. Natural capital, defined in that sense, can be measured by a NATURAL CAPITAL INDICATOR.

Natural Wealth. See NATURAL CAPITAL/NATIONAL ASSETS.

Social Capital: the level of trust, public commitment, and capacity for cooperation of a community or group that provides a basis for effective functioning. Social capital is said to rest on the capacity of social institutions, ranging from political and legal institutions to clubs and voluntary groups, to foster cooperation and instill identification with broad societal units. Advocates of the social-capital concept, dating at least as far back as Max Weber's notion of moral behavior of individuals toward all other members of society as a component of the "Protestant ethic," have argued that it is essential for understanding why economic growth does or does not occur; thus it should be considered as an "input" in the production process just like other forms of capital. They argue that economic development and democracy require the strengthening of micro-level social institutions as much as the provision of money, development expertise, or formal macrolevel democratic practices. Some theories of social capital (such as those of Edward Banfield and Robert Putnam) argue that social capital is so deeply imbedded in social institutions that its level persists over many generations. The maintenance of social capital is considered to be one of the prerequisites of STRONG SUSTAINABILITY.

See Edward Banfield, *The Moral Basis of a Backward Society* (Chicago: Free Press, 1958); Max Weber, *The Protestant Ethic and the Spirit of Capitalism* (New York: Charles Scribner's Sons, 1958); Robert D. Putnam, *Making Democracy Work* (Princeton, NJ: Princeton University Press, 1993).

COMMENTARIES ON SOCIAL CAPITAL

The term *social capital* is attributed to James Coleman, who used it to describe the trust, norms, and networks that allow social institutions to function. The term has been repopularized with Robert Putnam's work "Bowling Alone," which details trends of declining social capital in American society. Putnam asserts that while the number of bowling leagues has diminished, the number of individuals who bowl alone has increased. Social capital has been a touchstone concept since the 1990s. Many researchers, authors, and practitioners have used the term to describe or explain the alienation or apathy of various publics with respect to involvement in political life. Consequently, the term *social capital* has become so overused that it risks losing any real sense of meaning. *Toddi Steelman*

It is always good to review existing social institutions and study their origins, since they often respond to barely perceptible barriers to sustainability. Local provision of public goods tends to be underestimated in the management of natural resources, resulting in the imposition of official institutions that can take away local ownership of a resource. In Central America, local arrangements of the commons are often replaced by natural-resource management entities that look at the community as miners of a resource, which may not be the case. *Gustavo J. Arcia*

Undercapitalization: the lack of adequate investment capital to operate a firm, industry, or sector at high levels of efficiency and societal benefit. While many economies suffer from excess capital that is drawn without productive purposes into an industry or sector, the opposite problem of undercapitalization hampers the adoption of optimal scale and technologies. Undercapitalization may reflect price distortions that make other industries or sectors more attractive for investment, or it may reflect government decisions to drain capital away from a particular firm, industry, or sector. For many decades the governments of developing countries undercapitalized the agricultural sector through pricing and taxing policies that drew capital into industry. In some cases, governments have also undercapitalized their own state enterprises in the resource sectors, in order to bring the resource surplus into the central treasury or to weaken the enterprises when their power seemed threatening to the government.

1.1.2 CLASSIFICATION OF NATIONS AND MAJOR POLITICAL UNITS BY DEVELOPMENT CHARACTERISTICS

Countries in Transition: formerly centrally planned economies making the transition to a market economy. The designation covers Albania, Armenia, Azerbaijan, Belarus, Bosnia and Herzegovina, Bulgaria, Croatia, the Czech Republic, Estonia, Georgia, Hungary, Kazakhstan, Kyrgyzstan, Latvia, Lithuania, the Republic of Moldova, Poland, Romania, the Russian Federation, Slovakia, Slovenia, Tajikistan, Turkmenistan, Ukraine, the former Yugoslav Republic of Macedonia, and Yugoslavia. Major international environmental conventions acknowledge special circumstances of these countries. For instance, within the FRAMEWORK CONVENTION ON CLIMATE CHANGE, countries in transition as well as ORGANIZATION FOR ECONOMIC COOPERATION AND DEVELOPMENT (OECD) countries, agreed to voluntarily reduce their greenhouse emissions by the year 2000 to 1990 levels. However, in contrast to the OECD countries, these nations are not obligated to provide technological, financial, and intellectual assistance for reducing greenhouse emissions from DEVELOPING COUNTRIES.

Developed Countries: nations with relatively high levels of economic infrastructure and production so as to yield high national income. Although the status of being a developed country is often determined by per capita income levels, not everyone agrees that national income per capita per se is equivalent to "development." Countries with high revenues from raw materials, such as certain oil-exporting countries, may not be "developed" in terms of long-range productive capacity. A similar controversy surrounds the tendency to equate development with industrialization, in light of several cases of nations that are prosperous and productive without heavy reliance on manufacturing (e.g., Norway).

5

How far can the United States qualify as a "developed" country when: with 5 percent of the world's population, it accounts for 23 percent of the world's carbon dioxide emissions; it has a higher proportion of massively overfed people than any other country; it has a per capita waste level higher than any other country; it produces proportionately more unintended pregnancies and unwanted babies than any other "developed" nation; it maintains a puritan spirit that berates its president for sexual peccadilloes while maintaining the largest pornography industry in the world; it complains about the high price of gasoline even though it costs far less than bottled water in the supermarket, is cheaper in real terms than when Americans started to pump it out of the ground, and features a "full social cost" of $6–8 per gallon; and it reneges on its agreed payments to the United Nations, even though the budget of the United Nations Environment Programme is smaller than that of the New York City Fire Department. Conversely, it is the nation that probably contains more people who remain determined to build that "shining citadel on the hilltop"; idealism forever! *Norman Myers*

Developing Countries (Less Developed Countries [LDCs]): nations with relatively low levels of economic development. Various international agencies define nation-states as developing if they fall below certain targets, for instance, in terms of GROSS NATIONAL PRODUCT (GNP), industrial production, literacy rates. The United Nations, UNITED NATIONS INDUSTRIAL DEVELOPMENT ORGANIZATION (UNIDO), the BRETTON WOODS INSTITUTIONS, and some other international organizations and development agencies have each produced their own lists of developing countries that are slightly different. There is an important debate whether countries with high GROSS DOMESTIC PRODUCT (GDP) per capita (like Kuwait or Saudi Arabia) ought to be considered "developing" because of economic or social characteristics not captured by aggregate income.

Most international environmental conventions give developing countries substantial concessions in terms of financial assistance, technology transfer, and extended time schedules for implementation of obligations. For instance, within the framework of the GLOBAL CLIMATE CONVENTION they are not supposed to match commitments by industrialized countries to cut GREENHOUSE GASES, and they are entitled to financial assistance and technology from the developed nations to initiate changes in their environmental policies.

East vs. West: an outdated but still widely used reference to the division of the world in terms of cultural and political matters. The "West" is equated with the developed countries, while the "East" refers in some cases to the former socialist bloc and more often to all countries other than the First World.

The East-West contrast maps a huge number of ostensible differences onto a geographically simplistic distinction. As Lewis and Wigen note, "its staple feature has historically been the linking of West with reason and progress and the East with spirituality and stagnation." In addition to being confusing from the geographical and political perspectives, this East-West binarism also has an implicit message that progress is achieved exclusively via a Western pattern.

See Martin Lewis and Karen Wigen, *The Myth of Continents. A Critique of Metageography* (Berkeley: University of California Press, 1997). *Natalia Mirovitskaya*

Economy in Transition. See COUNTRIES IN TRANSITION.

Fifth World: a label applied by some geographers in reference to the poorest and LEAST DEVELOPED COUNTRIES of the world. Some scholars use this term in relation to refugees.

First World. See INDUSTRIAL COUNTRIES.

Fourth World: 1) See LEAST DEVELOPED COUNTRIES (LLDS).

2) a label denoting indigenous minorities who lack political autonomy or adequate status (such as the Dayaks in Indonesian Kalimantan or the so-called pygmies of Central Africa), who have often been displaced onto marginal lands (highlands or remote forests), or end up in urban slums. The connotation of the term is that they remain culturally separate, even if their traditional lifestyles erode in the face of pressure from the more dominant cultures.

Group of 7 (G-7): 1) a grouping of seven of the wealthiest nations: the United States, France, the United Kingdom, Germany, Canada, Italy, and Japan. The leaders of these states annually gather at an economic summit to discuss most serious world concerns and to coordinate their actions. The Group was set up in 1975 at the peak of the world energy crisis triggered by the ORGANIZATION OF PETROLEUM EXPORTING COUNTRIES (OPEC) decision to raise the price of oil. Since 1988, environmental concerns have been included on the agenda of the summit along with other major political and economic problems. For the last few years, the president of the Russian Federation has attended the G-7 meetings; these meetings are typically labeled the "Group of 8."

2) also "Megadiversity 7"—a label for the countries that in combination contain more than a half of the world's biodiversity: Australia, Brazil, Colombia, Democratic Republic of Congo (formerly Zaire), Indonesia, Madagascar, and Mexico.

Group of 77 (G-77): an organization of DEVELOPING COUNTRIES that emerged during the preparatory stage for the first UNITED NATIONS CONFERENCE ON TRADE AND DEVELOPMENT (UNCTAD) in 1964 in response to the growing recognition of the North-South rift in the world economy. Since then, G-77 membership has increased to 132 in 1997 and its permanent institutional structure has been developed.

COMMENTARY ON G-77

The G-77 chapters in FOOD AND AGRICULTURE ORGANIZATION (FAO) (Rome), UNITED NATIONS INDUSTRIAL DEVELOPMENT ORGANIZATION (UNIDO) (Vienna), UNITED NATIONS ENVIRONMENTAL PROGRAMME (UNEP) (Nairobi), UNITED NATIONS EDUCATIONAL, SOCIAL, AND CULTURAL ORGANIZATION (UNESCO) (Paris), and INTERNATIONAL MONETARY FUND (IMF)/WORLD BANK (Washington) provide the means for the developing countries to articulate and promote their collective economic interests, to strengthen their joint bargaining leverage in the North-South dialogue, and to promote economic and technical cooperation among developing nations. As the world's largest coalition, the G-77 is one of the most influential actors in international environmental diplomacy. It has pursued reform in environmental problems of particular concern to developing nations such as desertification, drinking water, and sanitation. Usual requests of the G-77 during international environmental negotiations include technology transfer on preferential and concessional terms, development assistance, and future access to resources guaranteed by allocation (or a share in the proceeds that the industrial nations earn from the use of international commons). The G-77 succeeded in the official linking of environmental issues with development concerns and in securing substantially lower level of obligations by developing nations to international environmental conventions. For instance, the FRAMEWORK CONVENTION ON CLIMATE CHANGE (FCCC) specified that the extent of developing countries' commitments depends on the effective transfer of technology and financial resources from developed countries. The latter should also take into account that economic and social development and eradication of poverty are overriding priorities of the developing nations. However, the environmental clout gained by the G-77 during international negotiations of the 1990s has been largely negated by the low level of implementation of international obligations by the major Northern countries. *Natalia Mirovitskaya*

Industrial Countries (First World): term denoting the world's wealthier nations, but usually excluding the high-per-capita-income, raw-material-exporting countries such as Saudi Arabia or Brunei that have very little manufacturing. The term is somewhat of a throwback to the conception that economic wealth and development correspond to the degree to which a nation has industrialized, even though several "industrial countries," such as Norway, do not have large industrial sectors.

Least Developed Countries (LLDCs): countries identified by the United Nations as low-income countries encountering long-term constraints to economic growth, particularly low levels of human resource development and severe structural weaknesses in development: e.g., the lowest income (per capita gross domestic product [GDP] of $100 or less at 1970 prices), low level of industrialization (secondary or manufacturing industries contributing of 10 percent or less of GDP), and lowest literacy rates (20 percent or less of literate persons over age fifteen). The main purpose of this classification is to identify the countries in most need of foreign assistance. As of 1997, the number of LLDCs was roughly forty-eight. Ethiopia with its negative annual growth rates of per capita income, national poverty line of 86, HUMAN DEVELOPMENT INDEX (HDI) value of 0.244 and GENDER-RELATED DEVELOPMENT INDEX (GDI) value of 0.233 is an exemplar of this category. See also FOURTH WORLD.

See United Nations Development Programme, *Human Development Report* (New York: UNDP, 1997).

Less Developed Countries (LDCs): See DEVELOPING COUNTRIES.

Newly Industrializing Countries (NICs): countries that since the 1970s were able to achieve rapid (although not always consistent) growth of manufacturing output and to create a modern industrialized society. Principally located in Southeast Asia and Latin America, they may not be considered part of the THIRD WORLD, in the sense that they are not entitled to development assistance or other special prerogatives provided to the Southern nations within the frameworks of international environmental agreements. The UN list of NICs includes Thailand, Singapore, South Korea, Taiwan, Malaysia, Brazil, and Mexico.

North vs. South: a reference to the division of the world between more industrialized and wealthier countries, typically north of the equator, and the less-developed countries, many of which are south of the equator. The North-South division is loose and flawed from geographical, economic, and political perspectives; however, it is persistently used in political discourse. Economic and social differences between developed and developing countries are reflected in the so-called North-South gap, a set of national and regional estimates of different criteria of development (life expectancy at birth, infant mortality, adult literacy, per capita daily calorie supply, access to safe water, etc.). However, international agencies that use the North-South gap indicators in their work have different lists of countries included into these main categories.

Raw-Material-Intensive Economy: an economy that rests heavily on the extraction and processing of natural resources such as oil, hard minerals, or timber. Although in earlier eras it was assumed that the resource endowment gave an important economic advantage to raw-material-intensive economies, a contemporary issue in development theory is why such economies among developing countries apparently suffer from slow economic growth—the so-called resource curse (see also RESOURCE-CURSE HYPOTHESIS). Possible explanations include the difficulty of pricing raw materials appropriately when their abundance reinforces political pressures to maintain inefficiently low prices, and the possibility that relative raw-material dependence is a reflection of the failure to develop economically rather than a cause.

Second World: a label applied to the countries of the former socialist bloc (and ardently detested by them). See COUNTRIES IN TRANSITION.

Third World: a label applied to low-income countries. See also DEVELOPING COUNTRIES.

COMMENTARIES ON CLASSIFICATIONS OF COUNTRIES

The term *developing country* is considered more politically correct and less demeaning by most low-income nations than *Third World,* which was a popular term mostly used in the economic and political debates in developed countries during the 1960s and 1970s. *Elsa Chang*

The quaint nomenclature of "industrial countries" is clearly outdated if one considers that the service sector in the rich countries accounts for almost one-half of economic activity. *Gustavo J. Arcia*

Regarding raw-material-intensive economies, the inability to develop the rest of the economy when one has a dominant raw-material sector is also known as "Dutch disease," which describes the situation in Holland, when natural gas revenues were so dominant—overvaluing the exchange rate—that they did not allow other sectors to be competitive. To this category I would add the term "aid-intensive economy," like Haiti, Nicaragua, and some African countries, where the external sector—represented by massive inflows of external aid—helps maintain an overvalued exchange rate, making these countries poor and expensive. In this case, Dutch disease is caused by aid flows that are larger than the absorptive capacity of the aided economy. *Gustavo J. Arcia*

1.1.3 DEVELOPMENT

Bottom-up Development: often contrasted with top-down development, bottom-up development refers to a participatory or "grassroots" approach to making important decisions about development. It requires that the planning, design, implementation, and evaluation of projects are to be done mainly by the project's beneficiaries. Local populations and indigenous knowledge provide the impetus for development projects. Priority is placed on self-determination and the right to develop in one's own way. Bottom-up development recognizes that local populations must be involved in their own development if such efforts are to be sustainable. Bottom-up or grassroots development efforts tend to be used more often in less developed countries than in more developed countries. In both cases bottom-up development is often a labor- and time-intensive process. See also PARTICIPATORY DEVELOPMENT. *T. S.*

COMMENTARY ON BOTTOM-UP DEVELOPMENT

Bottom-up development is a very good concept that is often abused. Bottom-up wisdom must always be challenged in order to make sure that we come up with a solution that it is optimal within a constrained environment—otherwise we may end up with a solution based on collective ignorance. Bottom-up development works fine when science and local interests coincide, but that coincidence sometimes goes unidentified. *Gustavo J. Arcia*

Capacity Building: efforts to increase an organization's ability to pursue development objectives and manage resources effectively and efficiently. Investments in capacity building often entail improvements in physical facilities, training, additional personnel, administrative reforms, and higher salaries to attract and retain more qualified personnel. Foreign assistance packages frequently include capacity-building components out of recognition that the effectiveness and sustainability of projects and programs depend on the quality of management. Natural-resource and environmental agencies have often been targeted for capacity building efforts because of their limited resources and capacities.

Some capacity-building efforts fail when they are more tailored to the project goals and agenda of the foreign assistance organization than the needs of the recipient country or organization. Often aid or donor agencies lack coordination amongst themselves and therefore duplicate efforts and waste resources. In some instances, capacity-building efforts have been abused by both the provider and the recipient group. The provider is more interested to getting visible results within a deadline. Recipient organizations sometimes accept all sorts of inconsistent capacity-building assistance without a specific plan, and send the same favored individuals for training. *Elsa Chang*

Core-Periphery Model: any of several analytical models that treat economic development as a set of linkages and interactions between centers (or core areas) and periphery areas, typically emphasizing how the people in peripheral areas become marginalized. Core-periphery models explain the persistence of pockets of underdevelopment within modernizing nations, though they may be used at a variety of scales, from local to regional to international. At different scales, of course, the definition of center and periphery will vary. In addition, areas may have multiple centers and/or peripheries. For example, at the national level there is an urban-industrial center region and a less industrialized, more rural, peripheral region; however, within the urban-industrial center region, there is also more urbanized center and less urbanized periphery. Center-periphery models have been developed by several authors and have been interpreted differently by those of different ideological persuasion. Core-periphery relationships are central to DEPENDENCY THEORIES, in emphasizing that developing countries are both peripheral to the world economic system and that they develop their own center-periphery distinctions in the development of economic activities geared to international markets (Cardoso and Faletto, 1979). Two of the first major statements on center-periphery relationships were by Gunnar Myrdal (1957) and Albert Hirshman (1958); both have remained influential.

See Fernando Henrique Cardoso and Enzo Faletto, *Dependency and Development in Latin America* (Berkeley: University of California Press, 1979); Terence K. Hopkins and Immanuel Wallerstein, eds., *Processes of the World-System* (Beverly Hills, CA: Sage Publications, 1980); Anthony de Souza and J. Brady Foust, *World Space-Economy* (Columbus, OH: Charles Merrill, 1979); Gunnar Myrdal, *Economic Theory and Underdeveloped Regions.* (London: Duckworth, 1957); Albert O. Hirschman, *The Strategy of Economic Development* (New Haven: Yale University Press, 1958).

Like two-toned shoes and lava lamps, this concept emerges in discussions once in while. So far, it has proved to be very useful for policy only when there has been a strong analysis of the structural characteristics of the links between the center and the periphery. For example, the traditional discrimination against agriculture—in favor of urban areas, to keep food cheap for urban consumers—did result in strong policy mandates favoring poverty reduction in rural areas. Because rural populations are generally less educated, more disperse, and with less voice than urban populations, they were less able to influence food price policy in their favor. To compensate for this inequity, governments are now making efforts to target rural areas with poverty alleviation programs. In cases such as this one the center-periphery model is quite relevant for analysis. *Gustavo Arcia*

Development: increased capability in the pursuit of wealth, well-being, or other values. Subject to many specific definitions, *economic development* is typically used to convey cumulative progress in economic institutions, human capabilities, economic policy, and physical and financial investment. Depending on whether development is seen as pertaining narrowly to economic aspects or more broadly to social progress in general, it can be seen as consistent with or antithetical to conservation. *Political development*, a concept popular in the 1960s as a label for progress in building more effective (and, usually, more democratic) political institutions, was heavily criticized as being Western-centered. Yet the term is experiencing a resurgence as world opinion has converged on the idea that governments must be more accountable, responsive, and effective.

Development Models: 1) broad interpretations of the nature of the development process and its timing and preconditions. Among the most prominent development models were W. W. ROSTOW'S STAGES OF ECONOMIC GROWTH, which predicted a "take-off" phase, after which economic growth is self-sustaining. Such development models often support or imply particular economic policy strategies.

2) broad strategies for economic development, varying greatly according to ideology and economic philosophy. Some development models emphasize increased investment through higher savings; they vary as to whether increased savings should come by concentrating wealth in the hands of the wealthy, who sometimes have a higher propensity to save and invest (the so-called TRICKLE-DOWN model), or by increasing demand by distributing wealth more evenly. Some development models call for greater insertion of the nation in the international financial and trade systems, in order to increase investment and benefit from the nation's comparative advantages; other models call for greater self-sufficiency to avoid disadvantages and distortions that some claim arise from interaction with stronger economies (the so-called dependency model; see DEPENDENCY THEORY). Some development models emphasize the need for developing particular economic sectors, such as the heavy-industrial sector (the so-called Mahalanobis model that was very influential in India in the 1960s and 1970s). In the early post–World War II era, most prevailing models emphasized industrialization; later models emphasized the need to reinvigorate the agricultural sector.

One must remember that the current consensus is on *human capital* accumulation—within a sound macroeconomic environment (no large fiscal deficits, no overvalued exchange rates)—as the key to development. There is also an emerging consensus that the economic troubles in Asia of the late 1990s was a market correction that would have nonfatal effects. Human capital investment is such a sound development base that the recuperation of Asian economies will take less time than expected. *Gustavo J. Arcia*

Economic Growth Debate: the controversy over the sustainability of economic growth in light of finite natural resources, environmental services, and carrying capacities. The antigrowth position rests on the reduced options for maintaining ecosystem functioning and continued economic expansion. Skeptics of this position point out that many resources predicted to come into short supply are more economically abundant than before (as indicated by their reduced prices) and that economic prosperity and technology can improve environmental services. The response to this argument has been that maintaining resource supplies and environmental services in the face of greater consumption and industrial activity has come at hidden costs of displaced stresses on ecosystems and on developing countries. Some antigrowth participants in this debate (e.g., Herman Daly) condemn economic growth, if defined in terms of growing material consumption, but not economic development, in the sense of improving human well-being. No standard definitions of these concepts prevail.

See Herman E. Daly, *Beyond Growth: The Economics of Sustainable Development* (Boston: Beacon Press, 1996).

COMMENTARIES ON THE ECONOMIC GROWTH DEBATE

The economic growth debate was epitomized by the bet between Julian Simon and Paul Ehrlich. Simon was a well-known proponent of population growth and believed in human ingenuity to overcome any adversities imposed by scarcity. Ehrlich, on the other hand, was known for his Malthusian views and predicted in 1968 that human populations would outstrip food production. In 1980 Simon bet Ehrlich that the prices of commodities would fall during the next ten years. The terms of the bet were to pay the difference on a $1,000 basket of minerals. In 1990, Simon won the bet. During the decade, mineral prices fell and Ehrlich paid Simon $576.06. Following the bet, critics have offered that instead of focusing on commodities, the bet should have focused on ecosystem functions, such as clean water, air, and fertile soil, as well as species, which seem to be more scarce over time. Nonetheless, when Simon offered a second bet based on commodities of Ehrlich's choice, Ehrlich declined. *Toddi Steelman*

As a citizen of a poor country I cannot but smile sardonically at antigrowth pundits who offer nonconsumption advice on grounds of world sustainability.

Sustainability for whom? My brain tells me they are right in the global sense, but my heart—undoubtedly fed by crass shopping impulses—tells me that they are wrong for the simple reason that it is easy to call for "sustainability" *after* you have done your share of consumption. *Gustavo J. Arcia*

Endogenous Development (Endogenous Growth): the economic development or growth due to internal (i.e., domestic) resources and activities within a national economy. For some, endogenous development or growth is viewed as a phenomenon to be explained: what characteristics account for greater or lesser development or growth in some countries, especially at the same point in time and with similar or different degrees of integration with the world economy? For others, endogenous development is a strategy of self-reliant economic growth, including such measures as protectionism, discouragement of foreign investment, and deemphasis of export-oriented industries. An endogenous growth strategy often appeals to nationalist sentiment and fears of interference by economic superpowers. The stagnation of nations that have attempted extremely self-reliant ("autarchic") development (e.g., North Korea, Myanmar, Albania), and the economic growth of some nations that have pursued export-oriented growth strategies (e.g., Chile and Taiwan), have dampened enthusiasm among economists and policymakers for such self-reliant strategies. However, the issue of whether partial self-reliance is better or worse than full exposure to the world economy is still hotly debated.

Exemplary Projects: in Russian urban planning, a wide range of community development projects claimed to demonstrate various options of alternative development. Their common feature is the pursuit of social values through grassroots democracy, avoidance of capital-intensive large-scale production, and attention to making the project appropriate to the specific context. One of the most interesting exemplary projects that emerged under the Soviet authoritarian regime was the Ecopolis program, which aimed to ensure a balance between the carrying capacity of several environmental sites near Moscow and different kinds of human pressure. The value of "exemplary projects" under the overwhelming pressure of globalization has been debated.

Growth, Economic: the overall increase in economic activity and the production of goods and services. Economic growth, typically measured by the growth in GROSS DOMESTIC PRODUCT (GDP) or GROSS NATIONAL PRODUCT (GNP) long dominated the conceptions of economic progress and development. An important branch of economics is devoted to understanding the conditions under which economic growth occurs and the limits of the rate of economic growth. Nations with low levels of economic growth often lack the resources to pursue noneconomic objectives, such as education, the development of health systems, etc. Nevertheless, there have been many critics of the tendency of the economic growth concept to distract attention from other aspects of human (and ecosystem) development. Critics like Herman Daly argue that if economic growth is taken to mean the physical expansion of production, it cannot be sustainable in a world of finite resources, although economic development, entailing the shift to resource-conserving production, may be sustainable.

See Herman E. Daly, *Beyond Growth: The Economics of Sustainable Development* (Boston: Beason Press, 1996).

Low-Level Equilibrium Trap: an explanation for economic stagnation first offered by U.S. development economist Harvey Leibenstein in the 1950s, proposing that producers in developing countries are trapped by the risks involved in moving up to higher productivity approaches. The theory weaves together risk aversion, economic inefficiency, and the logic of subsistence to account for stagnation without invoking external forces. It tries to demonstrate that underdevelopment, often attributed to the nonrational behavior of people in developing countries, may instead be the result of rational reactions to economic constraints.

See Harvey Leibenstein, *A Theory of Economic-Demographic Development* (Princeton: Princeton University Press, 1954).

COMMENTARY ON THE LOW-LEVEL EQUILIBRIUM TRAP

This is a big building block of development science that practitioners may take for granted (few people mention Leibenstein anymore). It helps explain why optimal farming systems may in fact be the optimization of poverty: not enough income to live decently, nor enough poverty to kick one out of farming altogether. It also explains the roots of technology adoption among low-income farmers and—indirectly—the need to invest in human capital as a way to increase labor productivity, since education is key for reducing the risk premium attached to investment and production uncertainty. *Gustavo Arcia*

Modernization: the processes of shifting from traditional social, economic, and political practices and institutions to those regarded as "modern." Although definitions of "modern" vary widely from one source or context to another, modernization typically refers to the transition to more efficient, instrumental, impersonal, and "rational" practices and institutions. Sociologists such as Talcott Parsons defined modern social relationships as those characterized by affectively neutral interactions in which individuals relate to one another without regard to their origins or their standing outside of the specific exchanges of the interactions, so that people are judged by their achievements rather than by characteristics unrelated to the transaction at hand. Economic modernization often entails a shift to technologically sophisticated, profit-maximizing production, with the consequent decline of low-yield activities, including, in many cases, small-scale agriculture and handicrafts. Modernization is often, though misleadingly, equated with industrialization, although economically advanced countries usually display a shift away from manufacturing to greater emphasis on the service sector. Modernization of management usually entails the shift from family firms to corporations run by professional managers. Political modernization is generally conceived as the rise of impersonal political organizations in place of kinship- or clique-based political arrangements, and from chaotic, personalist rule to regularized, predictable, "professional" politics.

See Talcott Parsons, *The Evolution of Societies* (Englewood Cliffs, NJ: Prentice-Hall, 1977).

Optimal Growth Theory: a range of theories that explore the conditions and policies for sustained economic growth. The relevant variables are typically the volumes and rates of capital increase and investment, the expansion of the money supply, the rate of population growth, and the broad choices in the allocation of investment. An important aspect of growth theories is their conclusions about the maximum sustainable rate of economic growth: beyond certain levels, growth-theory models predict that inflation and investment distortions will cut off future growth.

See Robert M. Solow, *Growth Theory: An Exposition* (New York: Oxford University Press, 1970).

Participatory Development: approaches to socioeconomic development that emphasize the active participation of affected actors in the planning and execution of development projects and programs. Participatory development calls for the involvement of STAKEHOLDERS in setting objectives of development activities and the control of their implementation, in contrast to the more common "top-down" development efforts characterized by control by government officials, particularly those of the central government. The thrust of participatory development may be to increase the typically low levels of influence of politically and economically marginal populations (i.e., to foster "popular participation") or to ensure the participation of the whole range of affected parties (i.e., "stakeholder negotiation"). The obstacles to participatory development include the difficulties of determining legitimate stakeholders (some argue that including already privileged groups as stakeholders often gives them effective veto power), the transactions costs of participation, and the resistance by government officials to the prospect of losing authority. Nevertheless, participatory development is increasing in many parts of the developing world, hand in hand with decentralization initiatives. See also BOTTOM-UP DEVELOPMENT.

COMMENTARIES ON PARTICIPATORY DEVELOPMENT

Participatory development was the catchword phrase in development lingo for the 1990s. In many places, the strategy has been employed indiscriminately with an eye toward pleasing funding organizations without understanding or appreciating the complexity and resource intensity of the process. Due to these half-hearted efforts, the twenty-first century is sure to see a backlash effect against this potentially useful, but misused tool of development. *Toddi Steelman*

Participatory development is, of course, sometimes subject to superficial implementation. Yet it is still a tremendous advance over the older perspective that people are merely the objects to be benefited—or transformed—by development programs. The significant advance is that genuine participation is now the standard by which participatory efforts are judged. A development agency, whether national, bilateral, or international, runs a risk of serious disapproval if its efforts to include local people in designing its programs are trivial. *William Ascher*

Participatory development has been a very popular term and the buzzword of the 1990s in the vocabulary of donor agencies, government officials, and nongovernmental organizations, who in good faith or in efforts of political and socioeconomic gains strongly support and advocate for a "BOTTOM-UP APPROACH." A number of participatory approaches and methodologies (participatory rural appraisal, participatory mapping) have been developed and put into practice for grassroots and community organization empowerment throughout Africa, Asia, and Latin America. Participation does not simply equate with just informing or consulting (seeking the approval of) stakeholders. Stakeholders often complain that their participation is included only for the implementation phase of a project, but the planning process ignores them in the earlier, crucial periods of project design and selection. *Elsa Chang*

There are many kinds of "participation": passive participation, participation in information giving, participation by consultation, participation for material incentives, functional participation, interactive participation, and self-mobilization. Organizing effective public involvement is not easy and has a mixed history. More case studies are needed that describe the type of participation used and its success. Too often, participation has been construed simply as "getting more citizen input," usually of a passive type. It is clear that natural resources management and sustainable development require more substantive ways to involve citizens.

See M. P. Pimbert and J. N. Pretty, *Parks, People and Professionals: Putting "Participation" into Protected Area Management,* United Nations Research Institute for Social Development Discussion paper DP 57:1–60, 1995. *Tim W. Clark*

Resource-Curse Hypothesis: 1) an assertion that the wealth of mineral resources might be a disadvantage for resource-dependent economies because it distorts the structure of national economies, decreases international competitiveness of other economic sectors, and discourages the adoption of policies designed to secure long-term development. Numerous examples of economic mismanagement with mineral revenues and comparatively poor performance of mineral-rich developing countries since the 1960s seem to support this thesis.

See R. M. Auty, *Sustaining Development in Mineral Economies: The Resource Curse Thesis* (Oxford: Clarendon Press, 1993).

2) the resource-curse thesis has another implication in terms of the CORE-PERIPHERY relationship. Numerous examples of development in the so-called remote resource regions attest that resource abundance of these areas is usually used by the center (whether the U.S federal government in relation to Alaska or the Russian government in relation to the Sakha Republic) without taking into consideration the needs and interests of native inhabitants of these regions, and often with severe consequences for their environmental health.

Rostow's Stages of Economic Growth: one of the most influential models of modernization, postulated by American scholar and politician Walt W. Rostow in 1960. According to this model, all societies develop from tradition to modernity in a linear pattern whose logic and direction are preestablished. This pattern of change includes four stages: a) traditional economic and social relations, with backward beliefs and technology; b) introduction to modern technology, like that experienced by eighteenth-century Europe; c) take-off toward modernization when new industries and entrepreneurial classes emerge; d) the stage of modernity when economic growth outstrips population growth and high mass consumption allows the emergence of social welfare. Growth and inequality were seen as temporary trade-offs; inequality was supposed to ensure capital accumulation and growth, so that eventually wealth could "trickle down" to the poor. The "advanced" countries had passed the stage of "take-off," and the developing nations might have followed them if obstacles, such as traditional forms of land ownership, some cultural and religious values, were removed. The model was criticized as being U.S.- or Eurocentric and imperialist; some of its opponents argued that the same pattern of development is not compatible with different cultures. Some feminists (e.g., Braidotti, Charkeiwicz, Hausler, and Wieringa, 1994) criticized it on the grounds that males tend to dominate the prescribed policy process. However, Rostow's model was an important pioneering effort to understand the sequencing of development stages, and shaped the practice of foreign assistance and development discourse for many years. See also DEVELOPMENT.

See Walt W. Rostow, *The Stages of Economic Growth: A Non-Communist Manifesto* (Cambridge: Cambridge University Press, 1960); Rosi Braidotti, Ewa Charkiewicz, Sabine Hausler, and Saskia Wieringa, *Women, the Environment and Sustainable Development: Towards a Theoretical Synthesis* (London: Zed Books, 1994).

Stakeholder: an individual or group that may be affected significantly by development projects or programs. A premise of PARTICIPATORY DEVELOPMENT is that stakeholders have a right to be included in the planning of such efforts, and that their exclusion often undermines the success of development because they will not participate with enthusiasm, or because project or program design is likely to be flawed if it is formulated without the input of key groups. One difficulty of including stakeholders in development planning (often called "stakeholder negotiation") is that the designation of stakeholders with sufficient standing to be included is often debatable.

Trickle-Down Development: the development model or strategy that is optimistic about economic growth when a large portion of national income is initially in the hands of the wealthier portion of the population. The expectation that greater production and growth would result from such a distribution relies on the assumption that wealthier people will invest a larger part of their savings in economically productive activities, thereby generating greater economic growth. Then expansion of employment opportunities will allow lower-income people to benefit from the expanding economy. As a strategy, trickle-down development calls for deemphasizing progressive redistribution (e.g., not tax the rich a larger proportion of their incomes) and perhaps making government credit amply available to the country's largest enterprises.

COMMENTARY ON TRICKLE-DOWN DEVELOPMENT

Some government officials, particularly in the 1960s and 1970s, invoked this reasoning to justify highly unequal income distributions, but few if any development economists who noted the fact that wealthier people tend to have higher savings actually advocated a trickle-down strategy. One problem is that even if the wealthy have higher savings, those savings may not be directed to productive investment within the national economy (e.g., they may opt for safer savings vehicles in developed countries). They also point out that higher wages for the lower-income population may come only after all of the unemployed find productive work; otherwise, wages will remain close to the subsistence level. *William Ascher*

Underdevelopment: an underlying concept of developmentalism (both as theory and practice) first coined by U.S. President Harry Truman in 1949 in a speech before Congress. Truman postulated the larger part of the world as "underdeveloped" while presenting the United States as an archetype of the "developed" world that embodied modernity, progress, and a superlative political model. All countries were supposed to follow the same path of success exemplified by the United States. The concept of "underdevelopment" set the frame for North-South development assistance for years to come. Despite the ideological differences in other realms, the Soviet concept of development also envisioned development as the passage from the state of "underdevelopment" to that of a bright future, albeit a socialist future typified by modernity and technological progress.

1.1.4 ECONOMIC PRODUCTION, INVESTMENT, AND CYCLES

Allocative Efficiency: efficiency in the assignment of resources for economic activities. The importance of the concept of allocative efficiency lies, first, in the insight that high volumes of resources, whether capital, land, other natural resources, technology, or labor, do not guarantee high levels of output unless the choices of resource allocation are made intelligently. This insight brought into question the simple formulations of "capital-output ratios" that were developed in the early post–World War II period to quantify how much capital was necessary to secure specified rates of economic growth. Second, the concept of allocative efficiency highlights the importance of efficient mechanisms for allocating resources across the economy.

COMMENTARY ON ALLOCATIVE EFFICIENCY

In principle, this is a sound concept. In practice it is an illusion fueled by relative prices. If taken lightly, it may result in environmental colonialism, since the underpricing of raw materials in an oligopolistic market is fairly common. If the price of minerals is controlled by a few importing firms to a level lower than it would be under competition, then allocative efficiency would induce poor exporting countries to deplete a resource or to exploit it without investing in environmental safeguards, since the lower price would make sound natural resource management unprofitable. In these cases, the principle of allocative efficiency would be invoked to legitimize unwise environmental exploitation. *Gustavo J. Arcia*

Decreasing Returns to Scale: declining rises in output for each additional unit of input. For some production processes, at a particular production level the efficiency begins to decline (for example, when more costly equipment is required). When returns to scale are decreasing, greater inputs still yield higher outputs, but the average costs per unit will increase. This principle is important for understanding why "scaling up" to bigger factories or farms sometimes reduces profitability.

Diminishing Returns (Law of Diminishing Returns): the phenomenon of smaller increases in outputs as inputs of one factor of production are increased. At certain points in the production process, under a given technology, adding more of one factor (e.g., the number of workers) without increasing other factors (e.g., the number of machines) will be less efficient in terms of output per unit of input. *R. P.*

Economic Efficiency: broad concept of achieving the greatest output per unit of input. This typically involves applying the best mix of inputs (ALLOCATIVE EFFICIENCY) and minimizing waste in productive activities, in order to achieve the returns on investments of capital, labor, etc. Economic efficiency can pertain to the entire economy or specific activities. For the entire economy, economic efficiency is approached when resources are devoted to efforts with the highest rates of returns; this requires examining the opportunity costs of seemingly good investments in denying investment for better opportunities. For specific productive activities, economic efficiency in the level of production is reached when the last unit produced (the "marginal" unit) has costs equal to its benefits. When lower production leaves marginal costs less than marginal benefits, then more production would provide greater efficiency; when higher production puts marginal costs above marginal benefits, then the loss detracts from economic efficiency. This marginal principle applies to natural-resource uses in explaining why excessive resource extraction reduces economic efficiency.

COMMENTARY ON ECONOMIC EFFICIENCY

Efficiency can also be understood as the driving moral value of utilitarianism since the Benthamite principle of the greatest happiness of the greatest number requires ensuring the largest possible pie. Any decrease in the size of the pie means less happiness to spread around. Unfortunately, however, utilitarians have never been able to articulate a consistent theory of distribution for the pie. This tends to exacerbate their inherent conflict with other moral principles such as the Kantian concern for individual rights. *Steve Rayner*

Industry: a term used to denote any set of firms producing similar products or services; or, more narrowly, economic activity producing physical outputs, particularly in manufacturing, mining, and energy production. For economic development theory, the perennial question is how much emphasis should be placed on promoting industry defined in this narrower sense, if this promotion must come at the expense of other sectors such as agriculture, or through inflation or indebtedness. A related question is whether industrialization per se is even a desirable trend, given that industrialization and economic growth sometimes do not go hand in hand. The environmental issues for a given industrial activity are whether it can be undertaken sustainably; and whether the ENVIRONMENTAL EXTERNALITIES are appropriately addressed in launching and regulating the industry.

Informal Economy: a label applied to businesses and workers operating outside the formal sector of the national economy. Informal-sector activity is usually beyond most government regulation, and its activities are not captured by national economic statistics. Despite the fact that it is often ignored or suppressed by many governments, the informal sector exists practically everywhere and is responsible for a significant share of national employment and production (especially in developing countries and societies in transition). Informal sector businesses traditionally avoid or evade governmental accounting and regulations, especially regarding taxes, workers' benefits, property titles, etc. On the other side, in some contexts it may be more economically efficient, flexible, and effective in promoting employment opportunities than the formal economy.

Kondratieff Cycles: temporal cycles of economic development first identified by Russian economist Nikolai Kondratieff (1892–1938). On the basis of historic fluctuations in commodity prices, Kondratieff distinguished long waves of economic expansion and contraction in the core areas of world economy, accompanied by substantial changes in the pattern of resource use. Each cycle of approximate duration of fifty to fifty-five years has involved different industrial products and resource demands, including food, raw materials, and energy. The first wave identified by Kondratieff coincided with the Industrial Revolution in Britain when wars and high prices for resources dampened economic growth and encouraged agricultural expansion. The material basis of the industry changed from organic and abundant materials to nonorganic and scarce materials, with substantial implications for the spatial patterns of resource use. This was followed by a trough in the 1840s when commodity prices declined while economic growth rates increased, with the cycle repeating itself later during the industrial expansion in Europe. Kondratieff cycles have traditionally been a subject of controversy surrounding their nature and especially their determinants. Most debate addresses whether the cycles are regular enough to be predicted, and determining the precise nature of the interactions among cycles, technological innovations that occur at their start (such as railroads or computers), and the dynamics of economic and political activities. The introduction of new technologies may induce new types and levels of resource demands, resulting in increasing scarcity of resources and higher prices, in turn reducing growth rates. Kondratieff waves may also be linked to cycles of environmental impact and fluctuations in environmental concern. While the latter is still debatable, there is little disagreement that successive waves of economic growth have occurred during the last two centuries and that each involved different types of environmental resources and affected spatial patterns of resource-use activities.

See Nikolai Kondratieff, "The Long Waves in Economic Life," trans. Wolfgang Stolper, *Review of Economics and Statistics*, vol. 17, part 2 (November 1935); James Shuman and David Rosenau, *The Kondratieff Wave* (New York: World, 1972); Michael J. Healey and W.Brian Illbery, *Location and Change: Perspectives on Economic Geography* (Oxford: Oxford University Press, 1990).

Market Failure: 1) the inability of market mechanisms to produce economically efficient outcomes. The existence of market failures under certain circumstances lends support for the view that the market does not provide a panacea for all economic problems. There are various ways in which an unregulated market may fail: imperfect information, unclear property rights, monopoly, and EXTERNALITIES (public goods and public bads for which private actors lack incentive to address). For instance, profit maximization under monopoly implies underprovision and overcharging for goods. Market failure provides for considering public regulation of some goods, public provision of others, and redistribution of incomes—the measures that are in turn exposed to potential failings.

2) In the environmental context, market failure usually refers to the fact that the economic values of environmental resources and services are not fully reflected by market prices. From that perspective a distinction is made between local and global market failure. The former relates to the inability of markets to capture some of the local or national benefits of biodiversity conservation or pollution control. The latter concept relates to the fact that biodiversity conservation yields external benefits to people outside the boundaries of the nation faced with the development/conservation choice. *R. P.*

COMMENTARY ON MARKET FAILURE

It is often forgotten that, as the economist Oliver Williamson observed, the failure of markets implies the success of something else. This implies that policies designed to make the world conform to the market model (correct market failure) may not be as effective as those that attempt to harness the existing structures to sustainable development goals. *Steve Rayner*

Opportunity Cost: the benefit forgone by using a scarce resource for one purpose instead of for its best alternative use. The concept of opportunity cost is the pivot of cost-benefit analysis (see BENEFIT-COST ANALYSIS) and project evaluation, because it permits the valuation of inputs and relates the pursuit of any particular investment to the range of possible alternative actions. The most prominent element is usually the opportunity cost of capital; short of assessing the return on capital from each potential alternative investment, the standard approach is to assume that the prevailing interest rate, insofar as it is a free-market rate, represents the opportunity cost. For natural resources and ecosystem elements, the opportunity cost of leaving the resource intact is the highest-yield use of the resource and the land; the opportunity cost of extracting the resource is the environmental service that it would have provided if left intact. *R. P.*

Overshooting: excessive investment in a particular economic activity, sometimes held responsible for boom-and-bust economic cycles in a particular industry or even in an entire economy. When a particular economic activity seems to have a high rate of return, investment may be forthcoming from many sources. If the profit and price signals are sluggish in revealing that so much production is reducing the rate of return, investment may continue beyond the point at which it is adequately productive. Investment may therefore be wasted, with possibly serious consequences for the economy as a whole, and the withdrawal of investment from that particular activity may then underserve its investment needs in the next time period. Resource sectors are particularly prone to overshooting because of uncertainty about the usable supply of natural resources and the length of time required to develop and extract them.

Price System: the prevailing set of prices for goods and services in an economy. The price system plays a central role in economic theory, particularly classical and neo-classical theory, because of the importance of prices in establishing the returns on particular investments. A crucial premise of mainstream economics is that when the price system sends accurate signals about the demand and scarcity of goods and services, investment will be directed into the most productive economic activities, because these activities will produce outputs that are priced to provide appropriate profits. A further premise is that the price system will send such signals if the economy is competitive and the government permits producers to set prices. However, the prices of some goods and services, in particular prices of public goods or for activities that produce positive or negative affects on others (positive or negative EXTERNALITIES), will not reach their optimal levels from market forces alone, and may require government intervention in setting or influencing their prices. In calculating the prices that should be used to evaluate the benefits and costs of particular goods and services, existing prices may reflect policy distortions that do not reflect the societal value of the goods and services. In such cases, analysts use "shadow prices" that correspond more closely to the prices that would prevail in a competitive market economy. This applies as well to environmental externalities that persist only because of an implicit subsidy to the polluter. *R. P.*

Production Function: model of the relationship between the volume of various inputs and the output of a particular good or service. The inputs (usually termed "factors of production") typically include capital, labor, land, natural resources, technology, etc. Production functions are frequently used to determine the optimal input mix and scale of production for a particular industry. Some production function models, including the widely used Cobb-Douglas production function, reflect constant returns to scale, while others reflect increasing or diminishing returns to scale. When applied to entire economies, production function models (often called aggregate production functions) try to gauge the impacts of total capital, labor, natural resources, etc., on total production. There is much debate on the meaningfulness of aggregate production functions, because of limitations in the measurement of aggregate inputs and the assumption that a particular form of the input-output relationship can hold for so many distinct industries.

Resource Cycles: a set of arguments that economic cycles of boom and recession are driven by investments and production in major resource sectors. When resource prices are high and supply is limited, investment is likely to boom in the resource sector, leading to eventual gluts, price collapses, and the flight of investment to other sectors. This pattern sets up another cycle of scarcity, investment, etc. See also KONDRATIEFF CYCLES.

Subsidies: direct transfers, favorable pricing policies, or favorable tax treatments for particular economic activities. Subsidies have been a key instrument for steering the economy by making particular types of investments attractive and by altering prices (e.g., low-priced, subsidized energy will make energy-intensive industry more attractive for investors). However, neoclassical economic theory argues that only a narrow range of circumstances justify subsidies, specifically when economic activities that would produce positive EXTERNALITIES are not attractive enough to induce these activities without additional incentive. Even under this circumstance, the subsidies must be of less value than the value of the positive externality, and the subsidy ought to be limited to the amount necessary to make the return on these activities marginally greater than alternative activities. Otherwise, subsidies distort prices and thereby lead to inefficient allocations of investment. *R. P.*

COMMENTARIES ON SUBSIDIES

While we need subsidies in certain circumstances, and will surely always do so, we could do without those many subsidies that are "perverse," that is, detrimental to both the economy and the environment. German coal miners, for instance, all 89,000 of them, are subsidized by the government to the extent of $76,000 each per year. It would be economically efficient for the government to close down all the mines and send the miners home on full pay for the rest of their lives. At the same time, Germany's environment would benefit through a massive decline in fossil-fuel pollution. Such perverse subsidies abound in several leading sectors, not only fossil fuels (and nuclear power), but road transportation, agriculture, water, fisheries, and forests. Worldwide these silly subsidies total almost $1.5 trillion per year. Thus they cause a great distortion in the global economy of $30 trillion, while causing environmental degradation on a spectacular scale.

Fortunately, a number of countries have started to phase out these subsidies. New Zealand (the developed country most dependent on agriculture) has eliminated virtually all its agricultural subsidies on the grounds that the subsidies had grown so large they were breaking the back of the national economy. Russia, China, India, Poland, Belgium, and Great Britain have all slashed their fossil fuel subsidies. South Africa is cutting back on its water subsidies. Several European countries are trying to get their car industries off what amounts to grand-scale welfare. Brazil no longer subsidizes the torching of Amazonia for cattle ranches.

But altogether these reductions amount to less than 5 percent of all perverse subsidies. Were governments to phase out even half of these subsidies, the funds released would enable governments to reorder their fiscal priorities in a fundamental fashion. At the same time, they would do more to restore and safeguard environments than through any other single measure. The subsidies represent funds that, by definition, foster unsustainable development. At the 1992 Rio Earth Summit, a budget was presented for sustainable development of $600 billion per year—which governments dismissed as unthinkably costly. Yet they could have found two and a half times this sum if they had looked at their perverse subsidies. *Norman Myers*

Subsidies are probably the most common cause of environmental damage, especially but not exclusively in developing countries. The main culprits are subsidies to polluting fuels such as coal, diesel, and gasoline or products such as electric power and water whose consumption or production is associated with pollution. The reduction or elimination of these subsidies offers significant scope for "win-win" actions whereby both environmental quality and economic efficiency can be simultaneously improved. In the most egregious cases, such as subsidies during the 1980s to coal use in the former Soviet Union and Eastern Europe, the environmental improvements that would have resulted from eliminating these subsidies far outweigh those that might have come from the improved application of targeted environmental policies. Since economic efficiency would also be enhanced, it is argued that where these distortions are serious, their removal should precede attempts to implement targeted environmental policies. *Sudhir Shetty*

1.1.5 HUMAN RESOURCE DEVELOPMENT

Basic Needs: the goods and services that people must have to meet some minimal standard of living. Basic needs are widely thought to include primary health care, basic education, family planning, nutrition, water, sanitation, and shelter, as well as emergency humanitarian assistance. Basic needs may be defined on a narrower basis as "those necessary for survival" or more broadly as "those reflecting the prevailing standard of living in the community." While the first criterion would apply to people who are practically near death, the second would cover those whose nutrition, housing, or access to services, though adequate to preserve life, do not enjoy a standard of living that measures up to the standards of the population. According to the United Nations Human Development Report in 1995, "basic social services—primary health care, basic education, safe drinking water and adequate nutrition—are not available to more than one billion people." The satisfaction of basic needs became the focus of development policy in the 1970s, most closely associated with Paul Streeten and Francis Stewart, following several decades of focus on macroeconomic indicators. While the basic needs approach puts primary emphasis on fundamental poverty alleviation, it also has been defended as a means to enhance human resources for long-term economic growth.

See Paul Streeten (with Shahid Javed Burki), *First Things First: Meeting Basic Human Needs in the Developing Countries* (New York: Oxford University Press, 1981). *T. S.*

COMMENTARIES ON BASIC NEEDS

This principle invariably overstates poverty. It confuses a *need* with the inability of local institutions to provide a basic good or service. It also confuses lack of access with inability to pay for a basic need or service. If a rural family lacks potable water, it may be because they are poor, or because they are not poor, but the rural water institution is totally inept. However, under the basic needs approach the only outcome would be that the family is poor. The concept is useful for determining public investment needs, for simple geographic targeting of public investment, and for dealing with the correlates of poverty, but not for poverty reduction. *Gustavo J. Arcia*

Basic needs of food, shelter, education, and a safe, healthy environment are essential to ensure sustainability and humans' well-being. Therefore, meeting these needs is now considered an undeniable human right. More programs and projects are now increasingly focused on addressing poverty alleviation and the satisfaction of immediate human basic needs, while at the same time working toward the long-term goal of safeguarding the environment. *Elsa Chang*

Human Capital. See under SECTION 1.1.1.

Human Development: the broad concept of progress in the material, intellectual, and spiritual development of people, often cast in terms of hierarchies (e.g., deference values and spiritual values such as respect, affection, and self-actualization rest on a base of physical well-being) and the expansion of choice. Thus, conceptions of human development often emphasize environmental concern as a relatively advanced stage of awareness that is more likely to arise once a modicum of material prosperity has been reached. See also HUMAN DEVELOPMENT INDEX (HDI).

Human Resource Development: the improvement of human skills and physical capacity. Human resource development is important both for increasing the contribution of labor to economic production, and for its own sake in terms of the intrinsic value of more enlightened, skillful, and healthy people. A human resource development strategy emphasizes investments in health, nutrition, education, and training that makes the workforce, particularly among those previously of low-income levels, more productive, while at the same time addressing BASIC NEEDS. Many East Asian countries, such as Japan and South Korea, heavily emphasized human resource development in their development strategies.

1.1.6 MEASUREMENTS OF ECONOMIC DEVELOPMENT AND GROWTH

Adjusted Income: income calculations modified to reflect cash transfers and in-kind transfers as well as market income (the actual amount of money earned by an individual or household). Adjusted income provides a more realistic measure of living standards. When income levels are to be compared across different countries, they can also be adjusted to reflect purchasing power. These purchasing-power adjustments for low-income countries like India reveal that ordinary per-capita income measures tend to underestimate purchasing power, especially when per-capita income comparisons are based on official statistics.

See Irving B. Kravis, Alan Heston, and Robert Summers, *International Comparisons of Real Product and Purchasing Power* (Baltimore: Johns Hopkins University Press, 1978). *R. P.*

Country Futures Indicators (CFI): a set of indicators of development proposed by economist Hazel Henderson in an attempt to correct deficiencies of GROSS NATIONAL PROD-

UCT (GNP). Now registered as the Calvert-Henderson Quality of Life Indicators (to develop her ideas further, Henderson formed a partnership with the Calvert Group, the U.S. asset-management firm), the CFI focuses on twelve very broad dimensions of quality of life to reflect how well a country is investing in its own people, whether it is investing in the resource base, and whether it is approaching sustainability. Among the criteria are social indicators (unpaid work, poverty gaps, provision of basic services, human rights observance), environmental indicators (depletion of nonrenewable resources, degradation of environmental resources, and quality of human habitat), and some measures of government effectiveness. The CFI try to meet the challenge of measuring development better than many other suggested criteria by being interdisciplinary, transparent, and understandable to the general public.

See Hazel Henderson, *Creating Alternative Futures: The End of Economics* (West Hartford, CT: Kumarian Press, 1996).

Gender-Empowerment Measure (GEM): one of the UNITED NATIONS DEVELOPMENT PROGRAMME (UNDP) indices of human development in different nations that indicates whether women are able to participate actively in economic and political life. It focuses on gender inequality in key issues of economic and political participation and decision making. In 1997 Nordic countries topped GEM rankings: these countries are successful both in strengthening basic capabilities for all their citizens and in opening opportunities for female participation in economic and political fields. Some developing countries outperform much richer industrial nations in terms of gender equality: Barbados is ahead of Belgium and Italy, while Japan ranks thirty-fourth in GEM behind many developing countries and countries in transition. Comparative analysis of data shows that there is a strong association between the extent of poverty and opportunities for women. However, gender disparity and income poverty are not necessarily connected.

See United Nations Development Programme, *Human Development Report 1997* (New York: Oxford University Press, 1997).

Gender-Related Development Index (GDI): one of the indices proposed by the UNITED NATIONS DEVELOPMENT PROGRAMME (UNDP) to measure achievement of basic capabilities by different countries from the gender perspective. GDI uses the same variables as the HUMAN DEVELOPMENT INDEX (HDI) but adjusts the average achievement of each country in life expectancy, educational attainment, and income for gender inequality.

COMMENTARY ON THE GENDER-RELATED DEVELOPMENT INDEX

Comparing HDI with GDI demonstrates that: a) no society treats its women as well as its men—the GDI value for every country is lower than its HDI value; b) gender inequality is strongly connected with human poverty; c) however, it is not always associated with income poverty; and d) gender equality can be achieved at different income levels and stages of development as well as across the range of cultures and political ideologies. In 1997 Canada and the Nordic countries topped the GDI rankings. Bulgaria, which was sixty-ninth in HDI, ranks forty-ninth in GDI and twenty-seventh in GEM—women from the former socialist bloc seem to show relatively high levels of resilience under the pressures of political and economic transition. The bottom five places in gender-related development are occupied by Sierra Leone, Niger, Burkina Faso, Mali, and Ethiopia: women in these countries face double deprivation.

See United Nations Development Programme, *Human Development Report 1997* (New York: Oxford University Press, 1997). *Natalia Mirovitskaya*

Genuine Progress Indicator (GPI): a measure of economic welfare based on the INDEX OF SUSTAINABLE ECONOMIC WELFARE (ISEW). It is thus an alternative to the GROSS DOMESTIC PRODUCT (GDP), which focuses on the value of goods and services without taking sustainability into account.

Genuine Saving: a measure of the new wealth created in the process of economic development. It is assessed as GROSS DOMESTIC PRODUCT (GDP) less consumption, depreciation of produced assets, and the costs of drawing down natural resources.

Gini Index of Inequality: a measure of the inequality in the distribution of any attribute, though most commonly applied to income or wealth. It is derived geometrically by measuring the area between the curve defining the actual cumulative distribution, beginning with poorest (LORENZ CURVE), and the line of complete equality. The Gini index is formulated such that a completely unequal distribution (i.e., one recipient obtains everything) will have a value of 1.0, and a completely equal distribution (i.e., everyone receives the same level of the benefit) will have a value of 0. Gini coefficients are often used to compare the income distributions within nations, with values typically ranging from around 0.2 to 0.5. However, the comparability is sometimes compromised by the fact that different coefficients will emerge depending on whether the recipients are defined as individual income earners, households, etc. Because the Gini index describes the overall difference between the Lorenz curve and the line of equality, systems with quite similar Gini coefficients may have different degrees of inequality over different segments of the distribution.

Gross Domestic Product (GDP): the monetary value of all goods and services produced in a country for a given year. "Real GDP" is assessed in constant—inflation-adjusted—currency units.

Gross National Product (GNP): the monetary value of all goods and services produced in a country for a given year, plus net property income from abroad. Traditionally, GDP and GNP per capita were regarded as the most important measures of a nation's economy. Economic policymakers indeed would be lost without a sense of the trends in GNP and GDP. However, such aggregate measures, even if averaged as per capita measures, do not convey anything about the distribution of income. In addition, GDP does not measure the contributions of uncompensated labor (e.g., women's household labor), and considers all expenditures as "contributing" even if they reflect economic or social losses, such as the recovery costs following a hurricane. Therefore, many alternative indicators of economic performance have been formulated.

COMMENTARY ON GROSS DOMESTIC PRODUCT (GDP)

A gross form indeed of measuring human welfare. To quote Robert Kennedy in 1968, "The gross national product does not allow for the health of our children, the quality of their education, or the joy of their play. It does not include the beauty of our poetry or the strength of our relationships; the intelligence of our public debate or the integrity of our public officials. It measures neither our wit nor our courage; neither our wisdom nor our learning; neither our compassion nor our devotion to country; it measures everything, in short, except that which makes life worthwhile." Moreover, it tends to highlight marketed goods and services, whereas the poorest one fifth of humankind, those living on a cash income of $1 per day, ostensibly account for only 1.5 percent of GROSS WORLD PRODUCT, whereas many of their goods and services simply do not register in the marketplace. By the same token, these people carry next to no economic clout, and have scant scope to register their basic needs through their dollar votes. Note too that the aggregate economic value of all environmental goods and services has recently been estimated (albeit a first-cut guesstimate) at $33 trillion, meaning that gross natural product is greater than gross national product, $30 trillion, worldwide. *Norman Myers*

Gross World Product: monetary value of all goods and services produced in the world. The concept is thus parallel to that of GROSS DOMESTIC PRODUCT (GDP) and GROSS NATIONAL PRODUCT (GNP) for nations, with similar limitations in its inability to speak to distributional issues or to distinguish benefit-enhancing expenditures from others.

Human Development Index (HDI): an aggregate index for measuring a country's human progress (rather than simply its economic performance), emphasizing health and education in addition to income. It was launched by the UNITED NATIONS DEVELOPMENT PROGRAMME (UNDP) in 1990 to "ensure that development planning is directed to people's needs." The HDI assesses comparative quality of life on the basis of weighted average of three indicators: life expectancy, educational accomplishment (combination of adult literacy and enrollment ratios), and standard of living (measured by per capita adjusted income). On the basis of HDI, nations are assigned to the categories high, medium, and low development. For instance, in 1997 Canada, Mexico, and Belarus were rated in the high category (64 nations); the Russian Federation, Turkey, and China in the medium category (66 nations); and India, Bangladesh, and Guinea in the low category (44 nations). Critics of this index argue that HDI does not take into consideration the environmental dimension (resource availability and pollution levels) as well as investments into social development, political freedoms, and human rights, which are important prerequisites of development. Therefore, they stress the need to find quantitative measures for such phenomena as free elections, multiparty political systems, absence of censorship, etc. However, the introduction of HDI has been instrumental in clarifying the distinction between means and ends in the process of development.

See United Nations Development Programme, *Human Development Report* (New York: Oxford University Press, 1997). *R. P.*

Human Poverty Index (HPI): a composite set of measures reflecting different features of deprivation in the quality of life. The components include the percentage of people expected to die before age forty, the percentage of adults who are illiterate, the percentage of people without access to health services and safe water, and the percentage of underweight children under age five. Like HDI, HPI does not reflect all critical dimensions of people's well-being, such as political freedom, ability to participate in decision making and the life of the community, personal security, and threats to sustainability and intergenerational equity. Nevertheless, like HDI, this index may serve as a useful complement to economic measures of poverty. Its authors advocate its use in at least three ways: as a tool for advocacy to mobilize public opinion and support; as a planning tool for identifying areas of concentrated poverty within a country; and a research tool in the assessment of the level of development.

See United Nations Development Programme, *Human Development Report 1997* (New York: Oxford University Press, 1997).

Index of Sustainable Economic Welfare (ISEW): a measure of broadly defined economic welfare proposed by Herman Daly and J. B. Cobb. The ISEW adjusts personal consumption by adding desirable services such as household production and subtracting regrettable expenditures, environmental pollution, and other welfare losses, for instance, from unemployment. In 1994 the name of this index was changed to the GENUINE PROGRESS INDICATOR.

See Herman E. Daly and John B. Cobb, *For the Common Good: Redirecting the Economy towards Community, the Environment and a Sustainable Future* (Boston: Beacon Press, 1989).

Indicators of Development: broad measures of economic development and growth. A large number of indicators of development emerged from the limitations of the original and more conventional measures of the growth rates of GROSS NATIONAL PRODUCT (GNP) and GROSS DOMESTIC PRODUCT (GDP). These measures do not capture distribution, sustainability, or a host of other aspects that can be considered as important aspects of development, such as environmental quality, institutional development, or greater equality for women. Many other measures have been created to capture these elements, yet each indicator, in emphasizing some aspects of development more than others, reflects a balance of values that some will reject. Despite the criticism leveled against any particular development indicator, the desire to evaluate overall economic performance keeps up the demand for development indicators.

Lorenz Curve: the geometric representation of the cumulative proportion of income, wealth, or other attribute across the distribution of recipients of that attribute, beginning with the least indulged to the most. In highly unequal distributions, the curve starts out flat and rises sharply toward the end of the distribution that represents the share gained by the wealthiest. The more equal the shares across recipients, the closer the Lorenz curve comes to a straight diagonal line (the so-called line of perfect equality). Therefore, the area between the Lorenz curve and the line of perfect equality is the basis for the aggregate measure of inequality, the GINI INDEX.

Measure of Economic Welfare (MEW; or Net Economic Welfare [NEW]): adjusted indicator of total national output that includes only the consumption and investment items directly contributing to economic well-being. MEW is calculated as additions to GROSS NATIONAL PRODUCT (GNP), including the value of leisure and the underground economy, and deductions like environmental damage.

See Paul Samuelson and William Nordhaus, *Economics*, 14th ed. (New York: McGraw-Hill, 1992).

Natural Capital Indicator: a measure of the noncommercial portion of natural assets of a country, suggested by a research team of the WORLD BANK. The natural capital indicator is designed to define a country's natural capital that is not already captured in conventional economic indicators for the country and defines each country's part of the world's total of remaining natural areas, adjusted for its biodiversity richness. This indicator can be used for several purposes, such as international analytical comparisons, policy formulation, calculation of financial burden sharing, and designing decision-making mechanisms. This indicator does not take into account the quality or condition of resources.

See E. Rodenburg, D. Tunstall, and F. van Bolhuis, *Environmental Indicators for Global Cooperation* (Washington, DC: Global Environment Facility Working Paper #11, 1995).

Net Domestic Product (NDP): the remaining value of domestic goods and services (GROSS DOMESTIC PRODUCT [GDP]) after deductions of the depreciation of humanmade capital. NDP does not, however, take into account the depletion or degradation of natural resources. Thus a nation may deplete its resource base while its NDP or GDP is growing. To capture the significance of resource depletion, measures of RESOURCE-ADJUSTED NET DOMESTIC PRODUCT (RANDP) can be used. *R. P.*

Pollution-Adjusted Economy Indicator: a measure of nonmonetized negative environmental services provided by a country to a global community (in terms of pollution of global commons). This indicator has been suggested by a research team of the WORLD BANK, as an attempt to define "shares" of different countries in global environmental services. In particular, the countries that are high emitters of carbon dioxide use more than their share of the limited sink capacity of the atmosphere, without paying for its use. The proposed indicator reflects a carbon dioxide pollutant factor to GROSS NATIONAL PRODUCT (GNP) and is calculated by multiplying current GNP by the ratio of average per capita carbon emissions divided by the actual per capita emissions for the same year. According to its designers, this indicator can be used in combination with a NATURAL CAPITAL INDICATOR and GNP as a classical measure of economic weight to determine environmentally adjusted country shares. This might be useful in developing global environmental cooperation.

See E. Rodenburg, D. Tunstall, and F. van Bolhuis, *Environmental Indicators for Global Cooperation* (Washington, DC: Global Environment Facility Working Paper #11, 1995).

1.2 Natural Resources and Environmental Services

1.2.1 CARRYING CAPACITY

Absorptive Capacity, Environmental (also Assimilative Capacity): the ability of the environment to accommodate waste products from human activities, within acceptable costs and risks to quality of life and economic opportunities. These activities include industrial and agricultural production and residential waste disposal. If the flow rate of wastes and residuals is in excess of assimilative capacity of the environment over time, the latter declines and may eventually go to zero. Some sites have greater absorptive capacity than others because of their physical characteristics (e.g., larger bodies of water such as oceans and seas can accommodate effluents more easily than smaller bodies of water). The limit of absorptive capacity as defined by health considerations often limits further economic expansion. Determining the level of absorptive capacity is not strictly technical, because it depends on the society's tolerance for pollution, health risks, deterioration of aesthetics, etc. Declining assimilative capacity of environmental media can be halted though not reversed by reducing the residual flow rate.

Assimilative Capacity. See ABSORPTIVE CAPACITY.

Carrying Capacity: the maximum population or levels of activity that an ecosystem can sustain and still retain particular desirable characteristics. Carrying capacity can be expanded by technology, but is limited ultimately by the system's capacity to renew itself or to absorb wastes safely. Carrying capacity is typically calculated by determining which of the necessary inputs will first emerge as the LIMITING FACTOR, and then determining the population or activity level possible given that constraint. Such constraints as food, water, and oxygen are typically limiting factors, although sheer space can also be a limiting factor due to the frequently deleterious effects of overcrowding (Calhoun, 1963; 1994). A system's carrying capacity—especially HUMAN CARRYING CAPACITY—is also a matter of preferences and priorities, in that the tolerance for particular ecosystem changes determines how much impact on the system will be regarded as acceptable (e.g., the accepted carrying capacity of an island where wild horses have been introduced depends on the public's [and the manager's] views on how much the vegetation of the island ought to be allowed to be degraded). See also ABSORPTIVE CAPACITY.

See David Price, "Carrying Capacity Reconsidered," *Population and Environment* 21 (1) (September 1999): 5; John B. Calhoun, *The Ecology and Sociology of the Norway Rat* (Bethesda, MD: U.S. Dept. of Health, Education, and Welfare, Public Health Service, 1963); John B. Calhoun, "Lemmings' Periodic Journeys Are Not Unique," *Focus: Carrying Capacity Selection* 4 (1) (1994): 78.

COMMENTARY ON CARRYING CAPACITY

The concept of carrying capacity was originally devised by ecologists to describe limits upon the size of deer herds arising from the availability of forage pasture. However, the capacity of animals for reflexive adaptation and collective action is generally considered to be more limited than that of humans. For instance, humans may seek to secure their food supply through agricultural innovation or trade. Hence, social scientists generally consider the notion of carrying capacity as underestimating the contribution of human ingenuity to sustainable development. Among the strongest opponents of the use of carrying capacity in social science was the late Julian Simon.

See Julian L. Simon and Herman Kahn, *The Resourceful Earth* (Oxford: Blackwell, 1984). *Steve Rayner*

Critical Environment: environment in which the extent or rate of environmental degradation precludes the maintenance of current resource-use systems or levels of human well-being, given feasible adaptations and the community's capacity to mount a response. The concept of environmental criticality, its definition, and assessments have been one of major point in environmental debate ever since Russian geographers began to produce a series of maps of critical environmental situations in the late 1980s.

See Jeanne Kasperson, Roger Kasperson, and B. Turner II. "Regions at Risk," *Environment* (December 1996): 4–15.

Critical Load: the level of exposure to one or more pollutants above which significant harmful effects on specifically sensitive elements of the environment occur. Determining the critical load establishes the critical levels of pollutants that must be taken into account in setting environmental standards. *R. P.*

Ecological Footprint: land and water area required to support the lifestyle of the particular population. The ecological footprint may be conceived as the inverse of the CARRYING CAPACITY of the territory.

See Mathias Wackernagel and William Rees, *Our Ecological Footprint* (Gabriola Island, BC and Philadelphia, PA: New Society Publishers, 1996.)

Ecological Integrity: an ability of a living system, when subjected to disturbance, to sustain and recover to a normal biomass end-state. Ecological integrity is conceptually similar to RESILIENCE and STABILITY.

Environment: the complex of natural objects and forces, within which humans live and which supports and limits human development. There is ongoing discourse on what components should be included into this notion (in particular whether natural resources and living things should be considered part of it) and whether human beings are part of nature or not. The British Environmental Protection Act of 1990, for instance, defines environment as "all of the following media, namely air, water, and land"; while the New Zealand Resource Management Act describes it as "ecosystems and their constituent parts, including people and communities; all natural and physical resources, amenity values, and the social, economic, aesthetic, and cultural conditions which affect the matters stated . . . above or which are affected by those matters." Natural and physical resources in this definition include land, water, air, soil, minerals, and energy, all forms of plants and animals (whether native or introduced), and all structures. In general, environment is a socially constructed idea, the precise definition of which depends upon the purposes of the author.

See John Adler and David Wilkinson, *Environmental Law and Ethics* (London: Macmillan, 1999), 8–9.

Environmental Space: a basic concept developed by the Dutch scholar Johannes Baptist Opschoor to denote the limits within which human beings can use the NATURAL ENVIRONMENT without doing lasting harm to its essential characteristics. The scope of environmental space depends upon the CARRYING CAPACITY of the ecosystem, the availability of natural resources, and their recuperative capacity. Therefore, environmental space depends on the abundance of resources and ABSORPTIVE CAPACITY. It has also been used (though in a slightly different form) by the South Centre in its presentation of the developing countries' concerns to the UNITED NATIONS CONFERENCE ON ENVIRONMENT AND DEVELOPMENT (UNCED) and is often found in the writings of Indian and Canadian scholars. The notion of environmental space is crucial for the elaboration of operational criteria of SUSTAINABLE DEVELOPMENT. It is often argued that each country has a right to the same amount of environmental space per capita.

See Johannes Baptist Opschoor, *Environment, Economics and Sustainable Development* (Groningen, 1992).

Limiting Factor: a factor or condition that limits or controls the abundance or distribution of a species. The concept is one of the oldest in ecology. Its origin is traced to the German chemist Justus Liebig, who in 1840 postulated that "growth of a plant is dependent on the amount of foodstuff which is presented to it in minimum quantity," the so-called law of the minimum. Limiting factors can be roughly divided into two groups: physical and biological. Physical factors that limit population growth include climate and availability of water. Biological factors involve competition, predators, disease, and other interactions between or within a species. The concept of limiting factors leads to an inevitable though often overlooked conclusion that no species, including humans, can expand its population or resource consumption indefinitely.

See Raymond F. Dasmann, *Environmental Conservation*, 5th ed. (New York: John Wiley, 1984), 76–78.

1.2.2 ECOSYSTEM

Biocenosis: a relatively self-contained association of different organisms in a well-defined natural system. The term was first coined at the end of the nineteenth century by German ecologist Karl Mobius and is still popular in European academia. *Biocenosis* is often used interchangeably with *ecosystem*. See also BIOGEOCENOSIS.

Biogeocenosis: in Russian ecological theory, a combination of flora, fauna, and microorganisms adjusted to each other as well as to the territory with its specific abiotic factors which they occupy. Biogeocenosis is close in its meaning to the contemporary notion of *ecosystem*. However, it is always connected with a particular part of the earth's surface, while an ecosystem might be applied to any system of living and nonliving components, including spacecraft.

Bioregion: large geographical areas sharing similar natural characteristics. Kirkpatrick Sale, a proponent of the term, defines it as "any part of the earth's surface whose rough boundaries are determined by natural characteristics rather than human dictates, distinguishable from other areas by particular attributes of flora, fauna, water, climate, soils and landforms, and by the human settlements and cultures those attributes have given rise to." Bioregions are further subdivided into ecoregions and georegions. Ecoregions encompass "the broadest distribution of nature, vegetation, and soil types" and are distinguished by their "spread of trees and grasses." The North American continent, for example, contains forty ecoregions, most of which cross various political boundaries. It is believed that prior to industrialization population resided mainly within the temporal and spatial parameters of ecoregions. Georegions are smaller ecosystems, such as the Klamath mountain range and Sierra foothills of northern California, that are tucked within ecoregions and share common characteristics as well as exhibit unique features.

See Kirkpatrick Sale, *Dwellers in the Land: The Bioregional Vision* (San Francisco: Sierra Club, 1985).

Biosphere: a term coined by Russian scientist Vladimir Vernadsky to define the earth as an integrated living and life-supporting system. Vernadsky described the biosphere as "the envelope of life, namely, the area of living matter. . . . The biosphere can be regarded as the area of the earth's crust occupied by transformers that convert cosmic radiation into effective terrestrial energy—electrical, chemical, mechanical, thermal, etc." More recently, it has been defined as a "thin layer of soil, rock, water, and air that surrounds the planet Earth along with the living organisms for which it provides support, and which in turn modify it in directions that either enhance or lessen its life-supporting capacity." By this definition, the biosphere extends from ocean depths to the upper stratosphere to an altitude of approximately forty miles.

See Vladimir Vernadsky, *Biosfera* (Moscow, 1926); United Nations Educational, Scientific, and Cultural Organization, *Use and Conservation of the Biosphere* (Paris: UNESCO, 1970).

Cultural Environment: human reshaping of the natural environment. The changes from "natural" environments to "artificial" environments reflect human life styles, resource uses, and waste disposal. Cultural environments, such as agricultural lands, usually are more dependent on external inputs, and have much less biodiversity than their predecessors.

Ecosystem: the set of interconnected living organisms (plants, animals, and microbes) in a given area and the nonliving physical and chemical components of their environment, linked through nutrient cycling and energy flows. For practical purposes, rough boundaries can be established for the ecosystem relevant to a particular area, species, or process, even though all of the earth's natural components are directly or indirectly connected with one another. The relevant ecosystem boundaries with respect to a particular species or process may partially overlap with the appropriate ecosystem boundaries for other species or processes; this is an important complication in defining ecosystem boundaries for research and management.

Natural Environments: ecosystems that are relatively unaffected by human creations, habitation, or influence. Virtually none of the environments on earth are completely natural in this sense; nevertheless, some ecosystems come much closer to being "natural" than others.

Noosphere: the heavily human-shaped biosphere, in which physical and biological conditions and processes are controlled or triggered by human intervention. Many believe that the major trend of the biosphere is evolving into a noosphere. The term was popularized by Russian scientist Vladimir Vernadsky (1863–1945) (see also BIOSPHERE), who defined it as a new geological phenomenon.

See Vladimir Vernadsky, *Biosfera* (Moscow, 1926).

Optimum Sustainable Population: the population size associated with the maximum sustainability, stability, or productivity of one or an assemblage of species. The optimum sustainable population depends on the CARRYING CAPACITY of the habitat, the interactions among species, and the health of the whole ecosystem. What is "optimal" also depends on the preferences of the observer. See also HUMAN CARRYING CAPACITY.

Resilience: ability of a natural system to recover some degree of ecosystem functioning following an outside disturbance. This is a function of several features of ecosystem dynamics, such as the degree of population fluctuations, the functioning of self-regulating mechanisms, the presence of diverse species that permit a regeneration of the assemblage of organisms necessary to restore the ecosystem, etc. An alternative definition of resilience is offered by C. S. Holling, who equates resilience to the capacity of species or ecosystems to survive in a wide variety of conditions (for example, varying climatic conditions). See also STABILITY.

See Stuart Pimm, *The Balance of Nature* (Chicago: University of Chicago Press, 1991); C. S. Holling, "Resilience and Stability of Ecological Systems," *Annual Review of Ecology and Systematics* 4 (1973): 1–23.

COMMENTARY ON RESILIENCE AND STABILITY

Resilience has been defined in two very different ways in the ecological literature. The first concentrates on stability at a presumed steady state, and stresses resistance to disturbance and the speed of return to the equilibrium point. The second definition of resilience concerns the ability of a system to *absorb* perturbations, or the magnitude of disturbance that can be absorbed before the structure of a system is changed. The second definition is based on a view of ecosystems as nonlinear, multiequilibrium, and self-organizing.

See C. S. Holling, "Resilience and Stability of Ecological Systems," *Annual Review of Ecology and Systematics* 4 (1973): 1–23; Lance Gunderson, C. S. Holling, and Stephen S. Light, eds., *Barriers and Bridges to the Renewal of Ecosystems and Institutions* (New York: Columbia University Press, 1995). *Fikret Berkes*

Seminatural Environment: a habitat where the basic type of vegetation differs from the natural environment, yet species composition has not been altered intentionally but occurred spontaneously. A seminatural environment may therefore be seen as an intermediate category of the degree of human impact on the environment. The most common modification is deforestation that has resulted in such new ecosystems as grasslands, such as the American prairies and African savanna, and grazing lands. Some seminatural environments can be rich in biodiversity and have high conservation value.

Stability (of Ecosystem): resistance to disturbances and the persistence of ecosystem composition over time.

Subnatural Environment: a habitat where human influence has caused some change to the ecosystem, but its structure has remained basically intact. It therefore denotes less human impact over the natural environment than the SEMINATURAL ENVIRONMENT. Many temperate forests exemplify the subnatural environment.

COMMENTARY ON THE SUBNATURAL ENVIRONMENT

Such environments are frequently the result of *highgrading*, the extermination of key fauna through overhunting or overfishing, or the introduction of exotic plants or animals into the system. The spread of human populations and activities over the world's surface has been so pervasive that a high proportion of landscapes that are apparently natural are actually subnatural. Sometimes aggressive management strategies must be employed (e.g., extermination of exotics) if original natural values are to be recovered. *Robert Healy*

1.2.3 ENVIRONMENTAL SERVICES

Amenity Services of the Environment: contributions of the environment to the enjoyment of life, often used to denote the nonproductive benefits such as appreciation of landscapes, enjoyment of very clear air, or sightings of wildlife. The subjective nature of amenity services raises particular challenges in measuring their value, which can only be determined by gauging indirectly what people would be willing to pay for the amenities, because they are generally not marketed separately. For example, the value of amenities can be estimated by examining the differences among prices for goods (such as housing or land) in areas with and without that amenity (the HEDONIC PRICING APPROACH to valuation) or by asking people to respond to the hypothetical question of what they would be willing to pay or forgo in order to have the amenity (the CONTINGENT VALUATION APPROACH).

Ecological Services. See ENVIRONMENTAL SERVICES.

Environmental Functions. See ENVIRONMENTAL SERVICES.

Environmental Needs: the concept based on the hierarchy of needs formulated by the American psychologist Abraham Maslow, who identified five levels of needs pursued by every individual. Physiological needs (food, shelter) are on the first level, followed by the needs for safety and security. The third level is acceptance by others, the fourth self-respect, and the fifth (highest) level is self-actualization. Translating this hierarchy to society helps to understand the evolution of environmental concerns. When people are fully preoccupied with meeting their basic needs (first level), environmental policy is not feasible. On the second level of safety and security, attention is given to threats to human health and environmental policy is oriented toward these threats. At the third level (acceptance by others), concern shifts to the well-being of community and therefore to the quality of shared habitat. At the fourth stage (self-respect), people rise beyond immediate concerns to their health and surroundings, and environmental conservation becomes an objective in itself. At the fifth level, people recognize the necessity of addressing environmental concerns as equal to socioeconomic and resource security. According to this approach, every phase can be reached only when the needs of the previous level are met.

See Pieter Winsemius, cited in N. van Lookeren Campagne, "Interviews with J. M. H. van Engelshoven, N. G. Ketting, H. H. van den Kroonenberg, E. van Lennep, O. H. A. van Royen, and P. Winsemius," in *Policy Making in an Era of Global Environmental Change*, eds. R. E. Munn, J. W. M. la Rivière, and N. van Lookeren (Boston and Dordrecht: Kluwer, 1996), 159–63.

COMMENTARIES ON ENVIRONMENTAL NEEDS

A very powerful concept, it helps explain why environmentalists have failed to get governments actively involved in resource management and conservation in developing countries. It also suggests why environmental policy wonks should change tactics and identify the subsectors of society where caring for the environment becomes an appropriate concern. *Gustavo Arcia*

Maslovian hierarchies of need are still taught in business school classes on motivation but have been largely discarded by contemporary social scientists. Empirical research finds that prepotency is not clear, thresholds where one need is satisfied and another emerge have eluded identification, many needs are complementary, and—where a most basic need can be identified—many researchers and policy analysts now consider self-actualization as a member of the community to be fundamental. How else can one explain the choice of a Somalian woman to starve rather than appear naked at a food distribution station? Ordering needs may be useful to provide a "framework in arguing intelligently about priorities," but should not be mistaken for social reality.

See United Nations Development Programme, *The Human Development Report 1998* (New York: Oxford University Press, 1998); Mary Douglas, Des Gasper, Steven Ney, and Michael Thompson, "Human Needs and Wants," in *Human Choice and Climate Change, Volume 1, The Societal Framework*, eds. Steve Rayner and Elizabeth L. Malone (Columbus, OH: Battelle Press, 1998). *Steve Rayner*

Environmental Resources: natural resources that not only contribute raw materials, but also ENVIRONMENTAL SERVICES.

Environmental Services: the benefits provided by natural systems apart from raw materials, as applied to both life support and productive processes. Elements of natural systems may stabilize conditions (such as climate, water levels, or soil conditions), provide aesthetic or other psychological benefits (such as knowledge of the existence of species—the so-called existence value invoked in resource valuation), and serve as absorptive sinks for human wastes, surpluses, and accidentally released substances. The cleanliness of air and water is one crucial aspect of the potential to provide environmental services; while these resources are virtually without limit, clean air and water may be scarce and therefore of limited value for health, production, or to absorb further wastes.

Environmental Space. See under SECTION 1.2.1.

Life-Support Services of the Environment. See ENVIRONMENTAL SERVICES.

1.2.4 OTHER BASIC ENVIRONMENTAL CONCEPTS AND PROCESSES

Biodiversity (or Biological Diversity): the variety and variability of the world's living organisms at all levels of organization, including genetic differences, species, habitats, communities, ECOSYSTEMS, landscapes, and the BIOSPHERE, as well as natural processes that maintain the functioning and stability of the system. The breadth of this concept reflects the interrelatedness of all its components. Biodiversity may be preserved for ecological reasons, its utilitarian, intrinsic, and aesthetic value, as well as NATURAL HERITAGE. The subject of biodiversity conservation is the current focus of many international efforts to reform natural-resource management and development practices, including several international treaties like the CONVENTION DESIGNED TO ENSURE THE CONSERVATION OF VARIOUS SPECIES OF WILD ANIMALS IN AFRICA, WHICH ARE USEFUL TO MAN OR INOFFENSIVE (1900 LONDON CONVENTION), CONVENTION RELATIVE TO THE PRESERVATION OF FAUNA AND FLORA IN THEIR NATURAL STATE (1933 LONDON CONVENTION), the CONVENTION ON THE CONSERVATION OF MIGRATORY SPECIES OF WILD ANIMALS (BONN CONVENTION), CONVENTION ON INTERNATIONAL TRADE IN ENDANGERED SPECIES (CITES), and the 1992 UNITED NATIONS CONFERENCE ON ENVIRONMENT AND DEVELOPMENT (UNCED; RIO CONFERENCE; EARTH SUMMIT).

See Reed E. Noss and Allaen Y. Cooperrider, *Saving the Nature's Legacy: Protecting and Restoring Biodiversity* (Washington, DC: Island Press, 1996); G. Meffe, C. Carroll, et al., *Principles of Conservation Biology* (Sunderland, MA: Sinauer Associates, 1997).

COMMENTARY ON BIODIVERSITY

A concept that has also become associated with agriculture and food production as there is a growing worldwide realization that biodiversity is fundamental to agricultural production and food security, as well as a valuable ingredient of ecological stability. "Agricultural biodiversity" or "agrobiodiversity" encompasses not only diversity among plant and animal genetic resources, soil organisms, insects, and other flora and fauna in managed ecosystems, but also diversity among elements of natural habitats that pertain to food production. The rapid decline and loss of diversity in crop varieties, soil and aquatic resources, insects, and the narrowing of agroecosystems jeopardize productivity, threaten food security, and result in high economic as well as social costs. Agricultural development and biodiversity conservation are sometimes perceived as opposing interests. Yet, with sustainable farming practices and changes in agricultural policies and institutions, conflicts can be overcome. Several major international bodies have recently focused on the issue of agricultural biodiversity: the Global Convention on Biodiversity and the 1996 World Summit Plan of Action for World Food Security.

See Lori Ann Thrupp, *Cultivating Diversity, Agrobiodiversity and Food Security* (Washington, DC: World Resources Institute, 1998). *Elsa Chang*

Biogeochemical Cycle: large-scale interactions among various living and nonliving components of the environment, including the atmosphere, water, soil, rock, and biotic systems. The biogeochemical cycles function at the global scale through processes of production and decomposition that involve such major elements as carbon, oxygen, hydrogen, nitrogen, phosphorus, and sulfur. Such contemporary environmental challenges as global climate change, ozone depletion, and eutrophication are believed to be manifestations of major changes in global biochemical cycle. This cycle has been a major focus of the INTERNATIONAL GEOSPHERE-BIOSPHERE PROGRAM, which, despite its undeniable advances, still was unable to produce a full understanding of the process.

Ecological Imperialism: a term coined to denote the dramatic and irreversible change of the native biota that occurred in the New World (North and South America, Australia, New Zealand) as a consequence of European expansion. European plants, animals, and diseases, which came with human migration, displaced native flora. In some areas, descendants of European species now form the great majority of the population. For an alternative meaning related to the imposition of environmental standards by one culture onto another, see under SECTION 4.1.

See A. W. Crosby, *Ecological Imperialism 900–1900* (Cambridge: Cambridge University Press, 1986.)

Ecological Entity: a general term referring to a particular species, ecosystem, habitat, or biome, usually used in ecological risk assessment and environmental management.

Entropy: 1) a measure of the randomness (disorder) of a system. The Second Law of Thermodynamics (the "entropy law") states that it is a natural tendency of the whole universe (or any thermodynamically closed system) to increase its entropy, or energy that is incapable of performing work. The notion of entropy has been linked to the concept of a nonequilibrium system, which requires a continuous inflow of energy and is characterized by "dissipative" structures, which may lead to self-organizing patterns. See also EVOLUTIONARY PARADIGM.

2) a quantitative measure proposed as an environmental indicator of the material limits to economic growth. Georgescu-Roegen originally applied the notion of entropy to the economic analysis of production processes, matter, and energy in agriculture and industry, though several ecological economists expanded the ideas of Georgescu-Roegen. The link between energy balances and economic growth remains highly controversial.

See Nicholas Georgescu-Roegen, *Energy and Economic Myths* (New York: Pergamon Press, 1976); Jeremy Rifkin with Ted Howard, *Entropy: Into the Greenhouse World* (Bantam, 1989).

Environmental Debt: the debt that human activity "owes" the environment (and future human generations) for environmental damage. According to proponents of the environmental debt, it must be taken into account in assessing the gains of development. One doctrine of sustainable environmental management calls for enough restoration of ecosystems to pay the environmental debt fully.

Closely related to the general idea of restoration ecology is an equally gran-
diose concept called the "environmental debt." The basic idea corresponds to
the cost of restoring the environment after the damage humans have caused
it over long periods of time. The concept is usually limited to "restorable"
damage, however that might be determined and costs assessed. Accounting
for damage that cannot be repaired and also determining how much restoration can and should be undertaken—and at what cost—are also problematic.
Pricing the "debt" or liability humanity has incurred, for instance in terms of
increased carbon dioxide created by the burning of fossil fuels for hundreds
of years, is the sort of problem one encounters in discussions of the environmental debt. *Garry D. Brewer*

Environmental debt is a useful (if not precisely measurable) concept because
it calls for keeping track of—and in principle addressing—the cumulative environmental deterioration caused by human activities. The valuation of environmental deterioration will always be disputable, but the discourse arising
from the dispute can be a healthy way to focus attention on the problem of
environmental degradation. *William Ascher*

Environmental Protection: any activity to maintain, restore, or increase the quality of
environment through preventing the emission of pollutants, reducing the presence of
polluting substances, preserving wildlife, etc. It may target a) changes in the quality
of good and services; b) changes in consumption patterns; c) changes in production
techniques; d) treatment or disposal of residuals in separate protection facilities; e) recycling; and f) prevention of degradation of the landscape and ecosystems. See also
CONSERVATION and PRESERVATION.

COMMENTARY ON ENVIRONMENTAL PROTECTION

Environmental protection is a reactive, remedial approach. It seeks to reverse
or prevent degradation of the environment. As such it seeks to restore an environmental condition to a satisfactory one. Environmental protection's reference point is some acceptable, stable condition of the environment with
energy exerted to preserve such a condition. This restorative approach is one
that is grounded in stasis as the objective of intervention. This can be contrasted with a view that intervention in the environment should be based on
improvement. Such an approach admits that change is inevitable and seeks to
take advantage of this change to benefit the environment. *Gerald Emison*

Equilibrium, Environmental: the stability of interactions and conditions within an environmental system. Several types of environmental equilibrium have been defined,
but the most important are "steady-state equilibrium" and "dynamic equilibrium."
Steady-state equilibrium is defined by the persistence of short-term changes around a

stable base state, requiring that inflow and outputs of energy and material are more or less matched. Dynamic equilibrium exists when the short-term changes are superimposed on a base state that has its own self-balancing dynamics. Environmental systems are in dynamic equilibrium if they can adapt to external change and maintain a degree of stability over long timescales. Many natural systems exist in dynamic equilibrium over long time periods. However, much of the concern over global environmental threats focuses on the rapid and unpredictable rate of change engendered by human activities, which may overwhelm the capacity of the natural system to maintain its long-term equilibrium.

COMMENTARY ON ENVIRONMENTAL EQUILIBRIUM

Perhaps ironically, the idea of environmental equilibrium seems to be based on the same enlightenment metaphysics as that of economic equilibrium since both notions emerged around the same time and for the same intellectual and social purpose. "These myths [of economic and environmental equilibrium] tend to downplay questions of instability (or the unpredictable) in favor of stability, control, and idealization against the backdrop of absolute time and space. . . . This need occurred with the gradual withering away of a belief in the correspondence of the world and the design of a God 'out there.'"

See P. Timmerman, "Myths and Paradigms of Interactions between Development and Environment," in *Sustainable Development of the Biosphere*, eds. W. C. Clark and R. Munn (Cambridge: Cambridge University Press, 1986). *Steve Rayner*

Evolution: change over time in the characteristics of living organisms or other systems, generally in response to changing or challenging conditions. The classic theory of evolution was developed to explain the emergence, disappearance, and change of plant and animal species by invoking the principle of "survival of the fittest," but the same logic can be applied to explaining the survival and changes in organizations, social practices, values, and so on. Insofar as the various aspects of natural systems have evolved together ("coevolved"), they tend toward mutual dependence. A crucial issue in the management of natural systems is whether the relevant social system is coevolving in ways that permit good management. The evolution of species or other systems is not necessarily adaptive in the long run, inasmuch as some adaptations are "evolutionary dead ends" that make future adaptations more difficult. The debates over evolution are not simply between evolutionists and creationists, but also among evolutionists over such issues as whether evolution is gradual or uneven in pace; whether evolutionary processes are responsible for altruism within species, etc.

COMMENTARY ON EVOLUTION

Humankind is conducting a planet-wide "experiment" with evolution, partly by eliminating species in unprecedented numbers, and partly by depleting evolution's capacity to generate replacement species. Given what we know of recovery periods following mass extinctions in the prehistoric past, the "bounce back" time could be at least five million years, or twenty times longer than humans have been a species themselves. Suppose that during those five million years the average world population is 2.5 billion, by contrast with today's 6 billion; the number of people living in an impoverished biosphere will be on the order of 500 trillion. Just one trillion is a large number; consider the length of time represented by one trillion seconds. *Norman Myers*

Feedback Mechanism: a mechanism that connects one component of the environmental system to another. Feedback usually occurs when a change in one element of the system results in a succession of changes in other elements, which eventually affect the element whose initial change triggered the systemic change in the first place. Due to the complex interrelation of a system's elements and overlapping connections, feedback phenomena are usually intricate and even a minor change in the system can have a radical impact. Feedback can either amplify (positive feedback) or dampen (negative feedback) the initial change within the environmental systems that was triggered by external factors. Most environmental systems are subject to negative feedback, which is why they either appear stable in the absence of external pressure or can usually adjust to external change.

Habitat: a place where a population (human, animal, plant, microorganism) lives and its surroundings, both living and nonliving. The latter include food sources, climate, water, soil and vegetation cover, and other resources and conditions the type, amount, and location of which are fit for breeding, escape, and breeding of a particular individual or species. There are different classifications of habitats, mainly based on the type of dominant vegetation and associated environmental conditions, for instance, tall-grass prairies in the United States and Australian coral reefs. Though humans have been altering habitats throughout history, during the last decades human-induced loss and fragmentation of natural habitat has become so significant as to be considered the main reason of global biodiversity loss.

COMMENTARY ON HABITAT

Habitat has been largely replaced in scientific ecology by the concept of *biome*, which emphasizes the interplay among species creating a characteristic landscape (e.g., grassland). However, *habitat* remains politically popular as it is often easier to mobilize popular support for landscape preservation around a single species. *Steve Rayner*

Media, Environmental: specific physical components—air, water, and soil—that are the subject of regulatory concern and activity. The traditional approach to solving environmental problems is the MEDIUM-BY-MEDIUM APPROACH, which targets each component of the environment independently from one another. *R. P.*

Natural Pollution: the production and emission by geological or nonhuman biological processes of substances not associated with human activities; for instance, natural oil seeps or toxins released by plants and animals.

Nonlinear Relationships: interactions among variables or factors that do not follow the straightforward linear relationship by which an increase in one is associated with a constant, proportional increase in the other. Nonlinear relationships are often characterized by threshold effects, ceilings, inflection points, and other more complicated patterns. The existence of nonlinear interactions makes the prediction of future patterns more difficult because the points at which thresholds, ceilings, etc. actually occur are often unknown, and even the nature of the relationship may be uncertain.

Pollution: in the general sense, the presence of matter or energy whose nature, location, or quantity produces undesirable effects. Though pollution is usually thought of as the humanmade or human-induced alteration of the physical, biological, or chemical parameters of the environment, it may also be of "natural" origin (see NATURAL POLLUTION). Throughout history, POLLUTION ABATEMENT focused on removing, altering, or dispersing pollution from where exposure to pollutants would cause harm (e.g., impounding toxic wastes into waste dumps rather than flushing them into rivers). Since the late 1980s, however, the emphasis in environmental policy has been shifted to POLLUTION PREVENTION, the active process of identifying pollutant-creating processes and activities for the purpose of reducing the use of toxics, substituting, altering, or eliminating the technological process to prevent the generation of pollutants.

Riparian Habitat: areas adjacent to rivers and streams. Riparian habitats often have high density, diversity, and productivity of plant and animal species relative to nearby uplands.

Species: a population of organisms that are able to interbreed freely under natural conditions. Species evolve distinct inheritable features, occupy distinct ecological niches (and often distinct geographical areas), and usually do not freely interbreed with other groups of organisms. The boundaries between species are nevertheless sometimes blurred. Breeding among related species does occur, although it typically results in infertile offspring.

COMMENTARY ON SPECIES

We are eliminating species in tropical forests alone at a rate of 50–150 per day, supposing that the global total is 10 million species (some biologists propose 100 million species). If allowed to proceed unchecked, this mass extinction of species will be the sixth such setback to life's abundance since the first flickerings of life 4 billion years ago, and during the 600 million years of major forms of life. *Norman Myers*

Synergism (or Synergy): the interaction of different components within an environmental system. As a result of synergy, the environmental system amounts to much more than the simple sum of its components. The phenomenon of synergy is often discounted when components of environmental systems are studied separately. See also REDUCTIONISM.

Thresholds, Environmental: critical points at which response and behavior of environmental systems change abruptly. Thresholds are therefore also important for establishing what levels of pollution (CRITICAL LOADS) or scarcity ought to trigger regulations or charges. The existence of thresholds is also important for understanding the non-linear behavior of environmental systems (see NONLINEAR RELATIONSHIPS) and their impacts (for example, a body of water may experience a sudden collapse of fish if the levels of oxygen fall below a threshold, or the concentration of a specific contaminant rises above a threshold).

COMMENTARY ON ENVIRONMENTAL THRESHOLDS

Thresholds are actually very hard to identify in nature or society (except some-times retrospectively). However, the idea is rhetorically very attractive for political mobilization. *Steve Rayner*

Wilderness: an area where the earth and its biological community are minimally trans-formed by humans, especially areas that have never been occupied by humans or inten-sively used by them. It is an environmental resource that is nonreplicable and irre-versible. Perceptions of wilderness and attitudes toward it differ substantially among various cultures, social groups, and across time. Once perceived as a hostile place of exile, wilderness has come to be seen as source of ENVIRONMENTAL SERVICES, valued for its aesthetic, spiritual, and recreational qualities. These areas are protected in many countries to preserve their natural wildlife and landscape.

1.2.5 RESOURCE USES AND TYPES

Commodity: a product that is essentially the same regardless of the producer. Most agricultural products and other raw materials are commodities; their uniformity often allows them to be sold in organized markets, including futures markets that permit hedging against supply-and-demand volatility. International commodities agreements have the same objective of stabilizing supply and prices. Manufactured products may verge on commodity status due to uniformity of technology and standards (e.g., IBM-compatible personal computers). Producers often try to differentiate their products so that they will not be regarded as commodities (e.g., Colombian coffee), so that the prod-ucts valued for their distinctive quality or image can be sold at higher prices.

COMMENTARY ON COMMODITIES

Historically, environmental and natural resources have tended to be treated as commodities. However, as income levels rise and leisure time increases, envi-ronmental and natural resources are valued less for their worth as economic commodities and more for their intrinsic existence. *Toddi Steelman*

Conditionally Renewable Resources. See NATURAL RESOURCES.

Ecological Rucksacks ("Forgotten Tonnes"): a label denoting natural assets merely used to make access to raw material possible and then returned to nature as wastes (e.g., the "overburden" or earth removed in mining; the usually contaminated water removed in oil pumping). These wastes represent an additional (and often enormous) ecological impact of the industries engaged in raw materials extraction.

COMMENTARY ON ECOLOGICAL RUCKSACKS

Reduction of ecological rucksacks would sometimes substantially increase the efficiency of resource use and, subsequently, effectiveness of nature protection. The phenomenon of ecological rucksacks is particularly important in the context of North-South relations. Because developing countries supply a major share of raw materials to the world market, they bear the brunt of additional ecological impact but are often not compensated for it. *Natalia Mirovitskaya*

Environmental Resources. See under SECTION 1.2.3.

Exhaustible Resources. See NATURAL RESOURCES.

Material Resources: NATURAL RESOURCES whose value is typically defined through their single use as inputs to economic activities, such as mineral or fossil fuel resources. Preservation of material resources does not help to increase non-market use such as recreation. Positive environmental effects through preservation occur mainly due to avoidance of pollution associated with their economic use, such as the burning of fossil fuels.

See Organization for Economic Cooperation and Development, *Natural Resource Accounts*, Environmental Monographs # 84 (Paris: OECD, 1994), 9. *R. P.*

Natural Resources: endowments or assets that come from nature rather than being humanmade. Natural resources include renewable resources (winds, tides, sunshine, forests, fish, crops, etc.) and NONRENEWABLE RESOURCES (oil, coal, iron ore, etc.). However, some renewable resources may be exploited in ways that for all practical purposes make them nonrenewable (e.g., "mining" trees species into virtual extinction). Therefore, the renewable resources can be classified as unconditionally renewable (such as sun, winds, tides, etc.) and conditionally renewable. Environmental services are natural resources in that they emerge from the quality of naturally occurring substances, such as clean air, clean water, beautiful vistas, etc. *R. P.*

Nonrenewable Resources: resources that cannot be naturally replenished or recycled. In the 1970s, nonrenewable resources were considered to be oil, gas, uranium, and various metals and minerals. The economics of nonrenewable resources is simpler than that of renewable resources, because the natural increase in the resource stock is not an issue. By the same token, the extraction of nonrenewable resources may require more careful planning because of its irreversibility. More recently, the media that hold pollutants (so-called sinks), such as air, water, and soil, have begun to be considered nonrenewable resources. As these resources have begun to reach their saturation points, their ability to continue to absorb pollutants and perform their ecological functions has been drawn into question. Some would even argue that resources technically defined as renewable, such as tropical timber, are in practice nonrenewable because their extraction rarely permits successful regeneration. *T. S.*

Resource Intensity: the relative weight of natural resources as inputs in a production function. The question of the optimal degree of resource intensity depends on a particular economy's "factor endowments" (the available inputs such as capital, labor, and natural resources), which in turn determine whether heavy use of natural resource inputs provide the greatest returns ("comparative advantage") for the producers and for the economy. The often erroneous belief that high resource intensity is a facet of underdevelopment prompts governments to induce lower resource intensity by providing cheaper capital or labor, which may result in inefficient use of inputs.

Tradable Commodity: goods that can be feasibly transported internationally, thereby capable of being exported and imported. The significance of tradability is that a good's value depends on the international market; its "free market" value can be estimated by the so-called border price of imports. In contrast to "nontradables" (e.g., land, cement, and most health services), tradable goods are directly affected by currency exchange rates. Overvalued exchange rates (i.e., local currency kept artificially high by government actions to limit how many units of the local currency can be purchased with a unit of another currency) will decrease the domestic producer's local currency income from selling tradables abroad at a given international price, or will force the domestic producer to increase the international price, although this risks making the commodity less competitive. Monetary policies therefore have a major impact on the relative costs of tradables and nontradables, and on the income distribution between those who are involved in producing each.

1.2.6 RESOURCE CONSERVATION: APPROACHES AND PRACTICES

Conservation: the most elaborate concept of environmental assets (see NATURAL ASSETS) protection. According to the WORLD CONSERVATION STRATEGY and subsequent WORLD CONSERVATION UNION (IUCN) publications, conservation includes both classic elements of nature PROTECTION and PRESERVATION, such as restoration and safeguarding of ecological processes and genetic diversity, as well as management of natural resources and ecosystems to ensure their sustainable use. For BIODIVERSITY, conservation is defined as the management of human interactions with genes, species, and ecosystems so as to provide the maximum benefit to the present generation while maintaining their potential to meet the needs and aspirations of future generations. Biodiversity conservation incorporates elements of research, protection, and use.

High Grading: the resource-extraction strategy of harvesting or mining the best and richest specimens or deposits. High grading is often a shortsighted extraction strategy; in forestry, it often degrades the forest quality because of the removal of the best genetic stock; in mining the rapid extraction of pockets of high ore grades frequently sacrifices nearby deposits of moderately good ore concentrations. Taxes on resource extraction that do not discriminate according to quality provoke high grading, because the extractor has an incentive to extract higher-valued output taxed at the same rate as lower-valued output.

Hotelling Rule: a nonrenewable resource extraction rule, expounded by economist Harold Hotelling, that specifies that the rate of extraction should be set such that the value of the remaining resource appreciates at the prevailing interest rate, which represents the return on alternative earning opportunities for the proceeds gained through extraction. This is because in a competitive-market context, the investment of receipts from more rapid resource extraction would yield a lower return than leaving part of the extracted resource in the ground; slower resource extraction would reduce the appreciation of in-ground resources below the interest rate. Harold Hotelling also demonstrated that the extraction rule optimizes for both the resource owner and for society. The Hotelling rule remains the basis of many refinements that take into account uncertainty about future prices and extraction opportunities.

See Harold Hotelling, "The Economics of Exhaustible Resources," *Journal of Political Economy* 39 (1931): 137–75.

COMMENTARY ON THE HOTELLING RULE

If rich countries are at a stage where environmental conservation is desirable and affordable, they could invoke this rule to compensate poor countries for leaving nature as is. However, in 1991 World Bank chief economist Lawrence Summers wrote that economic theory indicated that rich countries could pay poor "underpolluted" countries to accept their waste. The memo caused quite a stir but virtually no debate. *Gustavo Arcia*

Maximum Sustainable Yield (MSY): the largest harvest/catch of a certain renewable resource that allows the stock size of this resource to remain unchanged over time, or perpetually sustained. For instance, in fisheries the MSY is the maximum constant catch that allows all the relevant fish populations to be kept relatively constant over time and is equal to the growth of the stock. The concept was first developed for ocean fisheries and later was applied in forestry management. Although MSY might seem the most logical criterion for resource use on a sustainable basis, it does not take into account different compositions of resource endowment, market preferences, and additional costs associated with the effort. Therefore, in terms of efficiency the OPTIMAL SUSTAINABLE YIELD would seem to be a more appropriate criterion in renewable resource management.

COMMENTARY ON MAXIMUM SUSTAINABLE YIELD

Some ecologists and natural resource scientists see MSY as part of the problem of sustainability, rather than the solution. The reason for this is that MSY treats biological populations as discrete elements in space and time, in isolation from other elements in the ecosystem, and assumes away natural variability and complexity. The concept of MSY developed in the service of a utilitarian worldview in which nature was seen as a storehouse of raw materials and resources merely as commodities. The mathematical treatment of the supply and demand of resources ignored the fact that these "commodities" were embedded in ecosystems that were variable, complex, and unpredictable.

See P. Larkin, "An Epitaph for the Concept of Maximum Sustained Yield," *Transactions of the American Fisheries Society* 106 (1977): 1–11; Fikret Berkes and Carl Folke, eds., *Linking Social and Ecological Systems* (Cambridge: Cambridge University Press, 1998). *Fikret Berkes*

Multiple Resource Use: 1) in a general sense, use of land, bodies of water, or other natural resources for more than one purpose. For instance, forest can at once be used for recreation, timber harvesting, water supply, watershed and wildlife protection, and livestock grazing.

2) in U.S. environmental legislation, the management of the public lands and their various resource values so that they are utilized in the combination that will meet the present and future needs of the American people.

COMMENTARIES ON MULTIPLE RESOURCE USE

Like many environmental management concepts, multiple use management has tended to work better in theory than in practice (see SUSTAINABLE DEVELOPMENT). Coined as a term to give legitimacy to the variety of constituencies that valued and used public lands, multiple use management has come to mean all things to all people. As more people have sought to utilize public lands, there has been increased pressure on agencies, such as the United States Forest Service and the Bureau of Land Management, to balance the multiple and often conflicting multiple uses demanded from public lands. In many ways, multiple use management has been an abdication of authority by Congress and the land management agencies from making the tougher value-based decision of how our public lands should be managed and which values should be given priority. *Toddi Steelman*

Multiple use management, as a resource management principle, has one great virtue: the very term reminds resource managers of the need to *balance* objectives. Thus, it is an antidote to the common temptation to single out one objective—maximum profit, maximum environmental preservation, etc.—while dismissing the legitimacy of the others. Multiple use management, like any other management principle, should not be attacked just because it is open-ended. In any democratic context, the balance of uses will reflect the preferences of various STAKEHOLDERS, rather than any preset weighting that could be established technically through the application of the principle per se. *William Ascher*

Optimal Sustainable Yield: the harvesting principle that calls for an extraction rate that optimizes the full set of objectives of resource management, rather than relying on a simpler rule such as MAXIMUM SUSTAINABLE YIELD (MSY). The optimal sustainable yield depends on the value of the extracted resource, the growth rate of the unextracted resource endowment, the other goods and services that the unextracted resource endowment provides (e.g., a forest may provide habitat for valued species and protection from soil erosion; therefore greater extant forest may be preferable to the forest stock that would result from the maximum annual harvest).

Anything can be maintained at its optimal sustained yield. The trick is to figure out which resources and values should be optimized. *Toddi Steelman*

Preservation: an approach to conserving environmental assets that is aimed at maintaining their natural characteristics in a manner unaffected by human activities to the fullest extent possible (1991 Draft Covenant on Environmental Conservation and Sustainable Use of Natural Resources). In distinction to the more general terms of PROTECTION and CONSERVATION, *preservation* implies the intention to pursue durable measures of defending particular objects against external threats to their existence.

Protection: measures to reduce the degradation or reduction of natural resources and environmental services. It is the most general term for such efforts, both in academic discourse and legislation. Usually protection implies a reaction to some threat to the object that is to be protected. However, in distinction to CONSERVATION and PRESERVATION, it does not necessarily imply proactive or anticipatory practices or policies.

1.2.7 STANDARDS AND MEASUREMENTS

Ambient Standards: indicators of environmental quality specifying the amount of pollutants regarded as permissible per unit volume in different environmental media.

Biodiversity Indices: measures of species diversity expressed as ratios between the number of species and "importance values" (numbers, biomass, productivity, etc.) of individuals. The term may also refer to other levels of BIODIVERSITY (genetic, habitat, or ecosystem diversity).

Environmental Index: an aggregation of several specific indicators of environmental conditions and trends. Environmental indices are the apex of the environmental information pyramid, above indicators and analyzed data, with raw or primary data as its base. The goal of environmental indices is to communicate information about the environment—and about human activities that affect it—in ways that highlight emerging problems and the effectiveness of current policies. In summarizing several indicators (e.g., a composite pollution index aggregating indicators of the levels of most or all significant types of pollution), an index can provide a broad assessment of environmental conditions and trends. *R. P.*

COMMENTARY ON ENVIRONMENTAL INDEX

The weighting of the components in an environmental index is often controversial. The risk is that overly aggregated or poorly constructed indices will provide misleading information on environmental conditions and trends. *William Ascher*

Environmental Indicators: a set of quantitative measures relating to the environment and human impact. Environmental indicators can be subdivided into "state indicators," reflecting changes in the physical or biological state of the natural world, and "pressure indicators" measuring environmental consequences of human activities. Some authors also propose a set of "response indicators," reflecting policy measures adopted in response to environmental problems. State indicators, such as urban air quality or stratospheric ozone concentration, measure the "quality" or health of the earth's physical or biological systems. In contrast, pressure indicators, such as industrial-gas emissions or halocarbon production, reflect the causes of environmental problems. Response indicators, such as compliance with international obligations, budget commitments, or voluntary behavioral changes gauge the effort taken by the society, institutions, or social groups to mitigate human impact on the environment. For instance, in case of fishery stocks, the volume of catches might be presented as a pressure indicator, OPTIMAL SUSTAINABLE YIELD as a state indicator, and catch quotas as a response indicator.

See Allen H. Hammond, *Environmental Indicators: A Systematic Approach to Measuring and Reporting on Environmental Policy Performance in the Context of Sustainable Development* (Washington, DC: World Resources Institute, 1995).

COMMENTARY ON ENVIRONMENTAL INDICATORS

For environmental indicators to be useful, they must be consistently observed and reliably collected over a substantial period of time. These indicators often mirror a specific policy emphasis, with resources allocated to the "pollutant-of-the-month." The ability to maintain a scientifically sound and sustained measurement of environmental indicators requires a commitment to data collection that transcends immediate pollution control interests and looks instead at those properties of long-range importance. When resources chase the most recently discovered environmental threats, collecting and using environmental indicators presents real challenges. *Gerald Emison*

Environmental Standards: specifications of acceptable environmental quality, usually expressed in terms of limits on particular pollutants or conditions. Typically, the environmental standards for a given area or facility define the threshold for government restrictions. When environmental standards are exceeded, governments usually fine or prohibit further activity. When environmental standards are set on the basis of thresholds determined by health or aesthetic concerns, they may clash with economic efficiency. Heated debates occur over the appropriate levels of environmental standards, whether they should be uniform across different areas, and how they should be enforced. See also ECONOMIC INSTRUMENTS; EFFLUENT CHARGES.

COMMENTARIES ON ENVIRONMENTAL STANDARDS

Air quality standards may be of quite different forms: a) some require installation of a particular control technology; b) some require a specified performance characteristic; c) some specify an ambient concentration; d) some specify a harm threshold that shall not be exceeded. Most air quality standards are either performance- or ambient-oriented. For ambient standards to be effective, they must be connected to either the adverse health or welfare effects they seek to prevent. Establishing this degree of risk associated with an ambient standard presents three challenges. First, those with the technical knowledge to interpret the risk characterization may be either unwilling or unable to communicate to others the value choices inherent in setting an acceptable ambient standard. Second, for ambient standards to be effective there must exist a straightforward means of connecting emission reduction strategies with ambient concentrations. Usually this is done through imprecise mathematical models, with errors of +/- 100 percent or worse. Third, when the connection between controls and ambient conditions is well established, achieving the actual reductions depends upon credible enforcement by a legitimate authority or agency. If there is a failure of value articulation, practical strategy, results estimation, or enforcement, ambient standards can become little more than exhortatory. *Gerald Emison*

Environmental "standards" inevitably clash with the cost-benefit approaches that focus on the specific levels of pollution that correspond with maximizing overall benefits in the balance between economic activity and environmental degradation. The standards approach reflects a fundamentally different philosophy: that pollution beyond a particular level is simply not to be tolerated, regardless of the economic costs of staying at or below that level. However, in the practical world of inevitable concern over economic losses, the approach of trying to impose uniform environmental standards often deteriorates into a system of apparent standards and many exceptions or "variances." While such variances conceivably could be based on careful cost-benefit considerations, allowing exemptions from pollution limits only when the societal gain is clearly worth it, variance systems are prone to political pressures from both business interests and environmentalists, yielding outcomes that are often more reflective of political power than to direct balancing of societal costs and benefits. *William Ascher*

Index of Resource Depletion: one of the aggregate indicators proposed to measure the sustainability of natural resource use. It defines the value of the decline in natural resource stocks in a country as a ratio of the value of gross (or net) investment in humanmade capital during a given year. In other words, the index reflects whether the value of resource depletion exceeds or is matched by the creation of other assets. If resource depletion exceeds other value creation, then resource use is not sustainable according to the so-called WEAK SUSTAINABILITY criterion. An index of 1.0 indicates that the increase in national humanmade capital equals the depreciation of natural assets; an index greater than 1.0 signals unsustainable development, while a value of less than 1.0 indicates the enhancement of total assets. Advocates of the index of resource depletion assert that if a system of integrated environmental and economic accounts (SEEA) is implemented on a worldwide basis, this index of resource depletion could be routinely calculated in physical units and used at a national level for more effective natural resource policies and for international comparisons. See also ENVIRONMENTAL ACCOUNTING.

See Allen L. Hammond, *Environmental Indicators: A Systematic Approach to Measuring and Reporting on Environmental Policy Performance in the Context of Sustainable Development* (Washington, DC: World Resources Institute, 1995).

Living Standards Measurement Survey: a household survey methodology used by the WORLD BANK. It is based on the concept that poverty should be measured through family expenditures, since they tend to fluctuate less than income and, as such, are a better indicator of living standards. These surveys quantify household consumption from market purchases, as well as from home production and barter. From these data an analyst can develop a "Poverty Profile"to show the extent of poverty in a given country. The indicators used for assessing poverty are: a) "poverty line": the amount of money required by a person to be able to purchase a basic basket of goods and services. This basket includes nutrition and essential nonfood items; b) "head-count index": the percent of the population under the poverty line; c) "poverty gap": per capita expenditures expressed as a proportion of the poverty line—the poorer the person, the smaller the percentage, and the larger the gap; and d) "Foster-Geer-Thorbecke index": the poverty gap squared. This index places an extra emphasis on the poor in order to design compensatory policies.

See Martin Ravallion, *Poverty Comparisons* (Washington, DC: World Bank, 1992). G. J. A.

Pollution Index: one of the composite indicators aggregating a number of already highly aggregated sink indicators (Hammond et al., 1995). It is designed to represent the overall effect of emissions, nonrecycled wastes, and other products of human activity dissipated into the environment. Aggregating several different indicators (such as ozone depletion or acidification indicators) into one composite index is done by weighting each environmental issue on the basis of the gap between the current value of the indicator and the long-term policy target for sustainability: the greater the gap, the larger the weight assigned. The advantage of the pollution index is that it dramatically compresses and simplifies data on environmental pollution. Insofar as it is a valid reflection of overall pollution, its trends reflect the effectiveness of national environmental policies. Testing of the pollution index in the Netherlands has demonstrated its feasibility. Yet the pollution index remains controversial because of disagreement on how to weigh the importance of various types of pollution, as well as the sensitivity of the measure to the policy targets, which may reflect political motives more than balanced approaches for environmental protection.

Allen Hammond, Albert Adriaanse, Dirk Bryant, and R. Woodward, *Environmental Indicators: A Systematic Approach to Measuring and Reporting on Environmental Policy Performance in the Context of Sustainable Development* (Washington, DC: World Resources Institute, 1995).

Sustainability Indicators: broad measures of economic, environmental, and resource trends designed to facilitate analysis about the long-term possibilities for continual improvement in material well-being, equity, and ecosystem health. Sustainability indicators may encompass trends in the stock of key natural resources, reductions in various forms of pollution, more progressive income distribution, better ratios of resource consumption per unit of output, etc. Sustainability indicators are intrinsically controversial if only because the concept of sustainability is controversial, and the weighting of material well-being, equity, and ecosystem health is a highly subjective matter of values that cannot be determined technically.

1.3 Epistemology of the Field

1.3.1 APPROACHES TO KNOWLEDGE AND THEORY

Bounded Rationality: the principle, formulated by Herbert Simon, that the capacity of the human mind falls far short of the capacity necessary for objectively rational solutions to complex problems in the real world. Consequently, an actor must construct a simplified cognitive model of a situation in order to act upon it. Behavior that is rational with respect to the simplified model is not objectively rational, or even a close approximation. To predict the actor's behavior requires empirical inquiry into the actor's simplified model and the psychological processes by which it is modified. To predict behavior without such inquiry is an unattainable goal of classical economic theory. Similarly, in his proposal for a second environmental science, Paul C. Stern insists that consumer behavior must be understood from the consumer's perspective, and that interventions on behalf of environmental goals must be monitored and revised continuously because perspectives and behavior are dynamic, that is to say, they are responsive to interventions as well as other factors.

See Herbert A. Simon, *Models of Man* (New York: John Wiley and Sons, 1957); Paul C. Stern, "A Second Environmental Science: Human-Environment Interactions," *Science* 260 (25 June 1993): 1897–99. *R. D. B.*

Catastrophe Theory: mathematics that explores the way in which systems can change from one stable state to another by passing through a critical threshold or "catastrophe." Environmental scientists increasingly use the analytics of catastrophe theory to model and predict changes in the behavior of ecosystems subject to major, "nonlinear" transformations.

Chaos Theory: a set of mathematical and scientific constructs about systems that have erratic or "chaotic" behavior even if they are not subject to random disturbances. The discovery of how many different hypothetical and real-world systems are chaotic, rather than converging to a simple equilibrium, has led to much rethinking about the deterministic, mechanistic models that strongly dominated the natural sciences prior to the 1970s. Seemingly negligible differences in the initial conditions of different systems can result in dramatically different outcomes, thus reducing scientists' confidence in prediction.

COMMENTARY ON CHAOS THEORY

Chaos theory seemed to have burst on the scene in the 1980s as a new insight into discontinuous patterns in nature, providing an explanation for why prediction is so difficult. The fact that minuscule events (the proverbial flap of the butterfly's wings) can have escalating impacts has been lauded as a truly new perspective. In fact, in 1948 Warren Weaver published his classic article on "organized complexity," which recognized unstable equilibria and the limited applicability of the smoothing effects of the law of large numbers more than thirty years before "chaos theory" came into vogue. When elements of natural systems are interconnected rather than atomistic (e.g., the contrast between the subsystems of an organism and the molecules in a gas), much more complex and parameter-sensitive interactions occur, making prediction much more difficult. Ronald Brunner and Garry Brewer (1971) were pioneers in extending this idea into the social and policy sphere, and in dismissing the possibility of covering laws in the social sciences.

See Ronald D. Brunner and Garry D. Brewer, *Organized Complexity: Empirical Theories of Political Development* (New York: Free Press, 1971). *William Ascher*

Complex Adaptive Systems: systems distinguished by a large number of parts, each of which attempts to adapt to perpetually novel surroundings comprised of interactions among the many parts. The behavior of each part is based upon a simplified internal model that depends on previous interactions and is subject to change through further interactions. Computer simulation models of adaptive complex systems exhibit new forms of emergent behavior as the systems evolve; the evolutionary trajectories depend critically upon the specific history and context of the particular system. Such models cannot predict the behavior of any particular system, although their behaviors are reminiscent of real systems that include living forms. The variety of outcomes exhibited by such models call into question analytical approaches to adaptive complex systems that assume equilibrium, predictability, or replicability, or depend upon the hypothesis that the behavior of living forms can be reduced to stable and fundamental laws.

See John H. Holland, "Complex Adaptive Systems," *Daedalus* 121 (Winter 1992): 17–30; John Horgan, "From Complexity to Perplexity," *Scientific American* (June 1995): 104–9. *R. D. B.*

Complex Models: mathematical representations of how systems operate, consisting of interlinked equations. Some of these equations are "structural," in that they represent propositions about relationships among variables (e.g., more money in circulation leads to more spending); others are "identities" that sum up variables within a given category (e.g., total spending equals government spending plus nongovernment spending). In addition to the variables themselves, most complex models have constants (typically called "parameters") that convey how much of the change in one variable may affect others. These parameters are often estimated on the basis of previous empirical patterns (in economics, these are called "econometric models"), but are sometimes posited on the basis of theoretical considerations or even laboratory experiments. Complex models are used with great frequency to try to understand and predict the behaviors of complex systems, whether they are weather patterns, pollution effects, economic patterns, or population growth. The usefulness of complex models in almost all areas is highly controversial due to questions of how one knows whether the model is valid, and the concern that complex models will be used without due attention to their possible unreliability.

COMMENTARY ON COMPLEX MODELS

Complex problems may deserve complex analysis, but complex models often outstrip the knowledge mastery required for valid use. Mathematical models, no matter how intricate, are nothing more than expressions of assumptions about how systems work. The weaker the understanding of systems, the more likely the use of complex models would yield misleading results, without the model operators being able to recognize that it is misleading. The validity of complex models is often overestimated, because the parameters of the models are estimated on the basis of past performance, making the models seemingly consistent with past performance. Yet this consistency in itself does not mean that the model specifications are valid. Moreover, even a perfectly well-specified model developed at one point in time is not immune from the possibility that changes in the structure of the real world will render the model irrelevant. Thus the energy and economic models developed before the mid-1970s fell apart following the change to an economy in which high energy prices became a significant constraint. On the positive side, though, complex modeling is an exercise that usually forces the modelers to undertake systematic analysis, and make many more assumptions explicit than is usually the case with less formal analysis. *William Ascher*

Critical Theory: a body of theory developed by the members of the Frankfurt Institute of Social Research, or the Frankfurt School (notably Max Horkheimer, Theodor Adorno, and Herbert Marcuse) that has revised the Marxist heritage in several ways. One of the most important contributions of critical theory was to show that there are different levels and dimensions of exploitation beyond the economic sphere, including the subjugation of women and cruelty to animals. Members of the Frankfurt School replaced traditional Marxist critique of the allocation of the productive forces with a critique of technological civilization. In contrast to orthodox Marxism, early critical theorists saw increasing mastery of nature mainly in negative terms as giving rise to human domination over nonhuman nature and to a repressive division of labor and renunciation of the instinctual, aesthetic, and expressive aspects of the human psyche. Consequently, their quest was for the "resurrection of nature"—a new kind of mediation between society and the natural world. The Frankfurt School was particularly critical of instrumental rationality, that branch of Western intellectual tradition that understands human reason as effective adaptation and that accordingly apprehends only the instrumental (use) value of the phenomenon. The critique of instrumental reason in advanced industrial society was further developed by Jürgen Habermas, who argued that the advance of instrumental reason has led to the "scientization of politics," and subsequent manipulation and domination of the majority of the population by a technical and bureaucratic elite. Two central critical arguments of the Frankfurt School—the triumph of instrumental reason and domination of nature—link it to "green" political theory and practice. However, the political allegiance of critical theory proponents to the fortunes of democratic socialism has prevented its more direct impact on the green movement.

See Robyn Eckersley, *Environmentalism and Political Theory: Toward An Ecocentric Approach* (Albany: State University of New York Press, 1992), 97–117.

COMMENTARIES ON CRITICAL THEORY

Jürgen Habermas, one of the preeminent theorists associated with critical theory, argues that "systematically distorted communications" inhibit or prevent communicative rationality. To rectify this situation, Habermas proposes "communicative competence" and the "ideal speech situation" that emphasizes open, discursive, dialogic patterns of communication between parties. The intent is to broaden participation in decision making to foster more equitable outcomes. However, Habermas's writing has often been criticized for being too difficult for the lay person to understand.

Jürgen Habermas, *Between Facts and Norms: Contributions to a Discourse Theory of Law and Democracy* (Cambridge, MA: MIT Press). *Toddi Steelman*

A major feature of Habermas's views calls for the selection of public actions through discourse in which an ideal speech situation operates. This suggests that if communicative competence can be created among participants, whatever choice all accept is the rational choice. The approach rejects the positivist view that a truth exists outside of that reached through discourse. Habermas's approach is especially problematic when applied to environmental issues. Environmental protection depends on the ability of science to identify consequences of the physical world, yet Habermas's work suggests that if participants arrive under ideal speech conditions at a mutually agreeable decision that is at variance with what empirical science suggests, the decision is still rational. The incorporation of instrumental rationality into Habermas's discourse through ideal speech presents conflicts that have yet to be satisfactorily resolved. *Gerald Emison*

Cultural Determinism: an approach to explain different development levels among countries by focusing on cultural differentiation. Cultural differences among societies are reflected in systems of production and distribution, types of governance, lifestyle attitudes, and religious beliefs. Cultural determinism became an alternative to ENVIRONMENTAL DETERMINISM as an explanation of European dominance during the stage of imperialism often with racist overtones. On the one hand, some pro-Europeanists argue that certain aspects of European culture must be superior because people of European stock have expanded their economic and political domination around the world. However, according to some proponents of cultural determinism, European technology developed rapidly not because of cultural superiority, but rather because of the Western attitude toward the subjugation of nature.

See Clarence Ayres, *The Theory of Economic Progress* (Chapel Hill: University of North Carolina Press, 1944); Lewis Hill, "Cultural Determinism or Emergent Evolution: An Analysis of the Controversy Between Clarence Ayres and David Miller," *Journal of Economic Issues*, 23 (2) (June 1, 1989): 465.

Ecological Triad: analytical framework developed by American political scientists Margaret and Harold Sprout to examine an actor's behavior in the context of his or her environment. The "triad" comprises the actor, the environment, and the actor-environment relationship. This perspective presents a more complex vision of the human-nature interaction than the one assumed by ENVIRONMENTAL DETERMINISM. The Sprouts defined three main features as actor-environment linkages: environmental possibilism, probabilism, and cognitive behaviorism (reflecting the role of the actor's environmental perceptions).

See Harold Sprout and Margaret Sprout, *The Ecological Perspective on Human Affairs* (Princeton: Princeton University Press, 1965).

Environmental Determinism (Geographical School; Environmentalism): a school of thought that holds that the characteristics of the natural environment are decisive for social development. Various representatives of environmental determinism often distinguish a particular element of physical environment (climate, soil, territory, landscape, rivers, biota, etc.) as the main factor in effecting political and economic behavior, culture, or the settlement patterns of human populations. This school of thought became especially popular in the eighteenth and nineteenth centuries after the great geographical discoveries. Charles Montesquieu and Elisee Reclus (France), Friedrich Ratzel (Germany), Ellsworth Huntington (United States), L. I. Mechnikov, S. M. Soloviov, and V. O. Kluchevsky (Russia) were among its most prominent proponents. Russian historian Sergey Soloviov (1820–1879) argued that the emergence of Russian statehood and the most intensive land use in the center of the Russian plains were geographically predetermined. He also argued that particular natural conditions of the Russian central plains were a decisive factor in defining the character of economic activity, the form of population organization, and the comparative level of economic development in Russia. Karl Haushofer (Germany) borrowed some ideas from environmental determinism to "justify" his doctrine of GEOPOLITICS, which eventually became a rationale for Nazi Germany's expansionism.

See Lev Ilich Mechnikov, *Tsivilizatsiia i velikiie istoricheskiie reki* [*Civilization and Great Historical Rivers: Geographical Theory of Development of Modern Societies*] (Moscow: 1898); Mechnikov, *La civilisation et les grands fleuves historiques* (Paris: Hachette, 1889); S. M. Soloviov, *History of Russia*, vol. I (Moscow: Golos, 1993).

Game Theory: a method of applied mathematics that analyzes rational behavior and outcomes of simultaneous actions, often negotiations, of two or more players with divergent interests. Game theory is used to develop strategies when other actors' behaviors are not known in advance, and therefore must be anticipated by assessing the logic of how they can maximize outcomes from their own perspective of the "game." This approach is commonly used in the analysis of international environmental cooperation. For example, one of the applications of game theory suggests that the creation of regimes is more likely when a small group of actors is involved, because negotiators can then better understand the bargaining strategies of the other parties. Game theory can also be applied to the analysis of rational actions when faced with the uncertainties of nature, which may be modeled as an actor with random or semi-random behaviors. These analytical constructs are known as "games against nature."

See Carlo Carraro, ed., *Control and Game-Theoretic Models of the Environment* (Boston: Birkhauser, 1995).

General System Theory: an approach to analysis of complex systems based on the premise that different types of systems exhibit similar patterns of interaction and once these basic patterns are found, all systems, whether ecological, social, political, or economic, can be understood. The most notable proponent of this argument is Ludwig von Bertalanfi, who expressed the basics of systems analysis in his seminal work *General System Theory*. In some quarters the approach was considered highly promising; for example, one of the few cases of long-term cooperation between the Soviet Union, the United States, and European countries was the establishment of the International Institute of Applied Systems Analysis (IIASA). Though the basic promise of the general system theory (to find a magic bullet to apply to all systems) was never fulfilled, the general system approach remains a strong rival to classical reductionist science in the study of ecological and economic systems. From the historical perspective, general system theory develops ideas of Bogdanov's *tektology*. See also ECOLOGICAL ECONOMICS.

See Ludwig von Bertalanfi, *General System Theory* (New York: George Braziller, 1968).

Geopolitics: a method of political analysis that strives to explain and predict the behavior of states on the basis of their location, size, climate, natural resources, and other geographical variables. Geopolitical theories are often marked by very broad assertions that political leaders do or should follow particular doctrines. For example, leaders of dominant countries may be presumed to desire to expand their spheres of influence to the greatest possible extent; or to strive to protect sea lanes; or to protect their flanks from invasion or interference from rival powers. In one sense, geopolitical analysis has been an aspect of diplomatic and military strategic planning for thousands of years. More recently, though, the very term "geopolitics" has sometimes been associated with German political science of the 1930s and the territorial aspirations of the Third Reich. This association inhibited the development of geopolitics as an academic discipline in other countries. However, the potential merits of geopolitics have been recently recognized by experts in ECOPOLITICS as well as by military and political analysts. See also GEOGRAPHICAL DETERMINISM.

Indigenous Knowledge (Indigenous Technical Knowledge [ITK]; Traditional Knowledge): a body of knowledge maintained by a native people, often over a number of centuries in response to specific environmental and social stimuli. Common examples of ITK are herbal medicine, biological pest control, and natural systems of intercropping. Previously dismissed as "unscientific," indigenous knowledge is now considered productive, often environmentally sound, and crucial for the success of SUSTAINABLE DEVELOPMENT strategies. According to AGENDA 21, special efforts should be undertaken to preserve practices related to indigenous knowledge.

See Shelton H. Davis and Katrinka Ebbe, *Traditional Knowledge and Sustainable Development* (Washington, DC: World Bank, 1995); Fikret Berkes, *Sacred Ecology* (Philadelphia: Taylor and Francis, 1999).

COMMENTARIES ON INDIGENOUS KNOWLEDGE

Indigenous knowledge is a complex and dynamic phenomenon tied to a diversity of rich cultural beliefs, traditions, customs, and practices that can be passed down from generation to generation. Indigenous knowledge is threatened by global market pressures, widespread use of pervasive industrial agricultural technologies, and an intensive use of chemicals. The concept has been at the heart of the debate over intellectual property rights issues as local communities and indigenous people have been exploited, without compensation, for their traditional knowledge by private companies and institutions. *Elsa Chang*

The anthropologist Claude Lévi-Strauss notes that myths and rites preserve the results of observation and reflection on the requirements of adaptation to specific environmental and social settings. He calls such results "the science of the concrete," and considers them no less scientific or "genuine" than the results achieved by the exact natural sciences. "They were secured ten thousand years earlier and still remain at the basis of our own civilization."

See Claude Lévi-Strauss, *The Savage Mind* (Chicago: University of Chicago Press, 1966). *Ronald D. Brunner*

Systems of survival based on indigenous knowledge are very flexible in their strategies: interdependence and reciprocity, traditional ecological knowledge, and moral rules for upholding resource conservation. Indigenous systems that are still intact, their knowledge available, can accommodate environmental changes with little ecological damage. These systems are not compatible with market ideology and are therefore unlikely to survive the pressures of globalization. The Alaskan Pribilof Islands' monoeconomy and high-tech fisheries dependence is a case study of cultural incompatibility with global administration of fisheries. In the mid-1980s the Pribilof Aleuts fought for development of a small fishing harbor where they could transfer their skills and knowledge as subsistence halibut fishers to a limited day-boat fishery, as an economic alternative to the century-old fur seal peltry. Their small-scale proposals were overruled by the U.S. Army Corps of Engineers, which built a large port capable of accommodating large floating processors and factory trawlers. The Aleuts did get their halibut day-boat fishery, but their deep-sea port opened up the island and marine environment to ecological, social, and economic problems associated with servicing the richest bottom fishery in the world. Both the community and the environment have been irrevocably changed.

Many native people in the north are suspicious of ITK, which they suspect is another way for academics and governments to gain control of their episte-mologies while business continues as usual. As a result, many refuse to par-ticipate. The Bering Sea Coalition, for example, is setting up a protected Inter-net site where communities can exchange traditional knowledge, with the understanding that outsiders will not have access to it. *Susanne Swibold and Helen Corbett*

Maximization Postulate: the postulate, formulated by Harold D. Lasswell, that "living forms are predisposed to complete acts in ways that are perceived to leave the actor better off than if he had completed them differently. The postulate draws attention to the actor's own perception of the alternative act completions open to him in a given situation." As a logical tool for empirical inquiry, the postulate allows for the fact that an act may be spontaneous or deliberate; taken on behalf of an identified self ("we") or the primary ego ("I"); based on mistaken or valid expectations; and shaped by un-conscious demands, expectations, and identification, as well as conscious ones. The postulate also allows for the fact that these perspectives are subject to change through insight. Hence perspectives are to be inferred from patterns of acts observed in particu-lar situations, not taken as fixed or given or assumed. Like the esssentially equivalent principle of bounded rationality, the maximization postulate implies the need for a comprehensive conceptual framework to direct the attention of researchers systemati-cally to those aspects of the particular situation that are important for policy purposes. Otherwise, policy is likely to be better adapted to the limitations of the researcher's perspective than to policy purposes and the situation at hand. See also PRIGOGINE'S EVOLUTIONARY PARADIGM.

See Harold D. Lasswell, *A Pre-View of Policy Sciences* (New York: Elsevier, 1971), 16–18. *Ronald D. Brunner*

Positivism: a name originally given to the philosophical ideas of French sociologist Auguste Comte (1798–1857), but now applied to a range of views and scientific ap-proaches that share Comte's commitment to using induction to formulate generaliza-tions, which then must be tested through deduction. Comte's "positive method" was empirically based, in contrast with the metaphysical and religious argumentation of his day; today's positivists remain committed to the empirical demonstration of the principles they formulate. While Comte preached a holistic approach, in practice the current positivist search for general laws often fastens on a small number of variables that are believed to be related in consistent ways across very different contexts. Much of modern science is based on positivist premises, especially the primacy of testing empirical hypotheses as the essence of the scientific method. Thus, positivism now represents the approach of understanding the world through *general* laws that can be empirically validated. In this respect, modern ecology, insofar as it focuses on broad interrelatedness and the importance of contextual differences, is often antipositivist in rejecting the usefulness of searching for general laws and the dominance of hypothesis testing as the predominant means of scientific progress.

Prigogine's Evolutionary Paradigm: a paradigm that accounts for abrupt changes in sys-tems (both natural and human) that seemed to be in equilibrium. It is based on the discovery by Ilya Prigogine that, under certain conditions, a system, at the molecular scale, would cease to oscillate between one condition and another, and change its structure irreversibly. The same model can be applied to systems at more macro levels. Prigogine's evolutionary paradigm has been proposed as a fundamental dynamic of broad ranges of evolutionary patterns. Such changes of structure may occur suddenly, as a result of the combination of a multitude of NONLINEAR RELATIONSHIPS. Emergence of global environmental threats conform with the implication of Prigogine's evolutionary paradigm, which demonstrates how a new state of a system may be produced by relatively small, random variations within the process of its functioning.

See Ilya Prigogine and Isabelle Stenders, *Order out of Chaos: Man's New Dialogue with Nature* (New York: Bantam, 1984).

Rational Choice Theory: a framework for explaining human behavior that posits that individual behaviors reflect the efficient and knowledgeable choice of actions that will maximize the actor's interests. These interests may be selfish or other-regarding; the actor gets gratification from pursuing them regardless. The rational choice theory can be applied only if it is assumed that the analyst can also determine what course of action is optimal for pursuing the objectives; therefore, the theory presumes a very high level of knowledge of not only the actors' goals but also the workings of the systems in which they operate. Often the objectives are presumed to hold a priori for a class of actors. For example, rational choice theory may assume that all politicians are motivated to secure political office. Rational choice theory typically proceeds deductively, by specifying how rational actors would behave in different situations, rather than by empirical research on behavior. Decision rules are stipulated, usually framed according to uncertainty and risk, to define the universe of options on which rational actors can act. Rational choice theory is widely criticized for the unreasonableness of the assumption of how knowledgeable both the actors and the analysts can be of the truly optimal courses of action, and the frequently oversimplified depiction of the objectives to be maximized. *T. S.*

COMMENTARIES ON RATIONAL CHOICE THEORY

At the heart of rational choice theory is the assumption that it is possible to identify a stable set of actor's interests and objectives. Actors are well understood to possess a number of often competing objectives that may vary over time. Since it is not possible to consider all such objectives, we often simplify the situation by ignoring some objectives. Such a simplification can lead to a misspecification of an actor's interests. This simplification allows conceptual models to be constructed and deployed for understanding behavior. Yet we often overlook this simplification, assume that prediction is reality, and what starts as a small departure from reality ends up being substantially wide of the predictive mark. *Gerald Emison*

Rational choice theory is a very useful approach when applied judiciously. It tries to account for, and predict, behavior based on premises as to the objectives of the actors (e.g., firms approach environmental issues with a predominant aim to maximize profits), the incentive structure posed by the given situation (e.g., pollution fines make compliance the profit-maximizing strategy), and the assumption that the actors will behave knowledgeably and rationally (i.e., the firm's leaders will see the advantages of complying with environmental regulations and therefore comply). This is certainly a reasonable approach as a *first cut*, as long as the analyst is open to the possibilities that the firm's leaders may have more complicated objectives, or that they may not see the world as the analyst presumes, or that they may not respond as predicted (possibly because of the limitations of their information and analytic capacity). Often the rational choice analysis is useful primarily as a baseline to identify cases in which behavior actually deviates from the rational choice prediction. *William Ascher*

Reductionism: the analytic approach of dealing with complex systems by separating out components to be analyzed separately, and by searching for basic processes that may account for more complex behavior of the system as a whole. Some reductionist approaches try to identify one, or very few, factors that most account for the outcomes, with the implicit assumption that a focus on the dominant causal factors is sufficient for understanding the phenomenon.

COMMENTARIES ON REDUCTIONISM

Reductionism is an essentially mechanical and fragmented paradigm based on the assumption that progress is best attained by focusing on increasingly minute components of natural systems, while ignoring the interconnectedness of all things. This connotation is often used in criticizing Western approaches to environmental studies. Indeed, the system of research planning, financing, and administration in American academia encourages reductionism, even while the value of integral and interdisciplinary works is being verbally praised. *Natalia Mirovitskaya*

The significance of reductionism cannot be overstated. It has been an almost inconceivably potent strategy for extracting technological utility and control from nature, via traditional disciplinary physical science. The character of modern society owes much to reductionism. As a core aspect of modernity, reductionism is thus deeply internalized in the very idea of science and the very identity of scientists. Even those who willfully study complex systems often take an unapologetically reductionist approach, because to do otherwise requires an abandonment of the quest for causation and thus apparently rationality itself. The answer to reduction is therefore not interdisciplinary research—which itself can be reductionist—but a fearless acceptance that nature is saturated with indeterminacy. *Daniel Sarewitz*

Reductionism is expressed in computer models, such as the global circulation models that are prominent in global change research. Reductionism is often understood as the hypothesis that scientists can reduce the complex behaviors of living and nonliving forms to the same set of fundamental laws, which control these behaviors. Even if true, reductionism does not imply the "constructionism" hypothesis that scientists can start from those laws and reconstruct the behavior of the universe. The physicist Philip W. Anderson points out that the fundamental laws of particle physics do not explain the behavior of large, complex systems comprised of those particles. Instead, new and fundamental properties appear at each level of complexity. Thus, more complexity entails qualitatively different behavior, and "psychology is not applied biology, nor is biology applied chemistry." "More is different" became a rallying cry in the study of adaptive complex systems, which emerged in part from antireductionist sentiments.

See Philip E. Anderson, "More is Different," *Science* 177 (4 August 1972): 393–96. *Ronald D. Brunner*

Reflexive Modernization: a set of theoretical perspectives emphasizing the centrality of risk and trust in shaping a new form of modernity. Popular largely in Western Europe, reflexive modernization emphasizes the role of knowledge in the social and environmental spheres in creating new institutions sensitive to the negative consequences of wealth production. For such institutions to constrain the negative environmental and social consequences of capitalist industrialization requires critical reflection. How such critical reflection is achieved is a subject of some debate among the originators of this theory. Some variants highlight the importance of aesthetics and culture as a means of arriving at such a critical perspective (Scott Lash) while Anthony Giddens emphasizes individual ethics. Irrespective of these differences in emphasis, reflexive modernization underlines the need to address uncertainty and the unintended consequences of modernization. European responses to the reactor accident at Chernobyl, exposure to bovine spongiform encephalopathy (BSE, or "mad cow disease"), and public concerns over genetic modification have all underlined these concerns.

See Ulrich Beck, *Risk Society: Towards a New Modernity* (London: Sage, 1992); Ulrich Beck, Anthony Giddens, and Scott Lash, *Reflexive Modernization* (Oxford: Polity, 1994); Anthony Giddens, *The Consequences of Modernity* (Oxford: Polity, 1990); Anthony Giddens, *Modernity and Self-Identity* (Oxford: Polity, 1991). *I. W.*

Traditional Ecological Knowledge. See INDIGENOUS KNOWLEDGE.

1.3.2 ENVIRONMENTAL STUDIES AND ECONOMIC THOUGHT

Conservation Biology: a field of environmental studies that emerged as a recognized discipline in the 1980s. Some conservation biologists are trying to expand the interdisciplinary scope of the field to include a broad range of perspectives from genetics to ethics and law. A particular feature of this field is its explicit commitment to preserving sustainable assemblages of the earth's biodiversity to a maximum extent. Conservation biology relies on population genetics and biology, insofar as these fields explore populations' dynamics of flora and fauna and change in their genetic composition, as well as landscape ecology, which explores connections among species, natural communities, and landscapes from the conservation perspective.

Conservation biology is a new professional arena. The Society for Conservation Biology now has about five thousand members worldwide and continues to grow, reflecting a significant demand for focused biodiversity conservation and a desire to affect real world outcomes. The society was established to advance interdisciplinary collaboration and strongly influence policy for improved conservation. Unfortunately, its efforts have been largely "discipline-based," rather than "interdisciplinary," as called for in its mission statements. The society has had dubious effects on policy outcomes—locally, nationally, or internationally. The current form of the society is a reflection of its discipline-based membership, who operate from positivistic conceptions of science, society, and policy, a major barrier that must be overcome if the group is ever to achieve its goals of interdisciplinism and effective policy participation.

See T. W. Clark, "Conservation Biologists in the Policy Process: Learning How to Be Practical and Effective," in *Principles of Conservation Biology*, 2d ed., eds. G. K. Meffe and C. R. Carroll (Sunderland, MA: Sinauer Associates, 1997), 575–97. *Tim W. Clark*

Ecological Economics: 1) a recent approach to reformulating economics by emphasizing the relationships between ecosystems and economic systems, broadly defined. Human consumers are considered to be one important component in the overall economy and ecosystem, rather than the dominant and central force, and human society is seen to coevolve (in terms of preferences, understanding, technology, and culture) with the biophysical world. Consumption is subjected to not only a monetary budget constraint, as in traditional economics, but also to the constraints of nature and to physical laws. Sustainable management of the economic and ecological system is the focus, and the time frames considered are typically longer than in traditional economics. The normative commitment of ecological economics is striving for a "frugal society" (i.e., one based on a definition of thrift in terms of economic efficiency, and on achieving environmentally sound economic development). Ecological economics rejects the assumption that market forces will ensure the appropriate supplies of raw materials and environmental services, because of constraints in natural systems that ecological economists claim are ignored by neoclassical resource and environmental economics. Ecological economics emphasizes the importance of throughput (from basic inputs of water and soil to nature as a waste sink); carrying capacity (whose variability and unpredictability call for caution in order to minimize future regret); and entropy (the cost of any biological or economic enterprise is always greater than the product).

See Mostafa K. Tolba and Osama A. El-Kholy, eds., *World Environment, 1972–1992: Two Decades of Challenge* (New York: Chapman and Hall on behalf of the United Nations Environment Programme), 626–27. *R. P.*

2) a school of thought initiated in Russia by Sergey Podolinsky (1850–1891) to reformulate Marx's theory of surplus value in terms of appropriation of usable energy, thereby focusing attention on the limited CARRYING CAPACITY of the environment, the way

in which agricultural workers were being exploited, and how some regions were being exploited by others. This early trend of ecological economics (also sometimes labeled "ecological energetics") was further developed by Alexander Bogdanov (1873–1928) into a general theory of organization in *Tektology: The Universal Organizational Science* (St. Petersburg, 1913–1922). By contextualizing human beings as part of and within nature, existing only through their capacity to obtain and process usable energy, Bogdanov put the limitations of the natural environment at the heart of his theory.

Ecological Energetics Paradigm: a theory developed in the late 1920s by the Russian scientist Vladimir Stanchinsky, based on the premise that the quantity of living matter in the biosphere depends on the amount of solar energy that is transformed by natural communities at different trophic (i.e., food-chain levels). Encouraged by Vernadsky's ideas (see BIOSPHERE), Stanchinsky studied the "dynamic equilibrium" of natural communities and invoked the Second Law of Thermodynamics to explain decreasing biomass of the higher groups on the "trophic ladder." Each successive rung on the trophic ladder has less energy, as it depends on lower rungs for its energy supply (in the form of food) but cannot appropriate it all. The energy supply at the highest trophic levels (e.g., carnivorous animals) is most restricted of all, which severely limits its expansion. Stanchinsky's efforts to model the energetics role of different species within a theoretical natural community provided the first mathematical formulation of ecosystem dynamics. The ecological energetics paradigm was believed to be of high applied value: by studying the energy flows within natural communities, humans could both assess the productive capacities of these communities and structure their economic activity in conformity with the capacity of the environment. These goals inspired the efforts to create a system of ZAPOVEDNIKS.

See V. Stanchinsky, *Variability of Organisms and Its Importance for Evolution* (Smolensk, 1927).

2) from the contemporary perspective, a set of assumptions on the role of energy in connecting society with its natural environment. Several prominent Western scholars, such as sociologist William Frederick Cottrell (1955) and economists Georgescu-Roegen (1976) and Herman Daly (1977) promoted these ideas. The theory of ecological energetics was in turn popularized by American physicist Amory Lovins. He argued that most contemporary societies are on the so-called hard path of centralized energy production and high growth that discounts the Second Law of Thermodynamics (see ENTROPY) and the inherent limits to growth, eventually resulting in an irresolvable energy crisis. Lovins advocated a soft path emphasizing energy efficiency rather than growth, and a radical restructuring of the world energy supply system. From his perspective, such changes, in turn, would involve radical transformation of society itself toward more decentralized and sustainable system.

See William Frederick Cottrell, *Energy and Society: The Relation Between Energy, Social Changes, and Economic Development* (New York: McGraw-Hill, 1955); Nicholas Georgescu-Roegen, *Energy and Economic Myths* (New York: Pergamon Press, 1976); Herman E. Daly, *Steady-State Economics: The Economics of Biophysical Equilibrium and Moral Growth* (San Francisco: W. H. Freeman, 1977); Amory Lovins, *Soft Energy Paths* (Cambridge, MA: Ballinger, 1977.)

Ecology: the science concerned with interrelationships among organisms of different species with each other and with their nonliving environment. Though the word was first used in the mid-1800s, ecological studies were practiced even in ancient civilizations. Ecology is subdivided into autoecology, focusing on the study of the functioning of individual organisms in relation to the environment; population ecology; and community ecology (or "synecology"), which addresses how communities composed of different organisms interact.

Environmental Economics: the label generally applied to the NEOCLASSICAL ECONOMICS approach to addressing issues of pollution control, standard setting, waste management and recycling, externalities of private enterprise action, conservation, use of common property resources, etc., in order to provide guidance for efficient allocation and sound environmental policy. The externalities range from global warming to local problems such as soil erosion. The emphasis remains on mobilizing the market mechanism through adjustment of price signals to influence behavior of households and enterprises and so achieve environmental objectives in tandem with social and economic ones. *R. P.*

Environmental Studies: an academic field that implies an interdisciplinary approach to environmental problems. It emerged in the early 1970s both in the East and the West partially as a reaction to the challenges of the 1972 Stockholm Conference (see UNITED NATIONS CONFERENCE ON THE HUMAN ENVIRONMENT). The formidable task of environmental studies has been to change the traditional technology-oriented emphasis of applied science to a more comprehensive approach that would include exploring historical, political, social, and philosophical roots of environmental problems.

Green Economics. See ECOLOGICAL ECONOMICS.

Human Ecology: 1) a trend in social philosophy that strives to define how individuals or human groups relate to their environments. The Western tradition in human ecology probably dates back to Heraclitus, the Greek philosopher of the fifth century B.C., whose concept of reality was based on the notion of *logos*, the orderly process of constant transformation of world elements (fire, water, earth, and air) into one another. Eastern philosophies, such as Buddhism, Hinduism, and Taoism, are even more focused on the human-environment interaction than Western thought. See also BUDDHIST MODE OF DEVELOPMENT.

2) a paradigm developed in the early 1900s in the field of geography that is used to explore the relationship between natural environments and human activities from the standpoint of human adjustments to their surroundings. The organizing concept of that paradigm is the "region," which has been regarded by proponents of this approach as a living organism. One extension of this logic asserts that the state should be regarded as a living organism as well, thus supporting a philosophy of territorial expansion. Because of this "blood and soil" implication, few contemporary geographers support "human ecology" in this sense (see ECOFASCISM). However, the essence of this approach has been brought back by the proponents of BIOREGIONALISM.

3) the academic study of the interrelationship between the human species and its environment from an interdisciplinary perspective. Such social sciences as anthropology, geography, economics, sociology, and political science have indeed achieved an im-

pressive record of research into human-nature interaction, but human ecology aspires to establish more thorough transdisciplinary understandings, which to date are still rather modest. See also ENVIRONMENTAL STUDIES.

Neoclassical Economics: the contemporary school of economic thought developed out of "classical" liberal economic thought of the nineteenth century. Neoclassical economics emphasizes the importance of market equilibria and the price mechanism; its methodology is heavily quantitative, with strong reliance on econometrics. The policy prescriptions of neoclassical economics emphasize the importance of free-market prices, the inefficiencies of most forms of subsidies (except to encourage positive EXTERNALITIES), the importance of economic efficiency, and the risks of state interventions in the economy. Neoclassical economics regards the distribution of income and wealth as a value decision beyond economic theory, but a large body of neoclassical analysis does provide a framework for determining the impact of given distributions on economic productivity, and has posited "optimal distributions" to maximize productivity. The majority of economists from Western universities are trained in neoclassical economics. Therefore the economists of the staffs of most international development institutions can be characterized as neoclassical in their orientations and methods. In the field of environmental and resource economics, neoclassical economists also emphasize the importance of market prices for avoiding waste and excessive extraction of cheap natural resources and environmental services. However, neoclassical economics does recognize the need for government intervention to address imperfect information, monopolies, and unclear PROPERTY RIGHTS; as well as to discourage negative environmental externalities and encourage positive environmental externalities.

Policy-Relevant Research: research that directly addresses contemporary policy issues, and is used by policy advisers and policymakers in choosing their actions. The challenges to policy-relevant research lie in obtaining sufficient support to undertake the research, and ensuring that the results are understandable, credible, and relevant to policymakers. Often there is a gap between the research on conservation, environment, and development, and the focus of attention of pertinent policymakers. Policy-relevant research often requires intermediaries who can help identify relevant areas for research, translate research findings into digestible inputs, and certify the credibility of the research.

COMMENTARIES ON POLICY-RELEVANT RESEARCH

One of the ironies of academic research is that in an attempt to be theoretically relevant, the work produced is often policy-irrelevant. *Toddi Steelman*

Policymakers want solutions to politically difficult dilemmas; at the same time, scientists may be tempted to promise to provide these solutions. Such an arrangement can provide mutual benefit—policymakers can defer risky action until the "results" are in; scientists get funded in the meantime—as we saw with the National Acid Precipitation Assessment Program in the 1980s and much of the U.S. Global Change Research Program in the 1990s. In other words, research can be politically useful without being policy-relevant. *Daniel Sarewitz*

Political Ecology: 1) an emerging approach to environmental studies that applies political economic analysis to the environment. It attempts to account for environmental problems by invoking the broadest patterns of political forces and economic structures, emphasizing how political and economic domination lead to unsound resource and environmental practices. Most adherents of political ecology apply leftist critiques to the issues of power, poverty, and environmental transformation, in some cases through Marxist analytical frameworks. Political ecology also designates, more generally, several movements of community-based environmental activism targeting pollution and wasteful resource depletion.

See Adrian Atkinson, *Principles of Political Ecology* (London: Belhaven Press, 1991); Ted Benton, *The Greening of Marxism* (New York: Guilford Press, 1996); Raymond L. Bryant, "Political Ecology: An Emerging Research Agenda in Third-World Studies," *Political Geography* 11 (1992): 12–36; Raymond L. Bryant and Sinéad Bailey, *Third World Political Ecology* (London: Routledge, 1997).

2) initially, a label for the school of thought also known as "neo-Malthusianism" and "eco-catastrophists" and associated with the works of Paul Ehrlich (1968), Donella Meadows et al. (1972), and William Ophuls (1977). These authors predicted looming environmental and social crises instigated by uncontrolled population growth in the South and exorbitant consumption in the West, and suggested drastic political prescriptions to prevent it. Though this school of thought was criticized across the political spectrum, it prompted an increasing academic and social interest in exploring the connections between politics and environmental change. See also CATASTROPHE THEORY and CATASTROPHIST MODEL.

See Paul Ehrlich, *The Population Bomb* (London: Ballantine, 1968); Donella H. Meadows, Dennis L. Meadows, Jorgen Randers, and William W. Behrens III, *The Limits to Growth: A Report for the Club of Rome's Project on the Predicament of Mankind* (New York: Universe, 1972); William Ophuls, *Ecology and the Politics of Scarcity: Prologue to a Political Theory of the Steady State* (San Francisco: W. H. Freeman, 1977).

Resource Economics: the branch of economics mainly concerned with the efficient management of natural resources, and with the government policies necessary to achieve this efficiency. Within neoclassical economics, resource economics has tried to broaden conventional neoclassical economic concepts to include the true costs of natural resources, particularly common goods such as air and water, in the determination of optimal allocation of resources. Nonmarket valuation methods have been developed that enable economists to determine the costs or benefits attributable to nonpriced natural resources. Insofar as environmental services can be considered as natural resources, resource economics can encompass environmental economics. The prevailing controversies over resource economics concern the treatment of the intrinsic value of ecosystem elements (valuation in the neoclassical framework can only measure what people consider as valuable), the treatment of sustainability within frameworks that discount future benefits, and the treatment of inter- and intragenerational equity. See also ENVIRONMENTAL ECONOMICS. *R. P.*

Restoration Ecology: a newly developing field that attempts to recreate or revive lost or severely damaged natural ecosystems. The basic idea is not simply to return lost species but to ensure that the reconstructed system is self-maintaining and integrated into the larger ecological landscape. Restoration projects have been implemented in several countries and targeted areas of wetlands and forests, some water bodies, and coral reefs. Classic examples of such projects are the clean-up of the tidal Thames River by the Thames Water Authority (Great Britain) and reconstruction of an oak-and-grassland savanna in Northbrook, Illinois, by the Nature Conservancy (United States). However, restoration ecology is subject to criticism on the grounds that reconstruction of "pre-disturbance" ecological conditions may not be feasible or socially acceptable; the reconstructed ecosystems are only rather expensive shadows of the originals.

COMMENTARY ON RESTORATION ECOLOGY

A curious stalemate exists between two distinctive professional groups when it comes to cleaning up the environment. On the one side are conservation biologists and some kinds of ecologists, while on the other are mainline economists. In the purest form, those who support environmental restoration often argue for a return to pristine conditions—before human intrusion in the extreme case. Matters of cost or other "unnatural" consequences are simply not as important as returning an ecosystem to an often-romanticized state of nature. Likewise, in its purest neoclassical economic form, those who question the romantic arguments point out that items not ordinarily accounted for in traditional economic theory and models have little standing: "externalities" is the common catch-all phrase for these awkward details. Because there is so little common ground to be discovered between these two extreme perspectives and groups, there has been little productive discussion and even less useful or informed activity in the important realm of environmental restoration. A more useful concept, and one intended to close the gap in the interest of making some progress, is the idea of reconstitution—in which a damaged ecosystem is viewed as having various higher-order potentials than the state in which it currently exists. Thus an out-of-use factory site, or "brownfield," would not be subjected to 100 percent clean-up and restoration to its pre-human condition but rather would be improved and made useful for better human purposes. Better human purposes in this case becomes a matter for meaningful discussion, debate, and decision—all taken with respect to sensible economic realities. *Garry D. Brewer*

Welfare Economics: the branch of economics that assesses the social desirability of alternative economic activities and resource allocations in terms of their efficiency in improving social welfare. In asserting normative (i.e., value) positions, welfare economics can lead to policy recommendations, unlike "positive economics," which focuses exclusively on how economies operate. One application of welfare economics is BENEFIT-COST ANALYSIS, which attempts to compare benefits and losses from different policy options. The challenge of welfare economics is to determine which outcomes are to be considered as improvements over others, given the difficulties of comparing alternatives when people differ in the utility they would assign to different outcomes.

2

Sustainability

2.1 Basic Sustainability Concepts

Agricultural Sustainability. See SUSTAINABLE AGRICULTURE.

Coevolution: related or coordinated changes over time of two or more components of systems. The concept of coevolution has been used by environmental theorists such as Richard Norgaard to emphasize that social systems involved in resource and environmental management must evolve along with the nature of the ecosystem and available technologies. Coevolution has also been used by evolutionary theorists such as Stuart Kauffman to account for the mutual dependence of humans and other species and structures of the broader ecosystem.

COMMENTARY ON COEVOLUTION

"Don't mess with Mother Nature" takes on a theoretical foundation when we recognize that human beings have coevolved with other species and overall ecosystem patterns. It means that the preponderance of ecosystem elements has supported human existence, or we would not have been among the survivors. In concrete terms this means that Nature has undergirded our nutrient cycling, climate moderation, control of pests, etc. To be sure, Nature has her nastier elements—poisonous snakes, malaria-bearing mosquitoes, tapeworms, and so on. But unplanned and unanticipated impacts on Nature risk disrupting the supportive environmental services more than disrupting the patterns that pose risks to human well-being. If such actions as clear-cutting a natural forest or introducing an exotic species have unexpected consequences, they are more likely to threaten valued wildlife, initiate soil erosion, or unleash damaging pests than to eradicate the few noxious species. This imbalance—negative surprises are likely to dominate positive surprises—runs contrary to the usual agnostic view prevalent in project evaluation that the unknown should be treated neutrally.

See Richard Norgaard, "Coevolutionary Development Potential," *Land Economics* 60(2) (1984): 160–73; Stuart Kauffman, *At Home in the Universe: The Search for Laws of Self-Organization and Complexity* (New York: Oxford University Press, 1995). *William Ascher*

Community Sustainability: an aspect of social sustainability that specifically relates to communities of people. It has dimensions of equity (fairness, social justice), empowerment (ability of people to exert a degree of control over decisions affecting their lives), capacity building (the ability of a group to solve its own problems), and sustainable livelihoods (the capacity to generate and maintain one's means of living). Community sustainability involves shared values and commitment to a community and is a foundation of social order; this concept is sometimes termed "social cohesion." See also SUSTAINABLE SOCIETY. *F. B.*

Sensible Sustainability: a level of sustainability that would require not only maintaining the total level of capital intact (a condition of WEAK SUSTAINABILITY) but also maintaining critical levels of each type of CAPITAL (humanmade, natural, human, and social). It is assumed that while humanmade and natural capital are substitutable (at least over some margin), the full functioning of the system requires a mix of the different kinds of capital.

Steady State: 1) in environmental laboratory sciences, the equilibrium at which the amounts of test substance being taken up and shed by the test organism are equal.

2) in environmental and socioeconomic systems, the state in which all system elements are fully operational and their consequences fully felt, as opposed to the start-up of systems under new dynamics. For example, if property rights are altered, the steady-state, long-term consequences may be very different from the initial consequences.

Steady-State Economy: 1) one of the most interesting economic ideas of the nineteenth century, expressed by John Stuart Mill in his classic *Principles of Political Economy*. Mill argued against the mere increase of production and accumulation and for a steady state in which people would be able to enjoy the fruits of their earlier savings, or material abstinence, which had been necessary for the accumulation of industrial capital. Mill presented this idea under the label of "stationary state," by which he meant a condition of zero growth in population and physical capital stock, but with continued improvement in technology and ethics. These arguments, with various modifications, are still invoked today.

See John Stuart Mill, *Principles of Political Economy with Some of their Applications to Social* (London: J. W. Parker, 1848).

2) a concept further developed by Herman E. Daly to define a sustainable economic development strategy. It defines the state of affairs in which flows of resources into production and of pollutants back to the environment are kept at a steady level. The steady-state economy, as well as steady-state environment, is not "frozen"; it exists in the process of constant change, adapting itself to external and internal factors and keeping its resilience. Daly's concept of steady-state economy has been instrumental in the development of ECOLOGICAL ECONOMICS.

See Herman E. Daly, *Steady-State Economics: The Economics of Biophysical Equilibrium and Moral Growth* (San Francisco: W. H. Freeman, 1977).

Strong Sustainability: a level of sustainability that requires maintaining different kinds of capital intact separately. From this perspective, given the limits to substitution between NATURAL CAPITAL and other economic assets, and the problems of irreversibility, uncertainty of threshold effects, and the potential scale of social costs associated with the loss of certain environmental assets, SUSTAINABLE DEVELOPMENT cannot be assured without complementary functioning of natural and humanmade capital. Therefore, preventing depletion of both is necessary. *R. P.*

Sustainability: the potential for a system to maintain or improve its functioning and the benefits derived from it. To be meaningful, the types of benefits (e.g., preservation of nature, economic growth, economic equity) have to be specified for particular applications. The meteoric rise of the notion of sustainability in the past several decades reflects a new perspective on how to assess human impacts on the natural environment and resources. The term originally came from the realm of natural-resource management, where it refers to a regimen for renewable resource use that would maintain specific levels of harvesting over time. This initial meaning was first broadened by ecologists who applied it to the task of preservation of the status and function of the ecosystems. Later, representatives of other disciplines made this term a part of their lexicon, which eventually resulted in the extreme ambiguity of its current use and a wide range of options to achieve it.

Sustainable Development: a development path that meets the major needs of the present without endangering subsequent needs and aspirations of future generations, allowing for the conservation of nature to be part of this path. The concept was popularized by the WORLD COMMISSION ON ENVIRONMENT AND DEVELOPMENT (WCED) in its 1987 report, *Our Common Future,* and has a number of important antecedents. From a political perspective, the notion of sustainable development emerged from the "North-South" dialogue on development and environmental protection, ongoing since the 1970s. Like other intentionally broad and politically driven concepts, sustainable development is subject to many interpretations and is difficult to present in terms of operational criteria. However, despite the ambiguity of the notion and the multiplicity of definitions and interpretations both at theoretical and practical levels, there is a relative consensus that sustainable development includes three main components: integration of environmental, economic, and social concerns, intergenerational equity, and international justice. See also ECODEVELOPMENT, WORLD CONSERVATION UNION (IUCN), BUDDHIST MODEL OF DEVELOPMENT, and GAIASOPHY.

COMMENTARIES ON SUSTAINABILITY AND SUSTAINABLE DEVELOPMENT

Sustainability has been criticized as a woolly, ambiguous concept that is resistant to precise definition, fraught with internal inconsistencies, and difficult to apply in practice. It shares these difficulties with other core societal values, such as freedom, equality, and justice. *Daniel Sarewitz*

Sustainable development, like a cloud, is difficult to get your hands around, but you know it when you see it. *Toddi Steelman*

"Sustainable development" is a zone of creative ambiguity in environmental policy discourse that enables participants to agree to live together while continuing to pursue incompatible agendas. *Steve Rayner*

The reason why some love the concept of sustainable development and others hate it is that it is used for quite different purposes. When "sustainable development" is used as an expression of the common interest of people to balance long-term economic growth, conservation, and equity among people, it can be a useful rallying cry and signal of *somewhat* common ground. In other words, it has great potential for the *promotion* of policies and practices that strive for some balance among these objectives. This is reflected in Rayner's insight that people can act as if they have agreement about the concept even if they are pursuing incompatible agendas. Yet the question is the degree of incompatibility, which may be less than that among those who subscribe to "sustainable development" as an objective versus those who do not. This is true even if there is no scientific way to use the concept to define the balance that would be agreeable to all people, inasmuch as no two people are likely to hold precisely the same preferences for this balance. If people expect "sustainable development" to be a technical guide to optimal policy, they will surely be disappointed—therefore the concept has been roundly derided as a "scientific" term to be used in the *intelligence* function. However, while it may be difficult or impossible to state definitively what would constitute the most optimal set of policies and practices for sustainable development, it is often painfully clear when these policies and practices fall far short. *William Ascher*

Sustainable Natural Resource Management: resource management that provides for the continuity of progressively beneficial and equitable resource uses with minimal damage to the ecosystem. Sustainable natural resource management techniques encompass careful attention to extraction rates; appropriate pricing, including sufficiently high royalties on the extraction of natural resources from government-controlled land and the elimination of most subsidies; the encouragement of positive externalities and discouragement of negative EXTERNALITIES; and the strengthening of open, competitive markets for natural resources.

Weak Sustainability: a term used to denote a modest criterion for judging the sustainability of economic progress, stipulating that the value of savings must at least equal the depreciation of manufactured capital plus the depletion of natural capital. That is, weak sustainability is achieved if the society's total capital does not decline. As a standard for evaluating economic, resource, and environmental management, weak sustainability allows for the increase in one type of capital at the expense of others. *R. P.*

Bellagio Principles: a set of principles developed and endorsed in Bellagio, Italy, by an international group of practitioners and researchers from fifteen countries (1996) as guidelines to assess progress toward sustainability. These principles deal with four main aspects of such an assessment: a) establishing a vision and goals of SUSTAINABLE DEVELOPMENT that are meaningful for decision makers; b) defining essential elements, scope, and practical focus of assessment; c) ensuring key process issues, such as openness, effective communication, and participation of all STAKEHOLDERS; and d) establishing an adequate capacity for continuity of assessment. The Bellagio principles are intended for use by stakeholders at all levels: grassroots and other NONGOVERNMENTAL ORGANIZATIONS (NGOS), businesses, governments, and international organizations.

Constant Utility Path. See DISCOUNTED UTILITARIAN INTERPRETATION.

Discounted Utilitarian Interpretation: one of the major approaches developed by Graciela Chichilnisky and her collaborators to develop sustainability as a measurable definition. In this approach, SUSTAINABLE DEVELOPMENT policies are those for which future generations can enjoy nondeclining consumption (utility) paths. This interpretation implies a premium on the consumption of future generations and in particular that: a) the present and the long-term future are treated symmetrically with a positive value of welfare in the very long run, and b) the intrinsic value of environmental assets is explicitly recognized. The sustainability path can then be obtained by maximizing a criterion function defined over the discounted utilities over an infinite time horizon, where utility depends on the flow of consumption and the level of environmental pollution.

A criterion for sustainable development compatible with this approach is that of a set (or "vector") of social goals proposed by Pearce et al. (1990). The vector of social goals includes real per capita income, health and nutrition, education, access to resources, fair income distribution, and democracy. Sustainability is then defined as a state in which the development vector does not decrease over time.

See David W. Pearce, Edward Barbier, and Anil Markandya, *Sustainable Development and Environment in the Third World* (Worchester: Edward Elgar, 1990); Graciela Chichilnisky, Geoffrey Heal, and Alessandro Vercelli, *Sustainability: Dynamics and Uncertainty* (Dordrecht: Kluwer, 1998).

European Sustainability Index (ESI): an aggregate measure of urban sustainability developed under the auspices of the International Institute for the Urban Environment and applied on an experimental basis in twenty European cities: Aalborg, Bath, Bessançon, Bilbao, Charleroi, Delft, Dublin, Espoo, Gent, Götenborg, Granada, Hamm, Leicester, Porto, Rennes, Suhl, Terni, Venice, Vienna, and Volos. It is intended to assess the progress toward local environmental and quality-of-life sustainability, and is suitable for use in international comparisons. The ESI is comprised of ten core indicators: healthy environment (number of days per year that local standards for air quality are not exceeded); green space (ratio of people with access to green within a certain distance); efficient use of resources (total energy and water consumption, production of waste for final disposal, ratio of renewable to nonrenewable energy sources); quality of the built environment (ratio of open space to the area used by cars); accessibility (distance traveled by transport per capita per annum); green economy (percentage of companies that have joined ecomanagement and audit schemes); vitality (number of social and cultural activities or facilities); community involvement (number of voluntary groups per 1,000 inhabitants); social justice (percentage of people living below the poverty line); and well-being (surveys of citizen satisfaction with quality of life). Despite concerns about the range of indicators and their measurement, the development of ESI has certainly been a valuable initiative to provide quantitative criteria for analyses, adjustment, and improvement of local policies.

See Tjeerd Deelstra, "The European Sustainability Index Project," in *A Sustainable World*, ed. Thaddeus Trzyna (Sacramento: International Center for Environment and Public Policy, 1995), 115–51.

Factor 10: the magnitude of reduction in natural resource consumption that certain experts claim is necessary for the industrialized countries to achieve SUSTAINABLE DEVELOPMENT. A group of internationally recognized scientists (Wuppertal Club; also Factor 10 Club) calculated that sustainability of world development can be ensured only if the West reduces its resource consumption by a factor of ten. Though quite popular in Europe and adopted by several businesses, the Factor 10 notion presents an enormous social, political, and economic challenge.

See F. Schmidtt-Bleek, ed., *Carnoules Declaration. Factor 10 Club* (Wuppertal: WIKUE, 1994).

Rawlsian Interpretation of Sustainability: one of the three major approaches developed to make the definition of sustainability operational. It is based on the Rawlsian principle (Rawls, 1971) that the general welfare is maximized by maximizing the welfare of the worst-off generation. Thus, sustainability in the above sense requires determining the level of welfare of the worst-off generation corresponding to all feasible consumption paths and then determining a feasible path that maximizes this minimal value.

See John Rawls, *A Theory of Justice* (Cambridge, MA: Harvard University Press, 1971); Geoffrey M. Heal, *Interpreting Sustainability*, ENI Enrico Mattei Discussion Paper 1.95 (Milan: Fondazione, ENI, 1995).

Sustainability Indicators. See under SECTION 1.2.7.

Sustainable National Income (SNI). See ACCOUNTING.

2.3 Philosophical Approaches and Social Movements

Alternative Normativism: a label applied to a school of thought critical of the dominant Western mode of development. Alternative normativists propagate a holistic understanding of human well-being and emphasize the role of self-reliance, alternative lifestyles, and cultural fulfillment, as well as material aspirations. Among the most prominent representatives of alternative normativism are Indian scholars Vandana Shiva and Shiv Visvanathan, who denounce developmentalism by questioning its epistemological foundations. They see Western science and its definitions of development, progress, and sustainability as the product of white, male, Western thinking, which is essentially REDUCTIONIST, serves the interests of the global market economy, and therefore is the root cause for the global ecological crisis. Proponents of this worldview criticize the now-dominant concept of SUSTAINABLE DEVELOPMENT as being based on an instrumental view of both people and nature, and propose a shift of values to simple lifestyles based on local cultures and human and nature's integrity.

See Vandan Shiva, *Staying Alive: Women, Ecology, and Development* (London: Zed Books, 1989); Shiv Visvanathan, "Mrs. Brundtland's Disenchanted Cosmos," *Alternatives* 16 (3) (1990): 377–84.

Alternative Structuralism: a label applied to a variety of theories that emphasize the negative impact of global economic structures on the developing nations. This orientation unites such different ideological streams as DEPENDENCY THEORY, CORE-PERIPHERY THEORY (as developed by Samir Amin), and world system theory promoted in the works of Immanuel Wallerstein. Alternative structuralism has significantly contributed to the understanding of the developing countries' lasting impoverishment and environmental degradation. Proponents of this line of thinking advocate a radical restructuring of the global economic system in favor of developing countries. However, along with the champions of the dominant development paradigm, alternative structuralists accept the notion of progress brought about by modernization with rapid economic growth and emphasize the primacy of state agencies and the role of Southern elites to counter the process of capital accumulation in the North. Therefore, alternative goals of development like the need for grassroots and intermediary levels of socio-economic changes are not explored.

See D. L. Sheth, "Alternative Development as Political Practice," *Alternatives* 12 (1987): 155–71.

Animal Rights: a philosophical perspective and social movement based on the sentiment that animals should live free from human interference or predation. The principle of animal rights implies that nonhuman species have absolute moral rights to be treated as ends, and not as means for some human purpose. Particular characteristics that endow animals with rights have not been specified, though it is assumed that to have rights animals should possess "inherent value." As a social movement, animal rights activism has been historically focused on eliminating unnecessary suffering, especially that inflicted by cruel and uneducated owners. Currently, this movement is particularly active in the United States and in Great Britain. Most of the major animal rights groups share basic principles with the proponents of similar humanitarian causes and strong environmentalism.

COMMENTARIES ON ANIMAL RIGHTS

Among radical animal rights activists it would appear that, as Orwell observed, "some animals are more equal than others." For example, clandestine releases of an exotic species from mink farms in Britain have ravaged the indigenous mole and wildfowl populations. *Steve Rayner*

The "animal rights" movement is the result of the systematic stifling of human satisfaction and self-reliance, and is most pronounced in industrialized countries. Economics, technology, culture, and politics restrict sufficiency, protection, creation, identity, and freedom—the real stuff of individual and community. Who feels the stifling first? The young and educated, who sense inside themselves an emotional amputation that they cannot clarify or verbalize. "Silent" animals become the cause onto which they project their frustrations. These activists rarely have direct experience or education in the dynamics of species interactions and communities, and thus they fixate on the individual or group of animals as symbols of their spiritual and psychological dislocation. *Susanne Swibold and Helen Corbett*

Anti-Environmentalism: coordinated opposition against the environmental movement that emerged in industrialized countries by the late 1970s and reached its peak in the late 1980s. Elements of the business community, especially in capital-intensive and extractive industries, use a variety of strategies: political lobbying, public relations campaigns, and research designed to refute the claims of environmentalists and regulatory agencies. For instance, some U.S. corporations and business associations increased their use of litigation (though not successfully), in particular "strategic lawsuits against public participation" (SLAPPS). SLAPPS charge environmental groups and grassroots activists with defamation of character, interference with business, or conspiracy. Another strategy has been to form coalitions for political lobbying and public relations, such as the National Wetlands Coalition (oil drillers and real estate developers) and the U.S. Council on Energy Awareness (the nuclear power industry).

See Kirkpatrick Sale, *The Green Revolution: The American Environmental Movement 1962-1992* (New York: Hill and Wang, 1993).

COMMENTARY ON ANTI-ENVIRONMENTALISM

The Reagan administration (1981–89) was remarkably unsuccessful in creating a public backlash against environmental regulations on the basis of their cost to the economy (Council on Competitiveness), their alleged destruction of jobs, or their creation of bureaucratic hurdles for individuals and small businesses. *Robert Healy*

Back-to-Nature Movement: a movement to create experimental communities in the countryside and wilderness. This movement, started mainly by young intellectuals, has been based on the premise that a subsistence economy ensures harmony with nature and community peace, providing salvation from disrupted environments and the social delinquencies of contemporary industrialized societies. Such movements have existed since the Enlightenment, but have acquired more scale and visibility in the last few decades. The movement derives some of its ideological inspiration from the rhetoric concerning indigenous societies (see also INDIGENOUS KNOWLEDGE).

Ironically, many "back to nature" exiles occupy the same sites where these indigenous peoples were situated—in the wildest and/or most beautiful parts of the colonized countries: Maine and Washington state in the United States, Wales in Great Britain, the Crimea in Russia. Questions to ponder: a) Is a subsistence economy always sustainable? b) Is it possible to reconstruct a subsistence economy? c) Is it possible to ensure true isolation? d) Are these sites indeed "the Garden of Eden"? *Natalia Mirovitskaya*

Bioregionalism: one of the most ecocentric branches of Green political thought. Bioregionalism seeks the integration of human communities with the nonhuman world at the level of particular ecosystem (BIOREGION) and argues that self-government by human communities within bioregional contours offers the best guarantee of social and ecological harmony. It draws heavily on the ECOANARCHIST tradition of PIOTR KROPOTKIN as well as the cultures of indigenous peoples and the more successful local communities, such as the Amish (see also INDIGENOUS KNOWLEDGE). Bioregionalism emphasizes decentralization, human-scale communities, cultural and biological diversity, self-reliance, cooperation, and community responsibility (both social and biotic). From a political perspective, proponents of bioregionalism promote a patchwork of anarchist polities connected by networking and information exchange rather than through a formal state apparatus. Bioregionalism is particularly popular within the North American Green movement and enjoys the general support of both SOCIAL ECOLOGY and DEEP ECOLOGY.

COMMENTARY ON BIOREGIONALISM

One of the key elements of identifying a bioregion is naming it. The name is often taken from the local river basin, mountain range, or name of the indigenous group that historically inhabited it. The bioregional name can be helpful in ascribing certain natural or cultural characteristics to the area and in distinguishing it from traditional geopolitical units, which usually have boundaries less dependent on natural features. The term "ecological address" is also sometimes used to identify a bioregion. *Robert Healy*

Core-Periphery Model (Center-Periphery Theory). See under SECTION 1.1.3.

Deep Ecology: a philosophical perspective that envisions humans as equal but not superior to other life forms. The term was coined and advanced first by Arne Naess, a Norwegian philosopher and social activist. He defined deep ecology as a normative movement inspired by human experience in nature and by ecological knowledge. The primary norms of deep ecology center around the notion that humans are just one of the many entities on earth that have intrinsic worth. Deep ecology is often contrasted with "shallow ecology," which is defined as a concern about pollution and resource depletion and has an ANTHROPOCENTRIC focus. *T. S.*

Dependency Theory: an approach toward development that emerged in the 1960s and 1970s, largely through the work of Latin American social scientists. Dependency theorists see the world in terms of a center and a periphery (see CORE-PERIPHERY MODEL). Development occurs in the center of the world capitalist system (the First World nations) and underdevelopment occurs in the periphery (the developing nations). Development at the center occurs as a consequence of underdevelopment in the periphery. The premise is that the global capitalist economy appropriates surplus labor and capital from peripheral countries for the benefit of industrial countries. According to dependency theory, in order to develop, developing nations must break the pattern of dependency and reject a subordinate role. Critics of dependency theory have argued that it cannot explain the rapid growth of some developing countries, and that its mechanisms have not been empirically demonstrated.

See Amiya Kumar Bagchi, *The Political Economy of Underdevelopment* (Cambridge: Cambridge University Press, 1982); Fernando Henrique Cardoso and Enzo Faletto, *Dependency and Development in Latin America* (Berkeley: University of California Press, 1979). *T. S.*

Ecoanarchism: a diverse range of positions representing various versions of anarchist political theory with the view that exploitation of nature predates and reflects exploitation of humanity. Ecoanarchism thus has individualistic elements (after Stirner), syndicalist elements (after Proudhon) and revolutionary or radical elements (after Bakunin). All share a common and total rejection of the state as political institution, emphasizing the self-organizing capacity of "the people." Most would regard the state as a key institution responsible for the domination of exploitation of both nature and human beings. Proponents of this approach argue that not only is anarchism as political philosophy most compatible with an ecological perspective, but also that it is grounded in ecology. Ecoanarchists oppose centralization, hierarchy, privilege, and domination. Instead, they advocate maximum political and economic autonomy of nonhierarchical, self-governed local communities as the best option for a harmonious human-nature relationship. From a historical perspective, ecoanarchism is linked to the works of PIOTR KROPOTKIN and Elisee Reclus. The contemporary body of ecoanarchism thought is rather diverse, though united through its links to the political theory of anarchism and the strong support within the grassroots and extraparliamentary activities of the worldwide Green movement. The two most visible branches of contemporary ecoanarchism are SOCIAL ECOLOGY and BIOREGIONALISM. *I. W.*

COMMENTARY ON ECOANARCHISM

Despite the idea that anarchism connotes chaos, ecoanarchists have been responsible for a wide range of innovative local initiatives contributing to the goals commonly associated with SUSTAINABILITY. Community recycling, transport, and housing schemes in the San Francisco Bay area and Manhattan serve as clear examples. The more constructive movements within ecoanarchism have been overshadowed by individualist and radical versions of ecoanarchism, which include the practice of ecological sabotage ("ecotage"), and tend to attract "ecowarriors" indulging in rather masculine, patriarchal acts of destruction. It would be a mistake to discount ecoanarchism as a fecund source of sustainable practices due to the high profile gained by the more destructive tendencies within the movement. *Ian Welsh*

Ecodevelopment: 1) a doctrine of environmentally, economically, and socially self-supported development first coined by Canadian scholars and put forward as a planning concept of international development by the UNITED NATIONS ENVIRONMENT PROGRAMME (UNEP) in 1975. The concept was intended to reorient prevailing strategies of economic development toward meeting basic needs of low-income populations in developing countries, and to strengthen economic independence. Major prerequisites of ecodevelopment were the formulation of economic goals consistent with the potentials of local and regional ecosystems, and the use of technology and organizational forms that relate to the natural ecosystems and local and cultural patterns. The UNEP also recommended elaborating an international law of ecodevelopment by integrating environmental and development law, which were traditionally regarded as separate legal regimes. The concept of ecodevelopment was criticized as being too "ideological" and was eventually supplanted by the concept of SUSTAINABLE DEVELOPMENT, some aspects of which overlap with those of ecodevelopment.

See United Nations Environmental Programme *The Proposed Programme* (Nairobi: UNEP, 1975).

2) the term is also used generically to describe an integrated approach to environment and development.

Ecofascism: a denotation given by their critics to the proponents of LIFEBOAT ETHICS, whose basic doctrine is that because of absolute constraints on available resources, equality of access to those resources is impossible unless the numbers wanting access are greatly reduced. According to Murray Bookchin, such a doctrine will inevitably lead to an "ethics of repression and totalitarian control."

See Murray Bookchin, *The Philosophy of Social Ecology: Essays on Dialectical Naturalism* (Montreal: Black Rose Books, 1995); David Pepper, *Political Roots of Ecological Environmentalism* (London: Croom Helm, 1984.)

Ecofeminism: 1) prominent stream within feminist theory that assumes crucial connections between the domination of nature and of women. Therefore, merging the critical and transformative potentials of ecology and feminism would "remake the planet around a totally new model." Ecofeminism contains a range of theoretical positions united by their critique of patriarchy and patriarchal conceptions of self, society, and cosmos that are perceived to be at the roots of this dual oppression. Beyond this common ground, ecofeminism is divided into at least three main theoretical streams: cultural, social, and Southern (so called because of its association with the developing world). Cultural (also nature) ecofeminism asserts a special association between women and nature, and claims that women's essential features, such as empathy, caring, and intuition, can help to develop more sustainable ways of living and social relations. Social ecofeminism, in contrast, recognizes gender as a social construct and strives to enrich the socialist-feminist tradition with environmental perspectives.

Southern ecofeminism challenges both the patriarchal character of society and the dominant reductionist and dualistic system of science. Southern ecofeminists stress the need for new scientific and philosophical approaches that would respect the plurality of different cultures and traditions and would make it possible to transform the existing exploitative social order and achieve greater human potentials.

2) ecofeminism has also become a rather diversified and decentralized social movement. Ecofeminist groups in different countries campaign for diverse issues, such as GREEN CONSUMERISM, conservation of wildlife, and women's social, economic, and reproductive rights.

See Rosi Braidotti, Ewa Charkiewicz, Sabine Hausler, and Saskia Wieringa, *Women, the Environment and Sustainable Development: Towards a Theoretical Synthesis* (London: Zed Books, 1994).

Eco-Marxism: ecopolitical theory injected with various Marxist perspectives. Recent efforts to address environmental issues from the Marxist perspective have been conducted in two main directions: the so-called orthodox and humanist eco-Marxisms. From the perspective of orthodox eco-Marxism, environmental degradation can be traced directly to the exploitative dynamics of private capital appropriation and indicates the inefficiency of the capitalist system. Therefore, the solution of the environmental crisis is through the radical transformation of the relations of production combined with a progressive development of knowledge and technology. In the orthodox eco-Marxist approach, the working class is best suited for assuming ultimate power and responsibility over the whole transformed system, both social and environmental. On the other side, the proponents of humanist eco-Marxism seek to address environmental issues through a creative development of philosophical writings of the young Marx. They admit that this challenge requires significant correction of at least three basic areas of Marxist heritage: a) the role of material progress; b) the relation of forces and relations of production; and c) survival versus abundance. Humanistic eco-Marxists argue that the traditional Marxist belief in material progress and potential use of capitalist technology must be reexamined in view of the visible danger this technology poses for society's ecological support system. Also, the relation between forces and relations of production (the main contradiction of capitalism in the classical Marxist account) needs to be supplemented by recognizing the second contradiction—between forces and relations of production and conditions of production, which can generate a crisis by undermining the viability of the social and "natural" environment and bring about a social movement that acts as a barrier to capital accumulation. Social movements engendered by two main contradictions of capitalism (labor and environmental in particular) are natural allies in the struggle to transform the modes of production in the direction of greater socialization and justice (O'Connor, 1996). The third major issue addressed (though never resolved satisfactorily) by humanistic eco-Marxism is the distinction between freedom and necessity, the former corresponding to the mastery of social and natural constraints and the latter corresponding to subservience to natural and social constraints. The Frankfurt School developed an original strand of Marxist ecological thought (see also CRITICAL THEORY).

See James O'Connor, "The Second Contradiction of Capitalism," in *The Greening of Marxism*, ed. Ted Benton (New York: Guilford Press, 1996), 197–221; Robyn Eckersley,

Environmentalism and Political Theory: Toward an Ecocentric Approach (Albany: SUNY Press, 1992).

COMMENTARY ON ECO-MARXISM

Eco-Marxism represents a heroic attempt by committed Marxists to realign Marxist works with an environmental agenda. While it is too early to write off Marxism completely, as advocates of neoliberalism have done to the cost of their societies, a green Marxism is an oxymoron due to the celebration of productive forces, sustained economic growth, and the transformative potential attributed to the "working class" within Marx's work. Marx's critique of classical political economy still has much to offer, but its model of social change and transformation is significantly eclipsed by changes within the means and relations of production. These changes render ideas of a unified and disciplined class seizing state power for their own special purposes anachronistic, given the rise of globalization. The work of David Harvey stands out as one of the most sophisticated attempts to demonstrate the contemporary relevance of Marxism in the context of the environment and sustainability.

See David Harvey, *Justice, Nature and the Geography of Difference* (Oxford: Blackwell, 1996). *Ian Welsh*

Ecosocialism: a body of theory that attempts to reconsider the democratic socialist case in light of ecological challenges. Ecosocialists argue that it is the competitive and expansionist dynamics of the capitalist system that have led to widespread environmental degradation and social insecurity. Therefore, capitalism must be replaced with a democratically controlled, nonmarket allocative system that ensures ecologically sustainable production for genuine human need. In distinction from ECO-MARXISTS, ecosocialists recognize that the industrial working class can no longer be considered a revolutionary force and that progressive social, cultural, and political change is more likely to emerge from new social movements. Ecosocialist analysis of societal contradictions expands beyond that based on class to incorporate human/nature and North/South contradictions. The resolution of these contradictions requires, from this perspective, a "new internationalism"—a model of development based on the redistribution of wealth not only within nations but also internationally so that all peoples may pursue the lifestyle that is within the earth's CARRYING CAPACITY. Ecosocialists regard the state's role as vital for the development of a socially just and ecologically sustainable society. The major theoretical positions of ecosocialism—the rejection of economic growth orientation, the emphasis on ecologically sustainable production for human need and collective control over the relationship with nature, the attempt to encourage a working partnership between social movements, and the new internationalism—together represent a major revision of emancipatory political thought. The basic ideas of ecosocialism were most comprehensively presented by Andre Gorz.

See Andre Gorz, *Ecology and Politics* (Boston: South End Press, 1980); Andre Gorz, *Farewell to the Working Class* (London: Pluto Press, 1982); Robyn Eckersley, *Environ-*

mentalism and Political Theory: Toward an Ecocentric Approach (Albany: SUNY Press, 1992), 119–44.

Gaiasophy: a theory asserting that the earth's biosphere, or "Gaia" (the Greek for "Mother Earth"), is a complex self-regulating entity, which like a living organism is capable of creating the environment that maintains its sustainability for life. The Gaia hypothesis was introduced in 1972 by the British scientist James Lovelock and from the historical perspective might be considered as a development of Vernadsky's concept of BIOSPHERE. Some results of scientific research in the 1980s mitigated the initial skepticism of natural scientists toward this hypothesis and an INTERNATIONAL GEOSPHERE-BIOSPHERE PROGRAM (IGBP) has been (at least partially) designed by its authors to test it. The Gaia theory, which stresses the relatively low level of human knowledge on the resisting, recuperative, and regenerative capacity of the biosphere, also speaks to humans' ability to reconstruct the biosphere and the necessity to harmonize human activities with the natural cycles of raw materials.

See James Lovelock, *Gaia: A New Look at Life on Earth* (New York: Oxford University Press, 1979).

Gandhian Concept of Development (Swaraj Doctrine; "Hindu Development"): a doctrine of development that was taught in India beginning in the 1930s by Mohandas K. Gandhi and which expounded the notions of BASIC NEEDS and self-reliance. The Swaraj Doctrine promoted an ideal of development of self-supporting rural economies, for which each village would be perceived as a complete republic, independent of its neighbors for its vital wants (food, clothes, clean water, basic education, and recreation). These "individual republics" are nevertheless connected with many others. Most activities should be conducted on a cooperative basis. A series of demonstration programs for village transformation was developed in the 1930s. Gandhian ideas of development were replaced by European-style industrialization in the 1940s, though they were revisited under Nehru's great experiment with community development after Indian independence and during the 1950s. Some basic elements of the Swaraj Doctrine (decentralized decision making, local resource-based production to meet local needs) are found in the contemporary concepts of ECODEVELOPMENT and SUSTAINABLE DEVELOPMENT.

Gender and Development Approach (GDA): the most recent approach toward gender issues within the framework of economic development. In distinction from the WOMEN IN DEVELOPMENT (WID) approach, GDA attempts not only to increase women's participation in development, but also asserts that development cooperation must encourage and support changes in attitudes, structures, and mechanisms at every level to reduce gender inequalities, to ensure political power sharing, women's economic empowerment, and women's equal access to and control over social development opportunities. The GDA approach has been recently adopted by major development organizations and the EUROPEAN UNION, which have begun training their development officials in "gender literacy," including how to conduct gender analysis in assessing policy documents, employment policies, project proposals, data collection, and so on. However, GDA still treats development in largely technical ways, and as such does not address some of the wider political issues.

Currently GDA is the most progressive framework for thinking about women's roles in development, gender equity, and equality. If implemented fully and in good faith, it might bring necessary changes within development institutions and the process of aid. However, like the previous attempts to bring gender perspective, women have simply been "added" to traditional frameworks of development. In the final analysis, GDA is weak on the issue of policy coherence between development cooperation and all other areas of macroeconomic policy. It does not analyze gender-based inequalities and injustices resulting from all aspects of the North-South relationship, including trade, environment, agriculture, foreign, and security policies. Therefore, the traditional paradigm of development, firmly embedded in the frameworks of modernization and economic growth—ultimately reinforcing the permanent subordinate position of women in the society—is not challenged. *Natalia Mirovitskaya*

Holistic Ecology. See DEEP ECOLOGY.

Kropotkin, Piotr Alekseyevich (1842–1921): Russian anarchist and noted geographer. His writings include *Mutual Aid* (1902), which holds that cooperation rather than competition is the norm in both animal and human life, and *Fields, Factories and Workshops* (1899, reprinted recently under the title of *Evolution and Environment*). From Kropotkin's perspective, an ideal society is a self-sufficient decentralized institution, organized in small political units where relationships of domination and hierarchy are replaced by mutual support and cooperation. Such a society satisfies its needs by the minimum use of energy and through diversification of human occupations that are aimed to provide the stability of a community rather than profits for different groups. Domination and exploitation of nature by humans is but an extension of the domination of humans over one another. If small-scale decentralized societies can avoid exploitative human relationships, these societies would also be the best for harmonious human-nature relations. Kropotkin envisaged future industrial organization to be shaped by small-scale autonomous industries federating into co-operatives to overcome the disadvantages of market weakness. The purpose of mechanization is not to create profits but to free the worker for alternative work in the fields and more creative occupations. Technology has to be appropriate and owned by the small community. Such economic organization would result in radically different human landscape: small, green cities, hamlets, and villages with small intensively farmed plots located everywhere. Kropotkin's ideas are believed to have greatly influenced the Garden City Movement in Britain and were very much repeated in the seminal *Blueprint for Survival* (though without reference). Kropotkin's writings pioneered ECOANARCHISM, which was further developed by Murray Bookchin.

See Piotr Kropotkin, *Evolution and Environment* (Montreal and New York: Black Rose Books, 1995).

Widely known as the "anarchist prince" due to his roots within the Russian aristocracy and his espousal of mutual aid and pacifism, Kropotkin's radical credentials were irreparably damaged in the eye of many when he publicly supported the First World War toward the end of his life. Compared to the more revolutionary teachings of other classical European anarchists such as Mikhail Bakunin (who maintained a belief in the revolutionary potential of the peasantry) and Pierre Proudhon (a syndicalist famous for the statement "property is theft"), Kropotkin's anarchism was predominantly reformist. *Ian Welsh*

Land Ethic: a philosophical approach that extended ethical considerations from a human-centered focus to encompass land and other aspects of the environment. The basic principles of this approach were elaborated by Aldo Leopold in his 1949 book *A Sand Country Almanac*. Leopold viewed land as a community of soils, water, plants, and animals worth preserving, while humans should be held responsible for land stewardship as an "ecological necessity." He argued that many environmental services (or "land services") are essential but have no commercial value; therefore land conservation decisions should not be based exclusively on economic criteria but rather should also preserve the land's integrity, stability, and beauty. One of the most striking points of this approach is its reference to predators that are to be spared not only for pragmatic reasons but because they are members of a community in which humans are just other members. Considerations of the land ethic prompted Leopold (along with Robert Marshall) to establish the Wilderness Society in 1935. Many environmentalists consider land ethic the first formulation of the modern ecocentric ethics.

See Aldo Leopold, *A Sand Country Almanac, and Sketches Here and There* (New York: Oxford University Press, 1949).

Lifeboat Ethics: a label coined by Garret Hardin in 1974 to describe the ethics of having to sacrifice some people in the condition of scarce resources, through the following analogy: ten people are safely provisioned in a lifeboat in a sea full of drowning people crying for help. If they pull one more person on board, there will not be enough provisions for all. Therefore, the correct course of action is to ignore the cries for help and to ensure the safety of those already on the lifeboat. This analogy was easily transposed to the global scale and to the North-South relationship. Under the limited CARRYING CAPACITY and rapid population growth, disbursing of food aid may be viewed as violating the principle of natural carrying capacity of the planet. The opponents of lifeboat ethics positions have often referred to it as ECOFASCISM.

See Garret Hardin, "Living on a Lifeboat," *BioScience* 24 (1974): 10.

<h2 style="text-align:center">COMMENTARY ON LIFEBOAT ETHICS</h2>

Southern feminists, in addressing the population issue, expanded Hardin's metaphor to the following example from Bangladesh: there are only nine cabins in the steamer launch that comes from Dhaka to Patuakhali. In the nine cabins only eighteen people can travel. The ticket is expensive, so only the rich people travel in the cabins. The rest of the common passengers travel on the deck. The latrine facility is provided only for cabin passengers. But sometimes the passengers from the deck want to use the latrines. The cabin passengers allow them to do that because they are afraid that if the poor deck passengers get angry then they might make a hole in the launch. Then the launch will sink, they will no doubt die, but the rich cabin passengers will not survive either. So, my dear sisters, do not give birth to more children as they cause problem for the cabin passengers.

See Akter, cited in Maria Mies and Vandana Shiva, eds. *Ecofeminism* (New Delhi: Kali for Women, 1993): 280. *Natalia Mirovitskaya*

Radical Environmentalism: a branch of contemporary environmentalism character-ized by an ideological opposition to Western economic worldviews and ANTHROPO-CENTRISM. The radical environmentalists maintain the need for a dramatic transfor-mation of human values and beliefs as well as the total reorientation of a Western culture that is environmentally destructive. The intellectual basis of the new radical environmentalism is extremely diverse and includes such different ideological trends as DEEP ECOLOGY, SOCIAL ECOLOGY, ECOFEMINISM, and GAIASOPHY. Action-oriented radical environmental groups range from the local to the international level, and from electronic computer networks to activist organizations and political parties. In gen-eral, they have more informal and decentralized bases than other nongovernmental environmental organizations and apply militant nonviolent direct action as their cen-tral tactics. GREENPEACE INTERNATIONAL, EARTH FIRST!, and RAINFOREST ACTION NETWORK represent distinct kinds of action-oriented radical environmental organiza-tions.

Social Ecology: 1) in academic environmental studies, an interdisciplinary field encom-passing the disciplines concerned with individuals and societies in their institutional, built, and natural environments. Most broadly the field is bounded by a) ergonomics, ar-chitecture, and planning; b) sociology, economics, and politics; and c) toxicology, public health, and ecology. *S. R.*

2) a trend in environmentalism based on the premise that ecological problems stem from social problems, such as hierarchy and domination. Like DEEP ECOLOGY and ECOANARCHISM, social ecology contains elements of anarchism and has been particu-larly influenced by the teaching of PIOTR KROPOTKIN. The solution to environmen-tal problems suggested by social ecologists is to eliminate hierarchy and patriarchy in order to recreate a "natural society."

Spaceship Economy: an alternative model of economic priorities, suggested by Ken-neth E. Boulding, that emphasizes natural-resource and human-resource sustainability rather than economic growth, which has traditionally been the most important objec-tive of development. Given population growth, exhaustion of natural resources, and restricted space for disposal of various wastes, the world economy, according to Bould-ing, should be treated as a closed system for which the conventional measures of eco-nomic success (gross national product [GNP], consumption, and income) are inappro-priate. Rather, the maintenance of NATURAL CAPITAL and the development of HUMAN

CAPITAL are fundamental to human welfare. Though the concept of spaceship economy has been criticized for exaggerating resource and environmental problems, it is an important benchmark in the history of environmental thought.

See Kenneth Boulding, "The Economics of the Coming Spaceship Earth," in *Environmental Quality in a Growing Economy*, ed. Henry Jarrett (Baltimore, MD: Johns Hopkins University Press, 1966).

Stewardship: 1) defined broadly, the individual's responsibility for managing his/her life and property with appropriate respect to the rights of others.

2) from the perspective of contemporary environmentalism, stewardship includes the notion of human responsibility for nature, the ethical perspective that the value of nature transcends human needs, and a notion of fairness in sharing environmental space and natural resources. Stewardship therefore implies obligations as well as rights in the exercise of ownership and the uses of property.

Transpersonal Ecology: one of the most recent philosophical trends reflected in the writings of Australian scholar Warwick Fox. From his perspective, philosophy should look upon all objects in the world as a part of a continuous, interconnected whole. The proper state is the experience of totality, inner sameness—something that cannot be understood and reflected by mechanistic, REDUCTIONIST science. The sense of continuity should come from a "new interactive science," and the sense of identity from mystical tradition. Therefore, transpersonal ecology combines the urge for an interdisciplinary approach and the basic essence of transcendentalist philosophy: the sense of complete loss of personal separateness and of spiritual confluence with the evolving universe. This emancipated state is believed to be the basis for sustainability. One can see very strong connections between this outlook and certain Eastern religions (see also BUDDHIST MODEL OF DEVELOPMENT).

See Warwick Fox, *Toward a Transpersonal Ecology: Developing New Foundations for Environmentalism* (Toronto: Shambhala Press, 1992).

Women and Development (WAD) **Approach:** a neo-Marxist feminist approach that emerged in the late 1970s and was a logical, though more radical extension of the WOMEN IN DEVELOPMENT (WID) position. In distinction to the WID approach, the WAD perspective focuses on the gender-development relationship rather than only on strategies of "bringing women in." The essential point of WAD is its argument that women have always been important economic actors in their societies. However, the projects of "integrating women into development" designed by international financial institutions under the pressure of the WID approach in practice serve primarily to sustain existing international structures of inequality. The WAD perspective has been eventually substituted in feminist discourse by the GENDER AND DEVELOPMENT APPROACH (GDA).

Women in Development (WID) **Approach:** an analytical perspective dealing with the differential impact of development and modernization strategies by gender. The term emerged from Ester Boserup's groundbreaking book, *Women's Role in Economic Development* (1970) and was publicized in the early 1970s by women in academia and American feminists. The basic argument of the WID approach is that women's experience of development, of societal changes, and of modernization in particular differs from that of men and that the conventional development strategies do not bring expected benefits for women, particularly in developing countries. Within the WID framework, specific strategies for "integrating women into development" were put forward. Though criticized by some for the validity of its basic assumption, and by others for not going far enough (see GENDER AND DEVELOPMENT [GDA]; WOMEN AND DEVELOPMENT [WAD]), the WID approach has been instrumental in developing research and discourse focused on gender specifics of development.

COMMENTARY ON WOMEN IN DEVELOPMENT

Problems with women in development stem from the deep-seated prejudice that women, African Americans, native people, and the poor are inferior and therefore perceived as objects (to perform the necessary tasks of reproduction, child rearing, keeping the house clean). As objects they become property, and property has no rights, can be bought, sold, or killed. Women are being psychologically and physically marginalized, still treated as property. Man has created the rule of property—man can change the rule. We are further creating our own suicide chamber by dismantling the human community of voices, intuitions, insights, and values, limiting our choices for living when *women's voices* are silenced. Imbalance will find balance somewhere and it may be without humans at all. The question becomes, why have children at all? *Susanne Swibold*

See Ester Boserup, *Women's Role in Economic Development* (London: Allen and Unwin, 1970).

2.4 Sustainable Development Outlooks and Models

Boserupian Thesis: the proposition that population growth stimulates agricultural production, development, and transfer of advanced technology from one society to another and gives impetus to new methods of production. On the basis of her research of the early development of agricultural communities, Ester Boserup, a Danish agricultural economist, demonstrated that population pressure brought about the transition from shifting cultivation to increasingly intensive forms of agriculture. In other words, the population's pressure on current resource limits resulted in pushing back these limits. She also demonstrated that although wet tropical environments may allow abundant plant growth and potentially high food production, these areas are also afflicted with bacteria and parasites that have impeded human populations throughout centuries. Thus environments in cooler or drier areas, though faced with more difficult conditions for food production, may support larger populations. Boserup's research emphasizes the complexity of human connections with environmental resources and the importance of the life-support function of the environment.

See Ester Boserup, *Economic and Demographic Relationships in Development* (Baltimore: Johns Hopkins University Press, 1990).

Buddhist Model of Development: an approach to development that places the individual human being, rather than maximization of economic growth or capital accumulation, as the central focus. It derives from the Buddhist reverence for life and its basic tenets of karma and dharma. Buddhist ethics emphasize the satisfaction of BASIC NEEDS, modesty of lifestyle and social justice. From a Buddhist perspective, an action undertaken for selfish or greedy motives will result in bad karma, misery, and eventually in collapse of the physical world. Insofar as the basic doctrine of Buddhism is the mutual penetration and infusion of all phenomena, destructive actions toward any one being threaten all. Therefore, compassion for all living beings is essential to the continued existence of humanity that is part of the constant cycle of birth, death, and rebirth. The consequences of the Buddhist approach to the environment are obvious, but they are reflected in the specific patterns of economic development as well. Such basic principles of the Buddhist model as self-reliance, small-scale organization of economy, and autonomy from the West were implemented in the development strategies of pre-Cultural Revolution China and in the early 1980s in Burma. Though neither has proven a success, the Buddhist model of development has been popularized in the West by E. F. Schumacher and Erich Fromm, while some European Green parties have incorporated elements of the Buddhist development model into their platforms.

See Martine Batchelor and Kerry Brown, eds., *Buddhism and Ecology* (London: Cassell, 1992); Ernst F. Schumacher, *Small is Beautiful: Economics as if People Mattered* (New York: Harper and Row, 1973); Erich Fromm, *To Have or to Be* (New York: Harper and Row, 1976).

Development Models. See under SECTION 1.1.3.

Ecological Energetics Paradigm. See under SECTION 1.3.2.

Environmental Convergence Theory: a theory maintaining that environmental disruption will tend to be uniform among countries at similar levels of economic development, regardless of their political or institutional forms. *R. P.*

Global 2000 Report: a comprehensive study of global trends in population, resources, and environment commissioned by U.S. President Jimmy Carter to serve as the basis for long-term government planning. It was completed by thirteen U.S. government agencies in 1980 and concluded that "if present trends continue, the world in 2000 will be more crowded, more polluted, and more vulnerable to disruption than the world we live in now." Moreover, by 2100 the world's population might approach 30 billion and would be close to the CARRYING CAPACITY of the earth. As the first global model used by the U.S. government, the *Global 2000 Report* significantly enhanced the impact of two other most influential publications of that time: *Limits to Growth* and *Blueprint for Survival*. However, although the *Global 2000 Report* foresaw severe stresses of agricultural land, forests, water, and BIODIVERSITY, as well as the growth of world poverty, the Global 2000 model in general was more optimistic than the CLUB OF ROME projections. Its authors emphasized that their gloomy projections might be prevented by a radical change in domestic policies and by strengthening international cooperation to address global issues. The United States, whose policies have a major impact on global trends, should cooperate "generously and justly" with other nations. One of the cornerstones of Carter's political stand on global issues, the *Global 2000 Report* was practically ignored by subsequent administrations. The report was severely criticized for its technical deficiencies (the efforts to create a truly interactive model of growth, environmental and resource-use relationships were stymied by a lack of cooperation between government agencies), and for the ambiguity as to what the assumption of "unchanged government policies" could mean in the face of environmental deterioration and resource scarcity.

See Gerald O. Barney, ed., *The Global 2000 Report to the President: Entering the Twenty-First Century* (New York: Pergamon, 1980).

COMMENTARY ON *THE GLOBAL 2000 REPORT*

The Global 2000 Report was promptly and angrily criticized for its overall pessimism and its general disregard of the benefits of progress. Its most ardent critics, the late Herman Kahn and Julian Simon, emphasized the role of economic progress and human adaptability. "If present trends continue, the world in 2000 will be less crowded (though more populated), less polluted, more stable ecologically, and less vulnerable to resource-supply disruption than the world we live in now," they wrote in 1984. Apparently Kahn and Simon's belief in human wisdom and technological capacity was overestimated. Having witnessed various heated disputes on the prospects of society-economy-environment interaction, whereas most discussants use the same data but come up with opposite predictions and projections, one may wonder whether any objectivity in these issues can be achieved.

See Julian Simon and Herbert Kahn, *The Resourceful Earth: A Response to Global 2000* (New York: Oxford University Press, 1984). *Natalia Mirovitskaya*

"Golden Billion" Development: the assertion that the current exploitation of natural resources and the world's environmental space benefits one billion people and their descendants (mainly from First World countries) at the expense of curbing the progress and human development of the other five billion. The implication is that if consumption levels in the West are not moderated, other nations cannot prosper. See also ENVIRONMENTAL COLONIALISM.

COMMENTARY ON "GOLDEN BILLION" DEVELOPMENT

Russian parliamentarian M. Lemeshev made the following comment in a Duma speech in 1998:

All our relationships with Nature are reduced to the sale of natural resources: oil, gas, timber (with logs, the cheapest sort, being the main export), nonferrous metals, gold, high-grade uranium, etc. Everybody says they are ready to cooperate with Russia, but nobody hastens to invest a single dollar, say, in our agriculture, housing construction, medicine, education. That is, a policy of turning Russia into a source of raw materials for developed countries is being pursued again. Therefore an extremely intensive public activity is required to overcome the criminal strategy expressed in the idea of "the golden billion"—the Earth's resources are sufficient for one billion people, and the other nations must vegetate, remaining colonies with natural resources. I think it is in conflict with the resolutions of Rio-92 and with the Agenda for the 21st Century, in particular. *Natalia Mirovitskaya*

IPAT **Model:** an attempt to conceptualize the joint impact of the human causes of environmental change in a formula $I = P \times A \times T$, or *I*mpact equals *P*opulation times *A*ffluence times *T*echnology. This formula—not intended to be taken literally as a mathematical formula—was originally put forth by ecologist Paul Ehrlich and physicist John Holdren. They argue that the impact (I) of any population group or nation on environment is a product of its population (P), its level of affluence (A), and the damage done by particular technologies (T) that support that affluence. While the relative weight of these components is subject to debate, and the true interactions are clearly more complicated than this somewhat oversimplified equation, the IPAT formula may be useful at least from a conceptual perspective. *R. P.*

Leontieff Report: a study commissioned by the United Nations to investigate the interrelationship between environment and future economic growth. Beginning with the assumption that pollution was manageable, the expert group under the direction of Russian-born U.S. economist Wassily Leontieff listed a range of prerequisites necessary for expanding economic growth in the developing countries. The report also put forward two general conditions for an essential reduction in the income gap between developing and developed countries: internal social, political, and economic changes in the developing countries, and radical restructuring of the world economic order.

See Wassily Leontieff et al., *A Study on the Impact of Prospective Economic Issues and Policies on the International Development Strategy* (New York: United Nations, 1977).

Limits to Growth Report: a very prominent analysis published in 1972 by Donella H. Meadows, Dennis L. Meadows, Jorgen Randers, and William W. Behrens III, which predicted devastation for the earth if it continued on its trajectory of population growth and consumption. Nine million copies of the book were sold in over twenty languages. The analysis was based on a computer model that emphasized five variables: population, food, industrialization, nonrenewable resources, and pollution. The conclusions —the most notable being that population and industrialization would increase sharply and then collapse—and methodology were harshly criticized, particularly because the model had little provision for incorporating the possibility of technological improvements in prolonging the availability of resources, finding substitutes, and mitigating the environmental consequences of economic growth. A storm of controversy followed the publication of *Limits to Growth*, about which there is still no consensus more than twenty-five years later. Subsequent models developed by the *Limits to Growth* team have attempted to address the limitations of the original model.

See Donella H. Meadows, Dennis L. Meadows, Jorgen Randers, and William W. Behrens III, *The Limits to Growth: A Report for the Club of Rome's Project on the Predicament of Mankind* (New York: Universe Books, 1972). *T. S.*

The methodology and pessimistic conclusions of the *Limits to Growth Report* have been criticized on several grounds. Among the most serious shortcomings mentioned by other analysts is the inadequacy of the empirical base of the research, its neglect of the impact of technological innovation and price fluctuations on the dynamics of resource use, and its disregard of substantial geographical differences in basic components of the model (population dynamics, level of economic development, food availability, resource depletion, and environmental degradation). Some scholars question the very idea of academic technical experiments in social forecasting and subsequent policy recommendations. However, given all these shortcomings, the *Limits to Growth Report*, as well as other publications of the CLUB OF ROME, played an important role in demonstrating linkages among population, human activities, economic and environmental trends, and in emphasizing the urgent need for population planning, resource-use control, and policies relating to exploitation of the global life-support system. This publication contributed to the development of global ecological thinking and to the emergence of environmentalism as a worldwide movement. In particular, the report gave an impetus to the formation of the "global problematique research" in the former Soviet Union, designed to identify emerging worldwide environmental problems, and later to Gorbachev's concept of *perestroika,* or "New Thinking." *Natalia Mirovitskaya*

Outbreak-Crash Model of the Future. See CATASTROPHISTS. See also under SECTIONS 2.3 and 3.7.

2.5 Sustainable Development Strategy and Agenda Statements

Agenda 21: the major document produced at the UNITED NATIONS CONFERENCE ON ENVIRONMENT AND DEVELOPMENT (UNCED), also known as the "Earth Summit," in Rio de Janeiro, Brazil, June 3–June 14, 1992. Agenda 21 was the major outcome of the conference and sets out an agenda for the twenty-first century to tackle global environmental and social issues. Six themes form the core of Agenda 21: a) quality of life on earth; b) efficient use of the earth's natural resources; c) the protection of our global commons; d) the management of human settlements; e) chemicals and the management of waste; and f) sustainable economic growth. Agenda 21 does not propose solutions to the problems confronting the world, but establishes a set of actions to be taken by local, regional, and global actors to implement change and work toward SUSTAINABLE DEVELOPMENT. *T. S.*

Blueprint for Survival: a groundbreaking work by Edward Goldsmith, published in 1972. On the basis of analysis of the English countryside, Goldsmith predicted "the irreversible disruption of the life-support systems on this planet possible by the end of this century." To avert this catastrophe, Goldsmith proposed a framework of a stable society based on four conditions: minimal ecological disruption, maximum conservation of resources, zero population growth, and the "social conditions" under which the first three circumstances could become feasible. Recommendations on these "social conditions" included introduction of NATURAL RESOURCE TAXES and the system of SOCIAL ACCOUNTING, application of the POLLUTER-PAYS PRINCIPLE, and development of wilderness protection legislation. The most notable recommendation of the blueprint was the decentralization of megalopolises into "human-scale communities" that was supposed to reduce capital costs (through community self-sufficiency) as well as to ensure social responsibility and cooperation. Controversial, to say the least, this book significantly contributed to the "first round" of the Western environmental awareness and launched some ideas of environmental management that are still very much in vogue.

See Edward Goldsmith, *A Blueprint for Survival* (Harmondsworth, UK: Penguin, 1972).

Caring for the Earth: the second WORLD CONSERVATION STRATEGY (WCS) published in 1991 by the same agencies that produced the first WCS: the WORLD CONSERVATION UNION (IUCN), the UNITED NATIONS ENVIRONMENT PROGRAMME (UNEP), and the WORLD WILDLIFE FUND (WWF). The distinctive feature of this document is its acknowledgment that sustainability cannot be achieved mainly through technical solutions, whether they stem from economics or applied sciences, but must embrace a new ethic in human-nature and human-society relationships. *Caring for the Earth* calls for an agenda of promoting this new ethic for sustainable living, and translating its general principles into practice. *Caring for the Earth* outlines a set of principles for building a sustainable society and specified actions for their implementation.

See David A. Munro and Martin W. Holdgate, eds., *Caring For The Earth: A Strategy for Sustainable Living* (Gland, Switzerland: IUCN, UNEP, WWF, 1991).

Declaration on the Human Environment: a statement containing twenty-six general principles for guiding international responses to environmental problems, adopted at the 1972 UNITED NATIONS CONFERENCE ON THE HUMAN ENVIRONMENT (STOCKHOLM CONFERENCE). The document is best known for its frequently cited article 21, which provides that "states have the sovereign right to exploit their own resources pursuant to their own environmental policies, and the responsibility to insure that activities within their jurisdiction or control do not cause damage to the environment of other states or areas beyond the limits of national jurisdiction." Articles 8 through 15 respond to the concerns of developing countries that efforts to address environmental problems not be at the expense of their aspirations for economic development.

See Patricia W. Birnie and Alan E. Boyle, *International Law and the Environment* (Oxford: Clarendon Press), 45–47. M. S. S.

Development Decades: the United Nations themes for development during each ten-year period starting in 1961. This term is used in UN parlance to focus attention on the problems of developing countries and to encourage international action in order to accelerate the development of their human and natural resources. The results of Development Decades are assessed both on a midterm and a final basis with subsequent recommendations of policy changes.

Earth Charter: a set of documents that attempted to specify universally agreed-upon principles guiding human interaction with the environment. Efforts to develop such a charter were initiated at the 1972 UNITED NATIONS CONFERENCE ON THE HUMAN ENVIRONMENT (STOCKHOLM CONFERENCE). Since then a number of NONGOVERNMENTAL ORGANIZATIONS (NGOS) have proposed various forms of the charter. However, none of these efforts resulted in a universally accepted agreement, which would have the same impact as, for instance, the United Nations Charter of Human Rights. Drafting the Earth Charter was one of the main challenges of the 1992 UNITED NATIONS CONFERENCE ON ENVIRONMENT AND DEVELOPMENT (UNCED), OR EARTH SUMMIT, that revealed fundamental divergence between North and South. Developed nations were looking for a set of principles that would guide economic and other policies of all nations to take account of ecological constraints, while from the perspective of developing nations the key principles should have stressed their right to pursue economic development and the responsibility of the North to assist them. The final UNCED document did not meet the expectations of either side and thus had to be renamed the "Rio Declaration" to indicate that the text was not the list of universally accepted, explicit, and binding principles for global environmental management. In 1993, the EARTH COUNCIL, in collaboration with GREEN CROSS INTERNATIONAL, decided to take up the unfinished work of the Earth Summit and begin the process of articulating the Earth Charter. The document elaborated thus far draws substantially from the WORLD CONSERVATION UNION (IUCN) draft *International Covenant on Environment and Development,* particularly by highlighting its ethical foundations. However, no process for establishing its endorsement as a universally accepted set of principles has been established.

Global Marshall Plan for the Environment: a package of strategic proposals put forward by then U.S. Senator Albert Gore (later U.S. vice president) to alleviate the destructive effect of contemporary industrial society on the world's ecological systems. Like the Marshall Plan for economic recovery of postwar Europe, Gore's Global Plan for Environment focused on a few strategic objectives: a) stabilizing world population by ensuring universal availability of contraception and by developing functional literacy programs; b) developing environmentally appropriate technologies and distributing them worldwide; c) new global "eco-nomics"; d) a new generation of treaties and agreements, with the United States assuming the role of a LEAD STATE; and e) a cooperative plan for educating the world's citizens on environmental issues. If adopted as policy, the Global Marshall Plan for the Environment could have far-reaching consequences. For example, introducing a new global eco-nomics would mean redefining traditional notions of GROSS NATIONAL PRODUCT (GNP) and productivity, elimination of government's subsidies and discount rates, full disclosure of corporate responsibility for environmental damage, introduction of a virgin materials fee based on the use of nonrenewable materials in production, and establishment of the Environmental Security Trust Fund financed by the revenues from carbon taxation. According to its author, the Global Marshall Plan for the Environment would help to establish conditions conducive for SUSTAINABLE DEVELOPMENT, such as social justice, human rights, adequate nutrition and health care, political freedom, and participation.

See Albert Gore, *Earth in the Balance: Ecology and the Human Spirit* (Boston: Hougton Mifflin, 1992).

New International Economic Order (NIEO): a broad model for restructuring the world economy first put forward by the leaders of the Nonaligned Movement in 1973, in the aftermath of the world energy crisis. The NIEO was supposed to include a new system of international commodity agreements, reforms in the major global economic institutions (the INTERNATIONAL MONETARY FUND [IMF] in particular), a unilateral reduction of barriers to imports from developing countries to industrialized states, increased Northern financing and technology transfer, and recognition of rights pertaining to the economic sovereignty of states, particularly relating to nationalization and the control of the activities of multinational corporations. The initiative was taken to the United Nations General Assembly, which in 1974 adopted a Charter on Economic Rights and the Program of Action for the establishment of the NIEO. However, the basic idea of an NIEO to regulate (on the basis of the principle of equity) the free play of market forces in the international arena was unacceptable to the developed countries. Most of the provisions of the NIEO have not been implemented, but it remains a potent model in North-South discourse.

Small is Beautiful: an influential book by E. F. Schumacher, first published in 1973, one of whose main ideas is that the modern discipline of economics operates in a moral vacuum and derives from a purely materialistic Western philosophy, though still containing some prospects for good stewardship and responsibility. However, in the process of economic assistance, recipient countries gained only Western demands rather than Western spiritual values. Therefore, the aspirations of even the most principled efforts at DEVELOPMENT AID fail, as economic assistance results in Western pop-culture patches on the otherwise unchanged poverty landscape. Development in poor countries should be promoted in accordance with indigenous philosophy, such as a BUDDHIST MODEL OF DEVELOPMENT that would be based on a nonmaterialist concept of the importance of work to both the individual and the community, and would embrace "the systematic study of how to gain given ends with the minimum means." Following the GANDHIAN CONCEPT OF DEVELOPMENT, Schumacher differentiated between mass production, which required a massive injection of capital to create one workplace, and production by the masses, which created many workplaces with little capital. In drawing further practical conclusions, Schumacher promoted an idea of intermediate (i.e., appropriate) technology that would raise productivity in poor countries by providing as many people as possible with productive and satisfying jobs. In practice, such technology should be scaled down from greater to smaller and should have a centrifugal effect on the centralized industrial state, making efficient use of resources and eventually reinvigorating economic activity on a local level. Many of Schumacher's ideas are milestones of contemporary discourse on SUSTAINABLE DEVELOPMENT.

See Ernst F. Schumacher, *Small is Beautiful: Economics As If People Mattered* (New York: Harper and Row, 1973); Ernst F. Schumacher, *A Guide to the Perplexed* (London: Jonathan Cape, 1979); Ernst F. Schumacher, *Good Work* (London: Jonathan Cape, 1979).

United Nations Declaration of a New International Economic Order. See NEW INTERNATIONAL ECONOMIC ORDER (NIEO).

United Nations Development Decades. See DEVELOPMENT DECADES.

United Nations World Charter for Nature: adopted by the UN General Assembly in 1982, the charter proclaims principles of CONSERVATION to guide and judge human performance affecting nature: a) nature and its essential processes should be respected; b) genetic viability of all life forms on earth should not be compromised and habitats should be safeguarded; c) special protection should be given to unique areas, habitats, and representative samples of ECOSYSTEMS; d) ecosystems, organisms, and resources should be managed to promote sustainable productivity; and e) nature should be secured against degradation by warfare. The charter also asserted the need for conservation to be taken into account in the planning and implementation of social and economic development, and for ecological education.

World Charter for Nature. See UNITED NATIONS WORLD CHARTER FOR NATURE.

World Conservation Strategy (WCS): See under SECTION 6.3.2.

2.6 Sustainable Society: Challenges, Institutions, and Mechanisms

Accountability: 1) in general sense, the means by which individuals, organizations, governments, or other actors report to a recognized authority and are held responsible for their actions.

2) in the environmental context, responsibility for the decline of the environment, implying the allocation of environmental costs to the economic activities that cause such decline. See also POLLUTER PAYS PRINCIPLE (PPP).

Alternative Dispute Resolution (ADR): a variety of alternatives to litigation available to parties to legal disputes. These alternatives include voluntary mediation using a third-party mediator, binding arbitration, and other alternatives to formal litigation. In the environmental policy context, ADR techniques are available to assist in the resolution of disputes between and among environmental groups, the government, and industry, and have been used successfully in the resolution of disputes over liability in the U.S. SUPERFUND program. *D. S.*

Civil Society: a term widely used in academia and the development community to define social and political actors (individuals, groups, or organizations) and institutions that are outside of government, often excluding the business sector as well. Typical representatives of the civil society are NONGOVERNMENTAL ORGANIZATIONS (NGOS) and grassroots and community-based organizations. Advocates of the expansion of civil society often argue that these actors and institutions have more progressive values and political practices than traditional political actors, represent society more equitably (especially if the organized business sector is excluded from the definition), and provide a crucial counterweight against domination by government elites. The emergence of civil society is widely considered to be a vital element of SUSTAINABLE DEVELOPMENT, though both terms are contested for several reasons. See also GLOBAL CIVIL SOCIETY.

Collective Action Theory: a body of theory exploring the implications of individually rational behavior on problems of cooperation. One important focus of collective action theory is the stalemate that occurs when one or more actors can veto or otherwise undermine collective action. Another key focus is on the dilemma that individuals often lack the incentives to participate in collective action that serves the general good; the challenge is to create institutions and incentives to reduce this problem. Often collective action theorists try to formalize the collective action dilemma in order to apply mathematical methods such as GAME THEORY to understand why these problems arise and how the situation can be restructured. Collective action theory is epitomized by the work of Mancur Olson, particularly his 1965 book *The Logic of Collective Action*. In the environmental policy context, this literature explores problems associated with the provision of environmental public goods, such as clean air and clean water, in the absence of regulation. One commonly cited illustration of the collective action problem is Garrett Hardin's "Tragedy of the Commons," in which Hardin demonstrates that individually rational herdsmen will deplete resources such as grazing lands available to them all. Collective action theory has also been applied to the study of the formation and persistence of environmental groups, as well as to the problems of securing international environmental agreements when one or more countries can be FREE RIDERS without making their own sacrifices.

See Mancur Olson, *The Logic of Collective Action* (Cambridge, MA: Harvard University Press, 1965); Garrett Hardin, "The Tragedy of the Commons," *Science* 162 (1968): 1241–48. *D. S.*

Epistemic Communities: transnational networks of experts sharing "a set of symbols and references," common values, and approaches to problems. Common professional background of such experts tends to foster a distinct solidarity across borders. The existence of epistemic communities is believed to be instrumental to the process of creating stronger international regimes and their implementation. The notion has been popularized by Peter Haas through his analysis of the factors of compliance with the Mediterranean Action Plan. However, other evidence of the effect of epistemic communities on the process of environmental policy formation is scarce.

See Peter Haas, "Do Regimes Matter? Epistemic Communities and Mediterranean Pollution Control," *International Organization* 43 (1989): 378–403.

COMMENTARIES ON EPISTEMIC COMMUNITIES

"Networks of experts" acting as an organized political force ought to be a scary concept. They will not always be our experts. If they share "common values," then expertise becomes a credential that confers legitimacy in public discourse aimed at promoting these values. Data and models become the currency of debate, superseding and camouflaging the underlying value dilemmas. This is the road to technocracy. *Daniel Sarewitz*

Epistemic communities have the remarkable potential to transcend national and group interests by creating international solidarity based on professional and ideological affinity. Witness the cooperation of Russian and American scientists during the Cold War, through interactions between the academies of science and through such institutions as the INTERNATIONAL INSTITUTE FOR APPLIED SYSTEMS ANALYSIS (IIASA). *William Ascher*

Free Rider (Free-Riding Problem): the problem of COLLECTIVE ACTION THEORY that arises when a potential cooperator realizes that the actions of others may achieve the collective benefit even if that individual does not act. If several individuals try to take a "free ride," collective action may break down to the detriment of all potential cooperators. Environmental clean-up efforts by communities are often plagued by the free-rider problem, as each community member may believe that his or her participation will make only a small difference to the outcome. Various institutional arrangements, ranging from formal penalties to social pressure, arise to try to prevent free riding. Collective action theory takes the free-rider problem as one of its central dilemmas.

Global Civil Society: an international social system that is allegedly growing around the states system and is comprised of a variety of nonstate actors, including businesses, nongovernmental organizations, and interest groups. These actors operate above the level of the individual but below or even above the level of the state. According to the proponents of this concept, global civil society becomes increasingly influential as a shaper of international environmental policy.

See Paul Wapner, *Environmental Activism and World Civic Politics* (Albany: SUNY Press, 1996).

Governance: 1) the exercise of formal authority; the actions of government. In this sense, the capacity and quality of governance has become a growing concern as development experts have recognized the importance of incompetence and corruption as impediments to both economic development and environmental protection.

2) a pattern of resolving COLLECTIVE ACTION problems in a world of interdependent actors through the establishment and operation of informal social institutions, rather than through formal organizations and governments. The research on COMMON PROPERTY RESOURCES has demonstrated that groups of interdependent actors often succeed in resolving collective-action problems in the absence of direct government intervention. This notion of governance has become central to the contemporary theory on environmental cooperation; researchers have been engaged in defining the conditions under which "governance without government" can succeed.

See Oran Young, *International Governance: Protecting the Environment in a Stateless Society* (Ithaca: Cornell University Press, 1994).

COMMENTARY ON GOVERNANCE

Governance is a catch-all label for the diverse processes that communities employ to decide what is in their common interest. Governance is a "process of balancing, accommodating, and integrating the rich diversity of culture, class, interest, and personality which characterizes all situations" (McDougal et al. 1980: 207). Governance is an endless search for an ever better approximation to the common interest. For a community of diverse people with conflicting special interests, clarifying and securing the common interest can be challenging. Common interests are shared interests whose fulfillment will benefit the entire community. But some governance processes clearly favor special interests, which are incompatible with comprehensive community interests.

They are made by only certain elites and will benefit only a small segment of the community with a corresponding deprivation to the rest. The kind of governance process required to secure common interests varies depending on context. Knowledgeable citizens and scholars can help people to clarify and secure their common interests in public decision processes.

See Myres McDougal, Harold D. Lasswell, and Lung-chou Chen, "The Basic Policies of a Comprehensive Public Order of Human Dignity," in *Human Rights and Public Order: The Basic Policies of an International Law of Human Dignity*, eds. Myres McDougal, Harold D. Lasswell, and Lung-chou Chen (New Haven: Yale University Press, 1980). *Tim W. Clark*

Interdependence: the state of affairs of any given social group or system of relations in which the actions of each actor affect the other actors. The contemporary notion of interdependence became prominent in political discourse during the 1970s and was first related mainly to military-security issue areas. Later, studies of interdependence were refocused upon wealth and welfare economic issues, while in the 1990s it became a pillar of the theory of environmental cooperation as well.

Pareto Improvement (Pareto Optimality): a change in actions or distribution that leaves some better off and no one worse off. When no change in the allocation of resources or benefits can be achieved without harming someone, the situation is said to be Pareto optimal. Italian economist Vilfredo Pareto (1848–1923) proposed the Pareto optimum as a weak condition of societal welfare, recognizing that society may be even better off (according to a given rule for aggregating utility and a given position on distributional equity) even if someone is harmed by an improvement for others. Economic theories differ as to the usefulness of the concepts of Pareto improvements and Pareto optimality. While in practical situations there is rarely a policy change that would not harm someone, one can conceive of side payments to the harmed parties in order to use Pareto optimality as a criterion of efficiency.

COMMENTARY ON PARETO IMPROVEMENT

The search for Pareto optimality is often thwarted in situations where one or more of the parties adheres to moral principles other than that of efficiency. For instance, in surveys conducted in the early 1990s, Austrians overwhelmingly repudiated proposals to export municipal waste to poorer neighboring countries in Eastern Europe, although these plans would have relieved pressure on Austrian landfills and provided much needed income for the recipients of the waste. In this case, moral judgments placed international trade in waste beyond the boundaries of utilitarian calculation altogether.

See Joanne Linnerooth-Bayer and Benjamin Davy, *Hazardous Waste Cleanup and Facility Siting in Central Europe: The Austrian Case, Report to the Bundesministerium für Wissenchaft und Forschung* (Laxenburg, Austria: International Institute for Applied Systems Analysis, 1994). *Steve Rayner*

Prisoner's Dilemma: an analogy to a broad range of COLLECTIVE ACTION problems conveyed by the story of two prisoners who have been charged with a crime. If only one testifies against the other, the confessor will receive no jail sentence, while the nonconfessor will receive the maximum possible sentence. If neither confesses, both will receive light sentences. If both confess, both will receive relatively heavy (but not the maximum) sentences. Under this set of incentives, the prisoners—especially if they are not permitted to communicate with one another—have an incentive not to cooperate with one another, but rather to confess regardless of what the other does, leaving them both worse off than if they had cooperated by refusing to confess. Some contend that the problem of providing environmental and public goods mirrors that of the prisoner's dilemma game, in that individuals, firms, or communities may fail to cooperate because they anticipate that others will not cooperate either. *D. S.*

Regime(s): agreed upon principles, norms, rules, decision-making procedures, and programs that govern the interactions of actors in particular issue areas. Existing regimes vary substantially in terms of membership, functional scope, geographical domain, degree of formalization, and stage of development. Regime analysts tend to define three main categories of regimes: international, transnational, and mixed. International regimes are created by nation-state governments and their operations center on issues arising in international interactions. Examples of state-organized regimes include the Antarctic regime that is based on several components of the ANTARCTIC TREATY SYSTEM and the global climate regime consisting of a FRAMEWORK CONVENTION ON CLIMATE CHANGE and a series of protocols. In distinction to that, members of transnational regimes are nonstate actors while the areas of operations are issues that arise in GLOBAL CIVIL SOCIETY. This type of regime is exemplified by the social practice of World Wide Web users. Real-world regimes are typically mixed types. The notion of regimes is basic to the contemporary theory of international environmental relations.

See Oran Young, ed., *Global Governance: Drawing Insights from the Environmental Experience* (Cambridge, MA: MIT Press, 1997).

Regime Effectiveness: the ability of a REGIME to eliminate or substantially ameliorate the problem that led to its creation. Students of international environmental regimes define several interrelated dimensions of regime effectiveness. These dimensions can be categorized broadly into two main groups: compliance effectiveness and result effectiveness. Compliance effectiveness is the degree to which involved actors follow regime prescriptions. Result effectiveness is the degree to which the changes caused by regime produce real environmental improvement. Studying regime effectiveness is a burgeoning area in the contemporary theory of cooperation.

See Oran Young, ed., *Global Governance: Drawing Insights from the Environmental Experience* (Cambridge, MA: MIT Press, 1997).

Regime Robustness (Regime Resilience): the capacity of international arrangements, rules, and expectation, whether formal or informal, to remain in force despite external changes, such as shifts of power among the regime's members and changes of interest in the REGIME by its most powerful members. Lack of resilience might result in either a radical change of the regime's norms or the degree of compliance to the regime's norms and rules by its members. The notion of regime robustness is used in addition to that of REGIME EFFECTIVENESS to characterize the significance and permanence of institutional arrangements.

Social Institutions: sets of rules or codes of conduct that serve to define social practices, assign roles to the participants in these practices, and guide the interactions of those participants. Different societies evolve different institutions (political systems, family, religion, etc.) in response to particular social concerns. Institutions can be formal (through explicit rules such as laws, or through organizations) or informal (such as shared expectations and values regarding appropriate behavior). From the political perspective, the most significant examples of institutions are those designed to resolve social conflicts, promote sustained cooperation in mixed-motive relationships, and, more generally, alleviate COLLECTIVE ACTION problems in a world of interdependent actors.

See Oran Young, *International Governance: Protecting the Environment in a Stateless Society* (Ithaca: Cornell University Press, 1994).

Sustainable Society: one of the dimensions of sustainability put forward by the CLUB OF ROME's 1991 report. It defines a society "that is based on a long-term vision in that it must foresee the consequences of its diverse activities to ensure that they do not break the cycles of renewal; it has to be a society of conservation and generational concern. It must avoid the adoption of mutually irreconcilable objectives. Equally, it must be a society of social justice because great disparities of wealth and privileges will breed destructive disharmony."

See Club of Rome, *The First Global Revolution. Report of the Council of the Club of Rome* (New York: Club of Rome, 1991).

3

Main Factors Behind Development and Environmental Change

3.1 Population, Poverty, and Underdevelopment

Demographic Transition: 1) in a general sense, any change in demographic parameters, usually connected with the dynamics of development.

2) a specific model characterizing the effect of income on population growth as it relates to the stages of economic development. Using the historical record of population growth in Europe and North America as an example, stage I represents the traditional agrarian society, stage II represents the beginning of industrialization, and stage III, the industrialized society. In the first stage, both birth rates and death rates are relatively high, but with the former being only marginally higher than the latter, the population grows slowly. As the society begins to develop into stage II, death rates fall due to improvements in public health while birthrates remain nearly constant (only falling slightly), and thus population growth accelerates. In stage III, as the countries become more fully developed, the demographic transition occurs and birthrates fall to more closely parallel death rates so, once again, population growth slows. It should be noted, however, that this model does not provide an explanation of the determinants of birth and death rates, but a way of organizing our past observations on population growth rates.

See James R. Kahn, *The Economic Approach to Environmental and Natural Resources* (New York: Dryden Press, 1995). *R. A. K.*

COMMENTARY ON DEMOGRAPHIC TRANSITION

The demographic transition does not always happen—for instance, birth rates in some Middle Eastern countries are higher than would be predicted by levels of income and industrialization. The key driving forces in moving from stage II to stage III (the decline in birth rates) appear to be the rising costs of having children associated with leaving the family farm, moving to urban areas, and rising opportunities for wage employment by women. *Robert Healy*

Human Carrying Capacity: the number of people who could occupy a particular area, ranging from small units to the entire world. As a parallel to the more general notion of CARRYING CAPACITY that is frequently applied to animals and plants (most often to wildlife), the concept of human carrying capacity attempts to specify the maximum human population that can survive within the unit by identifying limiting factors. However, the parallel breaks down because animal and plant populations are constrained by natural limitations, such as the supply of food and water or the area required for successful food gathering and reproduction, whereas human population density can be increased either by investing in infrastructure or by settling for lower standards of quality of life. Therefore the practical human carrying capacity depends on a combination of investment available to enable people to live more densely and the tolerance for living at greater densities. While carrying capacity for other species is essentially an empirical issue of the nature of the species involved and the natural-system elements required for that species, human carrying capacity is as much a matter of the value judgment on what constitutes tolerable living conditions. Nevertheless, overcrowding per se can have damaging sociological and psychological effects.

See Joel E. Cohen, *How Many People Can the Earth Support?* (New York: W. W. Norton, 1995); Philip M. Fearnside, *Human Carrying Capacity of the Brazilian Rain-Forest* (New York: Columbia University Press, 1986).

COMMENTARY ON HUMAN CARRYING CAPACITY

The rather solid concept of carrying capacity for a given species in a given area becomes an absurdity when it comes to human populations. The population limits for animals and plants depend on physical limits of food supply, soil composition, space, etc., which can be established through technical analysis. Given the characteristics of the ecosystem, one can estimate carrying capacity and develop policies (such as fishing or hunting quotas) to guide sustainable resource exploitation and to avert population collapses. Yet human population limits reflect not only physical constraints, but also human values such as tolerance for crowding. Has Mexico City, with roughly twenty million people, severe pollution, horrendous traffic, and serious water shortages, reached its "human carrying capacity"? This is no longer a technical question, but a value and political question; people are still migrating to Mexico City. Even the physical constraints that would be limiting factors under certain conditions may be overcome through the choices to invest in infrastructure and technology. *William Ascher*

Human Poverty Index (HPI). See under SECTION 1.1.6.

International Human Suffering Index. See under SECTION 1.1.6.

Microeconomic Theory of Fertility: a demographic theory that explains the decision to have children as a microeconomic utility maximizing decision, where the household (both parents) weighs the costs and benefits of having an additional child. The optimal number of children occurs where marginal costs are equal to marginal benefits. Any factors that serve to reduce the marginal benefits will reduce the number of children per family, and any factors that increase the marginal costs will also reduce the number of children per family. Some factors affecting the economic benefits of having children are that children are labor inputs to production processes and that they can provide for their parents in old age. Conversely, some factors that affect the cost of having children are the costs of education, food and housing, and the opportunity costs of the mother's and/or primary caretaker's time. It is informative to use this economic model to explain the economic behavior that may be underlying the DEMOGRAPHIC TRANSITION.

See James R. Kahn, *The Economic Approach to Environmental and Natural Resources* (New York: Dryden Press, 1995). *R. A. K.*

Poverty: deprivation of material well-being, often reflected or measured by low income and wealth, but applicable to low levels of access to such services as health care, education, etc., as well as deference benefits such as respect. The definitions and measurement of poverty are complicated by the breadth of the concept; for example, poverty can be measured solely by the income received from economic activities or can include goods and services provided by the government. When, for policy reasons, it is necessary to establish a cut-off level of income or wealth under which households are designated as living in poverty, the arbitrariness of such cut-offs can create great controversy. The concept of "absolute poverty" presumes that there is a minimum level of income or wealth necessary to meet BASIC NEEDS, although this latter concept is also subject to varying interpretations. The concept of "relative poverty" captures part of the psychological and social impact of receiving lower incomes than others. While it avoids the issue of where to establish the income level designated as separating the poor from the nonpoor, it faces a parallel issue of setting the cut-off points for relative poverty (e.g., should the bottom 30 percent of the population be considered as suffering from poverty, or the bottom 20 percent, or those with less than 25 percent of the income of the top half of the income distribution). Many multilateral and bilateral foreign assistance agencies have focused on POVERTY ALLEVIATION as their primary objective. On issues of ENVIRONMENTAL PROTECTION and CONSERVATION, the role of poverty is also highly controversial. Some argue that poverty causes people to engage in unsustainable economic and resource-use activities, inasmuch as income insecurity denies them the luxury to plan for the long term; others argue that affluence is a more important cause of environmental deterioration and resource depletion, because the wealthy make greater demands on NATURAL RESOURCES and ENVIRONMENTAL SERVICES.

COMMENTARY ON POVERTY

Among some social scientists and development activists, poverty is being viewed less in terms of relative material discomfort and more as exclusion from social participation in family, civic, economic, and political life. This insight is embodied in the HUMAN DEVELOPMENT INDEX (HDI) produced by the United Nations Development Programme since 1990, which measures poverty in terms of life expectancy, education, access to resources, political freedom, and personal self-respect rather than in terms of income alone. The development of this index has been hailed as a great achievement. Douglas et al. note that "the index is a highly sophisticated attempt to assess the infrastructure of the individual's life. The individual person is not left swinging in mid-air, without support or clues as to what might realistically be possible at that time or place. A person's chances of schooling, nutrition, life expectancy, and income say much more about well-being than straight comparisons of income."

See Mary Douglas, Des Gasper, Steven Ney, and Michael Thompson, "Human Needs and Wants," in *Human Choice and Climate Change, Volume 1, The Societal Framework*, eds. Steve Rayner and Elizabeth L. Malone (Columbus, OH: Battelle Press), 241. *Steve Rayner*

Poverty Alleviation: efforts directly targeted to improve the incomes and living conditions of the poor. The focus on poverty alleviation as a distinct and explicit goal of development emerged most forcefully in the 1970s, particularly from international development organizations such as the World Bank Group, as a more socially conscious goal than economic growth per se. While targeting the poor for development interventions obviously has distributive implications, poverty alleviation programs are less explicitly focused on the struggles among groups than are directly redistributive programs. Among poverty alleviation strategies, different approaches include: a) the BASIC NEEDS approach of providing or ensuring the most fundamental requirements of life such as nutrition, shelter, basic education, health, etc.; b) HUMAN DEVELOPMENT approaches that concentrate on increasing the productivity of the poor in order to increase their income-earning potential; c) micro-enterprise finance to provide the poor with access to credit; and d) direct government transfers to the poor.

Size Distribution of Income: the shares of income received by different segments of the population according to their relative income levels. Thus the size distribution of income may be expressed as the share of national income received by the wealthiest 10 percent of the population, the second wealthiest 10 percent, etc. It is often expressed according to deciles (10 percent units), quintiles (20 percent units), or quartiles (25 percent units). The size distribution of income is one of the most straightforward expressions of income equality or inequality. However, there is considerable variability in the income shares for each segment depending on how income is defined (e.g., before taxes or after taxes; including or excluding the cash value of benefits provided by the government) and the unit of analysis (e.g., individual income or household income), making cross-national comparisons of the size distribution of income somewhat difficult. The size distribution of income is sometimes summarized by an overall measure, the GINI INDEX, but any such overall measure cannot capture the details of the specific shares received by each segment.

Time Horizons: a loose term for the time frame that people take into account in making decisions, particularly economic and resource-exploitation decisions. People with short time horizons are presumed to invest less in sustainable initiatives than in actions that bring in immediate gains. Individuals and firms engaged in natural-resource exploitation are more likely to invest in resource development if they have longer time horizons. Government officials with longer time horizons are presumed to care more about the long-term consequences of their actions. In economics, the concept of time horizons is typically conveyed by the DISCOUNT RATE.

Underdevelopment. See under SECTION 1.1.3.

3.2 Affluence and Consumption

Consumer Groups: citizen groups organized to defend the general interests of consumers, often targeted to product safety, consumer information, and affordability. Increasingly, consumer groups have focused on products' environmental impacts. The possibility of mobilizing consumer groups to condemn particular products, because their production, use, or disposal is polluting or unsafe, or because they use up scarce resources, is emerging as a nonregulatory approach to forcing producers to internalize environmental consequences.

Environmental Kuznets Curve (EKC): an inverted U-curve that represents a posited link between environmental quality and income. It was so named because of the analogy to the earlier "Kuznets curve of inequality," a proposition formulated by economist Simon Kuznets in the 1950s that low-income nations initially show rather high equality of income distribution, become less equitable as they develop economically, and then regain a degree of income equity when they become developed. The argument underlying the environmental Kuznets curve is that as incomes within a nation grow, environmental degradation first gets worse but then improves as preferences for environmental quality receive higher priority and more resources and technologies become available. The existence and particular shapes of the environmental Kuznets curve have been tested in the context of deforestation and also in relation to some industrial emissions. Some of the data confirmed the existence of the environmental Kuznets curve; however, in general the evidence is still ambiguous.

See M. Cropper and C. Griffith, "The Interaction of Population Growth and Environmental Quality," *American Economic Review: Papers and Proceedings* 84 (2) (May 1994): 250–54; Gene Grossman and Alan Krueger, *Environmental Impacts of a North American Free Trade Agreement*, National Bureau of Economic Research, Working Paper no. 3914 (Cambridge, MA: National Bureau of Economic Research, 1991).

Environmental Perceptions: different attitudes toward nature and its components as well as the levels of tolerance toward environmental degradation taken by various individuals, groups, and societies. Different cultural backgrounds lead to different assumptions about the use and management of environmental resources. See also CARRYING CAPACITY, ENVIRONMENTAL DETERMINISM.

Ethical Consumption: the principle that consumers should only use products that are produced and distributed in an ethical fashion, that is, without exploiting the workers involved, the local people from the areas from which the products are derived, or the ecosystem. Aligned to GREEN CONSUMERISM, ethical consumption extends concerns for the environment per se to the social, economic, and cultural rights of indigenous peoples affected by the extraction of raw materials, the production of goods, etc. Ethical consumption highlights abuses such as child labor, sweatshop conditions, and labor camps, thereby relating human rights a to wider political and economic agenda. In Europe there is evidence to suggest that nation-states are beginning to take an "ethical turn"; the United Kingdom, for example, now has an "ethical foreign policy." *I. W.*

Green Consumerism: organized or spontaneous consumer behavior that favors products whose production is believed to entail less pollution or less depletion of scarce resources. For green consumerism to be effective, consumers must be well informed about the pollution profile and resource inputs of the products they buy. *R. P.*

COMMENTARY ON GREEN CONSUMERISM

The geographical distribution of green consumerism thus far has been very uneven: Scandinavian countries and Germany are leading the way while in the United States this trend is limited to the upper middle class. Scholars still disagree on the impact of green consumerism on environmental strategies and practices of businesses. While Welford and Gouldson (1993) argue that the impact of green consumerism on most businesses is marginal, Garrod and Chadwick (1996) provide statistical evidence that the pressure of green consumerism on companies can be already equated to the impact of governments and regulators. An ongoing convergence of consumer tastes might well contribute to further development of green consumerism around the world.

See Richard Welford and Andrew Gouldson, *Environmental Management and Business Strategy* (London: Pitman, 1993); Brian Garrod and Peter Chadwick, "Environmental Management and Business Strategy: Towards a New Strategic Paradigm?" *Futures* 28 (1) (1996): 37–50. *Natalia Mirovitskaya*

3.3 Technology

Appropriate Technology: technology chosen for its consistency of fit with local conditions and endowments, in order to pursue particular economic, social, or environmental objectives. Because the pursuit of possible objectives (e.g., maximum economic growth, greater employment, pollution avoidance, and minimal down time of specific devices) may involve tradeoffs, the imperative for technologies to be "appropriate" does not yield a simple, noncontroversial prescription. When the logic of appropriate technology is applied to developing countries, the prescriptions often emphasize avoidance of complicated technologies that require scarce technical expertise to operate and maintain. When appropriate technology is committed to ENVIRONMENTAL PROTECTION and CONSERVATION, the focus is often on technologies that minimize pollution, energy use, or raw material inputs. From an economic growth perspective, the appropriateness of a technology depends on whether the inputs it requires are abundant (e.g., labor-intensive technologies for labor-abundant economies).

Best Available Technology. See under SECTION 5.3.1.

Biotechnology: 1) the use of organic materials to create improved products and processes.

2) any technique that uses living organisms, parts of living organisms, or organic materials to make or modify products, improve plants or animals, or develop microorganisms for specific uses. Examples of biotechnology include such products as vaccines, gene therapy, and monoclonal antibodies in human health care; bioprocessing/fermentation and transgenic animals in agriculture; and improved microbial systems for control of water, soil, and air in environmental management. Biotechnology is one of the most controversial issues on the international environmental agenda.

Clean Technology: 1) a term usually used to define the most advanced available technology for reducing pollution, or the policies requiring such technology. Both BEST AVAILABLE TECHNOLOGY and BEST ENVIRONMENTAL PRACTICE fall under these provisions. Like SUSTAINABLE DEVELOPMENT, clean technology is still a rather vague concept and the definitions used by various international and national organizations, think tanks, and individual authors differ. Most definitions focus on strategies related mainly to production processes and waste minimization. However, there is also a trend of extending the application of clean technology from processes to product changes, such as product substitution, conservation, and change in composition.

2) equipment that has been adapted in such a way as to generate less or no pollution. The concept increasingly implies a preventive approach to pollution control. In contrast to traditional END-OF-PIPE REGULATIONS, clean technology is supposed to meet both environmental and economic goals.

COMMENTARY ON CLEAN TECHNOLOGY

Inherent in corporations' proactive environmental management practices has been the search for new clean technologies—processes, systems, equipment, and know-how—that reduce or control pollution more effectively. Entrepreneurial firms see business opportunities in the form of less polluting durable goods, especially cars and trucks, mass transit, cleaner fuels, and efficient materials handling. Diesel engine makers such as Varity Perkins in the United Kingdom, for example, are developing new sealing technologies and using computer simulations to market a zero-pollutant engine. Technological innovation is the key to the environmental progress required for sustainable development. Effective regulatory compliance depends on businesses having access to effective and affordable processes and equipment to meet environmental standards. Pollution prevention is based on process or technological changes that reduce and eliminate waste. By assessing its product and processing technologies, S. C. Johnson and Son, for example, was able to cut manufacturing waste nearly in half, reduce the use of virgin packaging materials by more than 25 percent, and reduce volatile organic compound (VOC) use by 16 percent between 1990 and 1995. Its product stewardship projects produced environmental benefits in its plants around the world.

See Dennis A. Rondinellli and Michael A. Berry, "Industry's Role in Air Quality Improvement: Environmental Management Opportunities for the 21st Century," *Environmental Quality Management* 7 (4) (1997): 31–44. *Dennis A. Rondinelli*

Technological Fix: an assumption that human-induced environmental problems can be solved by introducing more effective or more ecologically sound technologies.

The danger of depending on a technological fix is well demonstrated by the case of chlorofluorocarbons (CFCs). These chemicals, which for fifty years were regarded as technologically benign, and indeed significantly improved the quality of life for millions of people, turned out to be a cause of a major global environmental threat, ozone depletion. The substitution of CFCs by other more "ozone-friendly" chemicals, such as hydrochlorofluorocarbons (HCFCS), is a likely contributor to global warming. *Natalia Mirovitskaya*

Technology Transfer: a transfer of patents, machinery, equipment, or expertise from one country to another, generally from an industrialized nation to a developing nation. This is a contentious issue between the North and South in international environmental negotiations, because firms of developed nations are reluctant to give up their technology without compensation. However, developing nations assert that they will not be able to achieve economic stability or reach environmental goals without technology transfers from the North.

3.4 Political Economy

3.4.1 EXTERNALITIES

Coase Theorem: the demonstration by Nobel Prize-winning economist Ronald Coase that under assumptions of clear property rights, perfect information, and no transactions costs, private negotiations between polluters and those suffering from pollution will be efficient. If polluters have the right to pollute, those affected by the pollution may be able to offer the polluters enough compensation to induce them to eliminate or reduce the pollution. Conversely, if the affected population has the right to be free of the pollution, the polluter may offer the affected population sufficient compensation for them to permit the efficient degree of polluting economic activity. The Coase theorem shows that under these restrictive conditions, government action is unnecessary for securing economically efficient agreements. However, it also implies that uncertainties in property rights or information, as well as costly procedures for reaching private agreements, can undermine efficient resolutions of disputes over pollution.

Environmental Externality Costs: the costs of environmental damage that are not borne by the polluting producer unless POLLUTION TAXES are imposed. These costs include a) living condition deterioration; b) reduced production and employment; c) mitigation costs borne by government or other actors. Accurate estimates of environmental externality costs are essential for effective pollution charges or taxes. There is a debate as to whether to use the cost of damages imposed on society, the costs of controlling the pollutants, or the "implied valuation" costs of meeting legislatively established standards.

See R. Ottinger, *Incorporating Environmental Externalities through Pollution Taxes* (Geneva, Switzerland: World Clean Energy Conference, 4–7 November 1991), 531. *R. P.*

Externalities (External Costs): costs or benefits that are not captured by the standard market payments received or paid by the firm or individual who takes a particular action. Because this cost or benefit is not "internalized," or borne by that actor, it often will not enter that actor's calculation of utility and therefore may lead to societally deleterious outcomes. Even neoclassical economics, generally favoring minimal government economic intervention, recognizes that externalities, as market failures, justify taxes (or charges) on activities producing negative external effects, and the promotion or subsidization of activities producing positive external effects.

COMMENTARIES ON EXTERNALITIES

The identification of externalities and their accurate measurement are two quite different activities. It may be possible to identify benefits or harm not reflected in market prices that accrue to a party by another's action. Identifying the amount of such benefits or harm requires attribution and distribution that often exceed straightforward analyses. Due to this, such externalities are often not considered directly in environmental analysis even when all parties acknowledge that they exist. The ability to quantify externalities is essential to successful application of the concept to practical environmental decision making. *Gerald Emison*

The problem, in the words of Karl Polanyi, is that "nothing obscures our social vision as effectively as the economic prejudice." Of course, failure to internalize costs of environmentally harmful activities can promote "societally inferior outcomes." But, from another perspective, the quest to internalize environmental costs reflects the misbegotten compulsion to rationalize all policies through quantification. If preservation of nature or "the natural" is an intrinsically worthy human goal, it is also a goal that can become obscured or distorted when viewed through a filter that interprets everything as either internal or external to the marketplace. Old-growth forests provide less carbon sequestration than young forests; should cutting down old-growth forests be assessed as a negative or a positive externality? Is this a question that we even want to ask? In many cases, the environment should be *protected* from "the economic prejudice," much as we would protect various freedoms (of expression, speech, religion, association, etc.) from monetary valuation. *Daniel Sarewitz*

Externalities, Environmental: environmental costs or benefits that are not borne by the actor responsible for causing them. In environmental economics, EXTERNALITIES can be positive, such as watershed benefits that society receives from tree planting, or negative, such as health or visibility costs from industrial air pollution. Pollution charges and subsidies for environmental enhancement activities are therefore justified. However, one risk of such charges and subsidies is that their levels will not be set appropriately, thereby causing too much or too little economic activity and resource-stock development. *R. P.*

Internalization of Environmental/External Costs: the principle of requiring those responsible for causing others to bear costs (i.e., causing negative EXTERNALITIES) to bear these costs themselves. Thus, internalizing environmental or other external costs usually entails paying a fine or tax, paying to mitigate the damage, or compensating the damaged parties.

Transboundary Externalities: EXTERNALITIES that arise when activities that occur within the jurisdiction of one country or subnational jurisdiction affect the lives and well-being of residents in neighboring jurisdictions or the global commons.

Transboundary Natural Resources: resources that are present in two or more jurisdictions, or are potentially affected by actions taken in multiple jurisdictions. Some natural resources, such as fish and marine mammals, migrate between areas under the jurisdiction of different states or between an area of national jurisdiction and an area beyond the limits of national jurisdiction (e.g., the high seas). Other resources, such as rivers and aquifers, or oil fields near national boundaries, can be affected by actions taken on the other side of the border (such as irrigation or, in the case of oil fields, directional drilling). Thus, the use of transboundary natural resources on the territory of one state may affect its use in the area of another state or by the international community.

Transboundary Water Resources (International or Shared Waters): atmospheric, surface, or underground waters shared between two or more states. There are more than 240 international river basins and an uncertain number of aquifers whose waters are shared by two or more countries. Any substantial activities performed by one country on the transboundary basins may affect—positively or negatively—other sovereign states. Given the projected shortage of freshwater resources in the very near future, the number of conflicts over the use of transboundary water resources is likely to grow.

3.4.2 PROPERTY AND USER RIGHTS

Bundle of Entitlements (or Rights): rights and responsibilities for the use of a resource by the owner. The term *bundle* implies that the entitlements may be varied, including privileges (such as construction, exclusive use, or subsurface resource mining) and limitations (such as prohibited activities or obligation to sell to the government in cases of eminent domain). The concept of the bundle also implies that simply designating that someone is the owner of a resource, or has property rights, is inadequate for understanding what that actor can and cannot do. See also PROPERTY RIGHTS. *R. P.*

Collective Intellectual Property Rights (CIPRS): patent rights that dictate that communities and/or indigenous cultures should be compensated for the use of their collective expertise about genetic resource use. CIPRs are a relatively new idea actively promoted by environmentalists from developing countries. However, the GENERAL AGREEMENT ON TARIFFS AND TRADE (GATT) and the WORLD TRADE ORGANIZATION (WTO) do not yet recognize these patent rights.

See Vandana Shiva, "Piracy by Patent: The Case of the Neem Tree," in *The Case Against the Global Economy,* eds. Jerry Mander and Edward Goldsmith (San Francisco: Sierra Club Books, 1996).

Common Property Resources (Common Pool Resources): resources for which the user rights are shared by a collectivity of people, often termed the "community." Common property resources are distinguished from state property, private property, and "open access" for which no effective exclusion is enforced. In some cases, different uses of the same resource (e.g., a forest) may be held by different communities (e.g., timber harvesting by one; grazing livestock by another). The proponents of common property resources argue that under certain conditions, communities can manage natural resources more sustainably and more equitably than private or state owners, and that the apparent failure of resource management of the "commons" (see TRAGEDY OF THE COMMONS) is a problem of open-access resources rather than common property resources controlled by a well-defined community. Community control has the advantage of being able to apply social pressure to enforce commitment to conservation.

COMMENTARIES ON COMMON PROPERTY RESOURCES

Elinor Ostrom's *Governing the Commons* (1990) is often cited as the authoritative work that lays out the conditions under which a common property resource management regime would be expected to flourish. The eight principles that Ostrom proposes are derived from successful and unsuccessful case studies of community-based management. For the most part, these examples tend to come from less developed countries and subsistence communities that are reliant on natural-resource-based economies. Consequently, their usefulness in more developed countries and in communities that are less reliant on natural resources has been questioned. *Toddi Steelman*

It is important to recognize that common property requires the same thing as does private property—legitimacy and protection from the state. Private property would be meaningless if all of us had to lie awake nights defending, by whatever means at our disposal, the sanctity of our homes, farms, and factories. The authority system we know as the state agrees to protect private property and in doing so legitimizes that legal structure we know as private property. Explicit in that protection is our ability to call upon the state (in the form of the local police and the courts) to come to our defense in the face of threats against our private property. In essence, individuals have the awesome capacity to compel the state to act in their behalf against threats toward property. Common property requires no less of the state than does private property. The mere fact that common property is owned by a group should not be taken as an excuse for the state to refuse to accord it the same protection given to private property. With this in mind, it is easy to see that open-access resources *(res nullius)* are available to all precisely because there is no authority system to recognize and protect those resources against threats from would-be users or claimants. *Daniel Bromley*

Emission Rights (or Credits): a type of tradable emissions credit designed to put a ceiling on particular types of pollution ("emission standards"), reduce pollution-abatement costs, and encourage the most productive activities that create emissions. The government first establishes emission standards that set the ceilings for specific pollutants, and assigns emissions rights. Rights holders can pollute up to the ceiling on their emission rights, or sell or lease some or all of their rights. Pollution reduction and abatement are thus rewarded by the opportunity to sell or lease emission rights; rights are also likely to be purchased by producers who can benefit from lower abatement costs. Higher overall ceilings on the emission rights granted by the government in a particular area will allow higher levels of emissions in places where purification costs are highest. There are various forms of emissions trading systems, such as BUBBLE POLICY, NETTING, OFFSET, and BANKING. *R. P.*

COMMENTARY ON EMISSION RIGHTS

Successful use of tradable emission rights requires accurate measurement of such emissions. There must exist adequate technology to measure the emissions and emission types must be easily measurable. While point sources are relatively easy to measure, area sources from diffuse sources make establishing the baseline for trading difficult. Additionally, independent, honest accounting for such emission rights is essential for them to be reliably traded. *Gerald Emison*

Environmental Rights: claims that individuals or communities are entitled to environmental improvements or protection from environmental damage. Some would hold that such rights encompass protection from resource depletion as well as pollution. Such claims have been increasing dramatically over the post–World War II period, to the point where environmental rights are often included among the so-called positive human rights along with health and education. Environmental rights may be considered as a form of PROPERTY RIGHTS, in that they imply the right to enjoy property free of damage. Yet they can also be a challenge to property rights, if environmental rights are posed as superseding preexisting property rights.

COMMENTARY ON ENVIRONMENTAL RIGHTS

A cynic might define this term as follows: "The rights of government to seize control over natural resources without having to pay for them, while at the same time claiming to do good." This is accomplished by ignoring that existing natural resources under private or communal ownership have typically been providing benefits for other people (e.g., forests provide wildlife habitat protection, soil protection, etc.). In fact, the decision to liquidate such resources is not so much an injury to society as a choice to cease providing a free benefit. The government should pay for these benefits on behalf of society, by compensating those with resource rights to induce them to maintain the resource stocks voluntarily. Instead, governments often brand these people as resource despoilers who are violating the magically emergent "environmental rights" of the broad citizenry. We often applaud this hypocrisy when it is directed against the wealthy—owners of beach property who dare to build on their own property, only to face new zoning restrictions; or the private owners of stately redwood stands who suddenly find that they cannot harvest their trees. Yet we conveniently ignore the fact that the poor are more likely to have to eke out a living using natural resources, and they are the most vulnerable target of the onslaught of "environmental rights." *William Ascher*

Global Commons: regions or domains considered to be beyond the exclusive jurisdiction of nation-states. The principal examples of global commons are the oceans and seabeds outside the two-hundred-mile EXCLUSIVE ECONOMIC ZONES (EEZS) of coastal states, outer space, including the moon and other extraterrestrial bodies, Antarctica, and the atmosphere. While global commons as geographical regions or domains are not available for exclusive claims by states, the resources contained in the commons, such as fish in the oceans, may be appropriated by states or their operatives for their individual use and benefit. Use of the commons may entail extraction of resources, such as minerals, or disposal of waste substances. Numerous international treaties regulate activities affecting global commons. Nevertheless, global commons have been subjected to overuse or misuse, leading to the depletion or the degrading of the resources contained within them. See TRAGEDY OF THE COMMONS.

See Marvin S. Soroos, "The Tragedy of the Commons in Global Perspective," in *The Global Agenda: Issues and Perspectives*, eds. Charles W. Kegley Jr. and Eugene R. Wittkop (Boston: McGraw-Hill, 1998), 473–86. *M. S. S.*

Intellectual Property Rights (IPR): the rights of businesses, individuals, or nation-states to protect their discoveries, ideas, concepts, and inventions from use by outside parties without consent. Intellectual property rights came into play in the environmental arena during the late 1980s when the issue of biodiversity conservation became entangled in a North-South struggle over genetic resources. Developing nations had long been dissatisfied with the fact that the hybrid seeds based on the genetic resources obtained from their countries were protected by intellectual property rights and thus were available to them only at much higher costs.

International Commons. See GLOBAL COMMONS.

Open-Access Resources: resources that are available for exploitation by a wide range of potential users, owing either to the vagueness of property rights or to official declarations that they are available to many. It has been widely observed that open-access resources, more than any category of property, are subject to unsustainable exploitation. The logic is that any given potential resource user lacks the incentive to forgo immediate exploitation of the resource in order to conserve it, because an open-ended set of other potential users could take advantage of any resources that are left intact.

The open-access scenario is analyzed by Garrett Hardin in his well-known TRAGEDY OF THE COMMONS, which is often erroneously equated with a COMMON PROPERTY arrangement. In a common property (or common pool resource) situation, the relatively well-defined boundaries of the group may allow for the motivation and feasibility of self-discipline within the pool of users. See also ENVIRONMENTAL REGIMES.

Plant Breeder Rights (PBR): property rights extended to plant breeders (primarily seed and biotechnology companies) to afford them exclusive rights over new plant varieties. The issue of plant breeder rights is highly politicized and extremely complicated by legal uncertainties. Traditionally "wild" genetic resources and germplasm were considered "common heritage of mankind" and exchanged freely across the continents. In 1961, the United States, under the pressure of seed companies, passed a law granting plant breeders patent-like rights. Thus, the largest agribusiness in the world was provided with a legal basis to make exclusive profit from germplasm. European governments followed the U.S. example in supporting seed patenting. In response, developing countries argued that the germplasm for new varieties originated usually from their biological wealth and was a product of farmers' selection and breeding over the generations, and therefore Western agribusiness has been engaged in "biopiracy." In addition, the practical implementation of plant breeder rights is believed to have significant global environmental and social consequences, such as erosion of genetic diversity, expansion of monoculture, monopolization of the market, and suppression of local agriculture. The UNITED NATIONS EDUCATIONAL, SOCIAL, AND CULTURAL ORGANIZATION (UNESCO) and the UNITED NATIONS FOOD AND AGRICULTURE ORGANIZATION (FAO) support the concept of "farmers' rights" (FR), wherein breeders are supposed to pay royalties on varieties they create from traditional seeds or wild genetic material. The controversy between Western agribusiness and developing countries' farmers has been repeatedly addressed by different UN agencies and by the Conference of Parties to the CONVENTION ON BIODIVERSITY (CBD). However, no solution has been achieved thus far. See also INTELLECTUAL PROPERTY RIGHTS and TRADE-RELATED INTELLECTUAL PROPERTY RIGHTS.

Property Rights: the entitlement to particular uses of resources, goods, or services. This "bundle" of rights ranges from exploiting the asset through extraction or depletion, using the asset as collateral, and selling it. Clear and secure property rights are fundamental for the efficient operation of markets and for promoting long-term investment. Property rights therefore need to be well-defined, exclusive, secure, transferrable, and enforceable. Property rights are often classified as private property, state property, or COMMON PROPERTY (for which the rights are held by a specified community). In the absence of property rights, the situation of OPEN ACCESS prevails. The property rights school of economic thought argues that the success of a market economy in allocating resources efficiently depends on clearly vested property rights in all inputs to production, and in all goods and services produced. Otherwise, producers and consumers will not bear the full costs, or capture all the benefits of their actions, and distorted incentives will lead to socially undesirable decisions. *R. P.*

Scarce Resources. See RESOURCE SCARCITY.

Shared Natural Resources: physical or biological resources that extend into or across the jurisdictions of two or more nations or subnational jurisdictions. They may involve renewable and nonrenewable resources as well as complex ecosystems that transcend the boundaries of national jurisdictions (for instance, international watercourses). There may be significant asymmetries in the interests and needs of the states who share natural resources as well as in their opportunities to use them. Establishing a joint management regime for shared natural resources is a major challenge; many shared natural resources tend to be overused. See also TRANSBOUNDARY RESOURCES.

See Oran Young, ed., *Global Governance: Drawing Insights from the Environmental Experience* (Cambridge, MA: MIT Press, 1997).

3.4.3 SCARCITY

Absolute Scarcity: a state of resource availability in which the physical quantity of a resource is fixed and insufficient to meet the demand for it. Absolute scarcity occurs only with exhaustible resources such as fossil fuels. This notion is used in contrast to relative scarcity of a resource, which implies that demand exceeds supply over a given period of time (which can apply to humanmade, natural, renewable, and nonrenewable resources). See also RESOURCE SCARCITY. *R. P.*

Ecological Scarcity. See RESOURCE SCARCITY.

Geopolitical Scarcity: a type of RESOURCE SCARCITY that occurs when a particular resource is concentrated in a small number of countries that control its output and export it in accordance with their economic or political purposes, thus failing to meet the demand from other countries. An example of geopolitical scarcity is the oil crisis of the 1970s, when, following the 1973 Middle East conflict, Arab oil producers cut production and for a while banned sales to major Western countries. This type of scarcity is not related to the physical quantity of the resource, but to the political control over its use.

Hotelling Rent: a measure of natural resource value and depletion in ENVIRONMENTAL ACCOUNTING. It is the net return realized from the sale of a natural resource under particular conditions of long-term market equilibrium. The rent can be calculated as the residual of revenue received less all marginal costs of resource exploitation, exploration, and development, and less a normal return to fixed capital employed. See also HOTELLING RULE.

Resource Rent: the intrinsic or original value of a natural resource apart from any value added by its extraction, processing, or marketing. At least in theory, the resource rent can be determined by a fair, competitive auction of the resource, or it can be calculated by subtracting all of the processing costs (including a normal profit) from the final sale price of the commodities produced from the resource (see HOTELLING RENT). Secure ownership of the resource gives the owner this rent without having to make investments; therefore the rent (obtained, for example, by selling the resource) is equivalent to the basic economic definition of rent as "profit in excess to the normal return on capital." The concept of resource rent is crucial because unless resource managers charge close to the resource rent for others to extract the resource, the cheapness of the resource will induce excessive and reckless extraction. This is why low royalties on timber or mineral exploitation typically constitute a major natural resource policy failure.

Resource Scarcity: insufficient availability of natural resources and environmental space to meet human wants. Usually a distinction is made between ABSOLUTE SCARCITY when physical quantities of the resource are insufficient to meet the demand for it, and relative scarcity, when physical quantities of the resource are sufficient but its exploitation is limited either due to financial or institutional reasons. The difference between absolute and relative resource scarcity might be presented in terms of "Malthusian" versus "Ricardian" scarcity. With Malthusian scarcity, land (or any other basic resource) is fixed in absolute size and therefore the production is limited to whatever the extraction costs. With Ricardian scarcity, the quality of the resource deteriorates as population and economic growth require the exploitation of progressively poorer grades of the resource and subsequently to rising production costs. In practice, Malthusian and Ricardian concepts of scarcity complement each other. Yet other classifications of resource scarcity include "ecological scarcity," the limited capacity of the earth's systems to absorb pollution and to sustain life (Ophuls, 1977), and GEOPOLITICAL SCARCITY, when physical quantities of resource are sufficient but its supply to the world market is limited by a political control of a producer-nation. Resource scarcity has a wide range of consequences and a rich history of human responses. On the one hand, shortages of food, water, and energy resources throughout history have created famines, war, and other human tragedies; on the other hand, resource scarcity is also believed to have provided a powerful impetus for technological development and subsequent economic growth in many areas. See also BOSERUPIAN THESIS, KONDRATIEFF CYCLES, and RESOURCE-CURSE HYPOTHESIS.

See William Ophuls, *Ecology and the Politics of Scarcity: Prologue to the Political Theory of the Steady State* (San Francisco: W. H. Freeman, 1977).

COMMENTARY ON RESOURCE SCARCITY

Why doesn't anyone talk about the problem of "resource overabundance"? Most energy supply projections show that world oil production is not expected to peak until around 2025. Abundant supply and continued low price of oil create huge political and economic obstacles to meaningful action on greenhouse gas emissions and energy policy. We could use a little more scarcity.
Daniel Sarewitz

3.5 Worldviews and Social Traps

Anthropocentrism: a worldview that puts the welfare of humans as the highest priority. Philosophically, it may rest on a view of humankind at the center of creation. From the anthropocentric perspective, human beings exist independently and largely separate from the natural world. This assumption leads to the definition of the world as an object that "surrounds" humankind and whose value is determined by its usefulness to humans. Historically, anthropocentrism has been the most persistent and dominant assumption of Western society. Recently, its role in the development of ecological crisis has been increasingly questioned by environmentalists and ethicists.

Basic Pessimist Model (Catastrophism): a vision of an inevitable human-induced crash of the BIOSPHERE brought on by reckless development, population pressure, and unrestrained consumption. Typical examples of this approach are the LIMITS TO GROWTH study, BLUEPRINT FOR SURVIVAL, and the GLOBAL 2000 REPORT, which all deal with models exploring growth, ecocatastrophe, and ways to avoid it. All three publications can be regarded both as products of and stimuli for the second wave of global environmentalism.

Biocentrism: a point of view, in opposition to ANTHROPOCENTRISM, that regards nature as having intrinsic value and rights, irrespective of its practical usefulness to humans. In this approach, humanity is no more, but also no less, important than other life forms.

COMMENTARIES ON ANTHROPOCENTRISM VS. BIOCENTRISM

There is a growing worldwide demand for a healthy environment for ourselves and future generations: a polluted, degraded world impoverished in biological diversity is not in the common interest. Yet finding the common interest is difficult as humans tend to divide themselves into camps. A prime example of this is the "anthropocentrism vs. biocentrism" debate. These recently coined labels characterize two opposing worldviews about the relationship of humans to nature, especially biological entities. It is clear that the "we vs. they" phenomenon is a central leitmotif that holds groups of people together and seems to be a major component of all societies. Overcoming the dichotomy, however, would aid the common interest in securing a healthy environment for all. We need to find workable alternatives to the status quo to move us all closer to the shared goal of "sustainable conservation" of the planet.

See A. Flores and T. W. Clark, "Finding Common Ground in Biological Conservation: Beyond the Anthropocentric vs. Biocentric Controversy," in *Species and Ecosystem Conservation: An Interdisciplinary Approach* (New Haven: Yale University School of Forestry and Environmental Studies, Bulletin Series 105, 2001). *Tim W. Clark*

We are between the no-longer and the not-yet in our human evolution. What we have known and how we know it has been with our five senses, outer senses that experience matter outwardly. This has generated a pattern of dominance and exploitation, materialism, of body separated from mind. Modern physics and ancient forms of spirituality have tapped into different layers of energy within the self that include intention, intuition, power to love, and to be fully present and intimately engaged with nature. This evolution of new space within the human allows a deeper understanding of evolution, of the nature we are part of and can never be separate from, and movement toward becoming responsive, caring, intentional, trusting, fully engaged human beings. Our evolution is moving from external power that is corrupted, stolen, transferred, inherited, or changed to an inner power that is untouched by the outer structures of power. *Susanne Swibold*

Catastrophist Model. See BASIC PESSIMIST MODEL.

Contempocentrism: a label applied to the very short time horizons said to be characteristic of many individuals and policymakers. Critics of contempocentrism trace it to the elevation of individualism over the human community, arguing that the mortality of individual humans, in combination with various forms of uncertainty about the future, often leads to a preoccupation with "the here and now," or, in economic terms, with "net present value maximization." Contempocentrism traditionally was not only a cultural value but also a survival strategy of humans as a species. The emergence of global environmental threats, however, attest that contempocentrism is no longer working well as an adaptive strategy.

Cornucopian Model: a highly optimistic vision of the capacity of technology to address resource and environmental challenges. Associated with the ideas of Julian Simon and Herman Kahn, the cornucopian model has confidence in the human capacity to achieve ecological balance and economic prosperity. Proponents of this vision argue that humanity is still in the period of transition that began with the industrial revolution. That transition follows an s-shaped curve—that is, exponential trends in world population and economic growth that were typical since the 1880s began to level off in the mid-1970s, and even decline, albeit along with increasing affluence. Current "superindustrial" society will (probably in the middle of the twenty-first century) be replaced by a world of large scale systems dominated by high technology, managed by technocrats, and characterized by economic efficiency, stabilized world population, and globally sustained affluence. Cornucopians anticipate the prosperity of a free-market global village, tightly linked by communication, education, and trade activity and powered by technoscientific innovation. In this view, human development activities, including population growth, are positive forces in improving the quality of life, and humans are essentially exempt from natural constraints. Though development-induced environmental damage, social inequality, poverty, and related geopolitical tensions are problems in the short run, these transitional problems will recede as the tide of human progress manages to undo much of the damage inflicted by earlier, unsophisticated forms of economic development. Though branded as dangerously utopian by other alternative visions, such as the BASIC PESSIMIST MODEL, the cornucopian model, positing the possibility of universal affluence, environmental sustainability, and a leisure-oriented society, has many advocates, especially among social scientists. In response to the popularity of the cornucopian model, two of the most prestigious academic institutions, the U.S. National Academy of Sciences and the Royal Society of London, issued a joint statement (1992) that "advances in science and technology no longer could be counted on to avoid either irreversible environmental degradation or continued poverty for much of humanity."

See Julian Simon and Herman Kahn, *The Resourceful Earth* (Oxford: Basil Blackwell, 1984); John Naisbett, *Global Paradox: The Bigger the World Economy, the More Powerful Its Smallest Players* (New York: William Morrow, 1994); George Tyler Miller, *Living in the Environment* (Belmont, CA: Wadsworth, 1992); Royal Society of London and U.S.

National Academy of Sciences, "Population Growth, Resource Consumption, and a Sustainable World," Rio de Janeiro, Brazil, June 1992.

Ecocentrism: a term coined and developed by Timothy O'Riordan (1989) to define a radical trend in contemporary environmentalism. Ecocentrism views humankind as part of the global ecosystem. Ecological laws, along with ethical considerations and a strong sense of respect for nature in its own right, serve as limiting factors to population growth and economic development. In opposition to TECHNOCENTRISM, ecocentrists do not have unrestricted faith in modern science, technology, and its elites, but advocate alternative ("soft" or "appropriate") technologies that are more environmentally benign and democratic (see APPROPRIATE TECHNOLOGY). From a political perspective, ecocentrism stands for redistribution of power toward a decentralized, federated economy with more emphasis on informal economic and social transactions and the pursuit of participatory justice. According to O'Riordan, 5–13 percent of various opinion surveys in Europe demonstrated public support for this trend, while most ardent followers are intellectual environmentalists, politicians of radical-social and radical-liberal orientation, and "green" youth.

See Timothy O'Riordan, "The Challenge for Environmentalism," in *New Models in Geography*, eds. Richard Peet and Nigel Thrift (London: Unwin Hyman, 1989): 77–102.

Exclusionist Paradigm: a system of beliefs about economics that excludes human beings from the laws of nature. This social paradigm has dominated public understanding of environmental management for the major part of human history. This paradigm is also termed "frontier economics" since it suggests the sense of unlimited resources in a society with an open frontier. In the United States, the exclusionist paradigm came under attack in 1962 with the groundbreaking publication of Rachel Carson's *Silent Spring*, which exposed the high prices paid by society for policies based on this mindset.

Human-Needs Theory: a conceptual approach to environmental conflict analysis and resolution, rooted in hierarchical needs-based theory of self-actualization and research on the physiological and sociobiological causes of conflict. In this approach, environmental conflicts are seen as coming from disagreements among parties ascribing to different values regarding environmental goods and services. Resource disputes, in contrast, are regarded as conflicts of interest in which parties may disagree on the distribution of a resource or environmental goods, while basically agreeing on the value of these resources or goods. The distinction between value-based and distribution-based disputes may have important implications for conflict management strategies.

See John Burton, ed. *Conflict: Human Needs Theory* (London: Macmillan, 1990).

New Environmental Paradigm (NEP): a set of beliefs that nature is intrinsically valuable, that humans should behave with compassion toward other humans, species, and generations, and that limits to growth exist. Since the 1980s the NEP has been challenging the so-called dominant social paradigm in many parts of the world, most notably in the United States. Though held mainly by members of radical environmental groups and by small influential minorities of the total population, the NEP has potential significance from the standpoint of resource conservation.

See Lester W. Milbraith, "Culture and the Environment in the United States," *Environmental Management* 9 (1985): 161–72.

Technocentrism: a trend in ecophilosophical thinking that recognizes environmental problems but asserts that humans will be able to solve these problems through technology and eventually achieve unrestrained material growth (CORNUCOPIAN MODEL), or that the problems can be mitigated by careful economic and environmental management. In either case, possible solutions are seen within the limits of classical science, technology, and conventional economics.

Tragedy of the Commons: the degradation of natural resources subjected to overexploitation when they are available to multiple actors to use for individual, private gain. The phrase "tragedy of the commons" was popularized in a 1968 article by Garrett Hardin, which analogizes the "tragedy" in terms of an old English village whose residents continue adding privately owned cattle to a community pasture known as the commons until the pasture is badly overgrazed and its grasses destroyed. The residents persist in this behavior because the payoff they receive as individuals from adding each additional head of cattle to the overused pasture is greater than their shares of the costs of overgrazing, which are shared with the community as a whole. The depletion of high-seas fish stocks by boats from several countries is a real-world example of such a tragedy of the commons. However, it is important to distinguish between "commons" as a communally held resource that is restricted to members of the community, and the "open access" resources that are available to any would-be user. Tragedies of the commons may be avoided either by regulations that limit use of the commons, or by dividing the pasture or other resource domain into sections for the exclusive use of individual actors. See also GLOBAL COMMONS.

See Garrett Hardin, "The Tragedy of the Commons," *Science* 162 (1968): 1241–48. M. S. S.

4

International Political Economy of Environment and Development

4.1 Globalization

Biocolonialism: a market-driven process of increasing control of different uses of domestic biological resources by foreign actors, mainly transnational corporations. Examples include contractual arrangements requiring the use of hybrid seeds sold by the corporation that also controls the marketing channels. See also PLANT BREEDER RIGHTS.

Code of Conduct of Transnational Corporations. See UNITED NATIONS CODE OF CONDUCT OF TRANSNATIONAL CORPORATIONS.

Comparative Advantage Theory: a basic tenet of trade theory, proposing that patterns of trade should be and are defined by the particular resource endowments of each country. The argument, dating back to the work of early nineteenth-century economist David Ricardo, is that under conditions of free markets, two or more countries will be better off on an aggregate basis if each concentrates on producing and exporting what it produces with a higher return ("comparative advantage") over the other possible products, even if one country produces everything better than another. Thus to gain the highest profits, a country should produce and export products that use factors of production with which it is relatively well endowed, and import those items whose production in that country would depend on scarce factors. This logic is counterintuitive to many people, who assume that a country should concentrate on producing and exporting those items that have a "COMPETITIVE ADVANTAGE." Ricardo's key insight was that a country's welfare lies in doing what it does most productively, not in vanquishing other countries by killing their capacity to export—which would also destroy their ability to import. Comparative advantage theory is a bulwark of the justification of free international trade. One implication of the comparative advantage theory within the realm of the global environment has already resulted in a logical, though ethically shocking, suggestion made by then World Bank chief economist Lawrence Summers that developed nations should export their waste and dirty industries to the developing countries

(*The Economist*, May 30, 1992, Survey, 7). Though met with public indignation, this proposal reflects the ongoing practice of GARBAGE IMPERIALISM. See also REVEALED COMPARATIVE ADVANTAGE.

COMMENTARY ON COMPARATIVE ADVANTAGE THEORY

Herman Daly questions whether comparative advantage is a useful concept in today's globalized economy. In the nineteenth century David Ricardo observed that countries that made the same product incurred different costs. Consequently, one country would find a comparative advantage in making that product over another country. If all countries freely traded in those products that produced at a comparative advantage, then everyone would benefit. However, in Ricardo's world the factors of production were immobile—especially capital. In today's globalized economy, capital can migrate to the country where production costs are the lowest. Often the savings come from labor and environmental costs. Furthermore, free trade can encourage the externalization of costs. This brings into question whether everyone really benefits from comparative advantage and free trade. *Toddi Steelman*

Competitive Advantage: in international trade, the capacity of a firm, region, or nation to compete economically against others, through a combination of superior quality and lower prices of specific types of goods and services. Competitive advantage derives from strong endowments in particular factors (such as capital, labor, natural resources, or technology) that enables production at competitive prices. Goods and services may become noncompetitive because of government policies, such as above-market wage-setting. In such cases, a region or nation's COMPARATIVE ADVANTAGE may be undermined. By the same token, government policy may intervene to make goods and services more competitive than the intrinsic endowments would dictate—subsidies, taxes, wage policies, and lenient environmental regulations often play a large role in determining whether a firm or geographical unit has a competitive advantage with respect to a category of goods and services. When government policy creates competitive advantage through these means, other countries or regions may object on the grounds of unfair "dumping." NEOCLASSICAL ECONOMICS has long criticized the efforts to create competitive advantage when comparative advantage does not exist for a category of goods or services. In addition, having a competitive advantage over another country does not necessarily mean that export production to that country is wise; classical Ricardian trade theory argues that a nation should only produce what it does well (comparative advantage), even if it produces other goods and services better than do other countries.

Cultural Imperialism: a concept denoting the imposition of cultural (including environmental) values onto other cultures. In the current era, cultural imperialism is generally seen as the imposition of Western values onto the peoples of developing countries and countries in transition. This imposition occurs through such means as globalization of the media, introduction of new technologies and products, proselytizing, use of personnel and advisers, and, according to some critics, the funding policies of developing agencies and organizations.

Diversity and the different consequences of numerous projects that might be associated with cultural environmental imperialism—such as a recent GLOBAL ENVIRONMENT FUND project to save an endangered herd of monkeys at the expense of resettling several villages of indigenous peoples in Africa or the African Heritage project to introduce more efficient irrigation techniques in Tanzania—bring two major issues to the surface: the price of progress and the dichotomy of imposition and free choice. *Natalia Mirovitskaya*

Delinking: a term used in ALTERNATIVE STRUCTURALISM and in North-South negotiations to define the possible disassociation of the South from the North (see NORTH VS. SOUTH). Its expanded application has been connected with the development of mutual cooperation between developing nations in their quest for economic independence and self-reliance. However, the alternative structuralist approach does not provide any concrete plans as to how to bring delinking into practice on local, national, or regional levels, while during negotiations this term is used most for rhetorical purposes.

Ecodumping: the practice of taking advantage of lax environmental regulations within a country to make its exports more competitive by keeping down the costs of environmental compliance. Critics assert that some governments, particularly in developing countries, purposely enact weak environmental standards in order to promote competitiveness of their industries. Developed nations claim that they face competitive disadvantages while importing goods from developing countries under such conditions, and claim that a system of green COUNTERVAILING DUTIES or tariffs should be implemented to level the playing field. Because enforcement difficulties reduce effective environmental standards in many developing countries, whether or not the governments intend these limits to enforcement, it has been very difficult to demonstrate a deliberate strategy of ecodumping.

Ecological Imperialism: a term recently coined to condemn efforts either to have developing countries accept pollution from developed countries, or to pressure developing countries into protecting their ecosystems in ways that primarily serve the interests and values of the First World. The term "ecological" or "green" imperialism has been viewed increasingly in some quarters as a form of environmental alienation. Some critics in developing and transitional countries see the movement for global environmental/resource management as an attempt by industrialized countries to take away their control of resources and to turn them into suppliers of environmental repair or caretaker services aimed to mitigate the problems created by Western civilization.

For an alternative definition referring to the biological phenomenon of species from one area to invade another, see under SECTION 1.2.4.

Ecological Shadow: environmental resources that a country draws from other nations and the GLOBAL COMMONS. Few if any of the world's urban/industrial regions are ecologically self-contained. The ecological shadow is invoked by those who argue that economically powerful Western nations are able to drain the ecological capital of other nations (mainly developing countries) to provide food for their populations, energy and materials for their economies, and sink capacity for accumulated wastes. While subsequent overexploitation and depletion of the developing countries' resources can provide developing nations with short-term financial gains, the long-term consequences of ecological shadows can be a steady reduction of the developing countries' economic potential over the medium and long term. For developed nations, that would also mean destruction of ECOLOGICAL CAPITAL on which prospects for their economic prosperity and trade depend.

See Jim MacNeill, Piter Winsemius, and Taizo Yakushiji, *Beyond Interdependence. The Meshing of the World Economy and the Earth's Ecology* (New York: Oxford University Press, 1991).

Foreign Direct Investment (FDI): the establishment or purchase of businesses and properties in a country by individuals or corporations from another country. Investment of foreign capital has historically been a major engine of economic development; it increases the amount of capital in the host economy, which in turn raises labor productivity, income, and employment. In contrast with foreign borrowing, which imposes obvious future obligations to pay back the loans, foreign direct investment typically imposes a smaller burden on the target country's balance of payments. However, many corporations finance the expansion of foreign subsidiaries by raising capital in the country where the subsidiary operates. While North-to-North capital flows are predominant, the developing nations receive a significant and growing share of global foreign direct investment (38 percent in 1995). Many developing and transitional countries are cautious about opening their economy to foreign investment, because of issues of national pride and fears of foreign interference. One of the most recent arguments supported by environmentalists is the presumed negative role of foreign direct investment in the economy-environment interface. A counterargument is that MULTINATIONAL CORPORATIONS often have a better record of environmental compliance than do domestic businesses in developing and transitional nations.

Multinational Corporations (MNCs; Transnational Corporations, TNCs): firms with operations and often affiliates or subsidiaries in multiple countries. Vernon and Wells (1986) define multinational corporations as companies that are "made up of clusters of affiliated firms that, although in different countries, nevertheless share distinguishing characteristics: (1) they are linked by ties of common ownership; (2) they draw on a common pool of resources, such as money and credit, information systems, trade names and patents, (3) they respond to the same common strategy." A multinational corporation is a single business organization that accomplishes its objectives and has a physical presence in more than one country, even if it has subsidiaries that are formally chartered in different countries. This term is used interchangeably with "transnational corporations," a term traditionally used in United Nations documents, but often with the negative connotation that the firms are beyond accountability in any given country. Multinational corporations can be distinguished from firms that are international only in the sense of purchasing or selling across national boundaries. In terms of environmental protection, there is a long-standing debate over whether multinational corporations are pernicious because they have the capacity to locate polluting operations in countries where regulations are lax, or whether they are beneficial because their environmental standards are often higher than those of national companies, particularly in developing nations.

See Raymond Vernon and Louis Wells, *The Economic Environment of International Business* (Englewood Cliffs, NJ: Prentice-Hall, 1986).

Multilateral Agreement on Investment (MAI): a multilateral framework for international investments that was supposed to liberalize investment regimes, ensure investment protection, and create effective dispute settlement procedures. The MAI initiative was taken by the ORGANIZATION FOR ECONOMIC COOPERATION AND DEVELOPMENT (OECD) Council of Ministers in 1995. Like the NORTH AMERICAN FREE TRADE AGREEMENT (NAFTA) and WORLD TRADE ORGANIZATION (WTO) agreements, the MAI preamble links investment and environmental concerns. The MAI also discourages the creation of POLLUTION HAVENS. However, these provisions are considered inadequate by environmental activists, who argue for the inclusion of additional requirements that companies operating abroad apply the stricter of home or host country environmental standards. Thus far, public pressure has been able to stall the agreement.

Pollution Havens Hypothesis: an assumption that lax environmental standards in one country might create COMPARATIVE and COMPETITIVE ADVANTAGES and encourage firms from countries with higher standards to move their productive capacity to so-called pollution havens. Such a move would result in loss of jobs and tax revenues; therefore labor unions in developed countries promote the idea of a GREEN COUNTER-VAILING DUTY in order to compensate for the assumed negative effects on competitiveness. Critics of the pollution haven hypothesis argue that empirical studies over the past two decades have not found evidence of governments deliberately using lower environmental standards to gain a comparative economic advantage, and that the threat of industrial migration is a pretext used by firms and governments that wish to reduce the burden of environmental costs.

See Tom Tietenberg, *Environmental and Natural Resource Economics*, 4th ed. (New York: HarperCollins, 1996).

COMMENTARY ON POLLUTION HAVENS

Little evidence supports the wholesale condemnation of MULTINATIONAL CORPORATIONS' (MNCs) environmental performance in developing countries. Although critics can always find some multinationals that misuse human and natural resources in developing countries, a series of studies going back to the early 1980s confirm that highly polluting industries do not generally migrate to developing countries to take advantage of lax environmental regulations. MNCs seek locations in developing countries for reasons far more crucial to their competitive success than exploiting weaknesses in environmental laws. A Conservation Foundation study in the mid-1980s (Leonard 1984) concluded that most U.S.-based MNCs' FOREIGN DIRECT INVESTMENT in developing countries was not in pollution-intensive industries. A survey conducted by the United Nations in the late 1980s and early 1990s found that the share of "pollution-intensive" industries (chemicals, pulp and paper, petroleum, coal, and metal) in the total outward investment stock of industrialized countries was relatively low—17 percent of France's, 19 percent of Germany's, 8 percent of Japan's, 13 percent of the United Kingdom's, and 19 percent of the

United States'—and that many of these industries had begun to adopt more effective environmental management practices (United Nations 1993). Many corporations now adopt international standards of environmental management based on British, U.S., or ISO guidelines for their operations in developing countries and use environmental practices that are more stringent than local and national laws.

See H. Jeffrey Leonard, *Are Environmental Regulations Driving United States Industry Overseas? An Issue Report* (Washington, DC: Conservation Foundation, 1984); UN Conference on Trade and Development (UNCTAD), *Environmental Management in Transnational Corporations* (New York: UN Conference on Trade and Development, 1993). *Dennis A. Rondinelli*

Revealed Comparative Advantage (RCA): a measure used to define comparative advantage of a country in the production of particular product. RCA is defined as a ratio of a country's share of world's exports of the particular product to its share of world exports of all manufactured goods. A country is supposed to have a COMPARATIVE ADVANTAGE in the production of the product if its RCA is higher than one. The RCA might be used for the analysis of trade-environment interactions, though its methodological value is often decreased by price distortions, caused, for example, by inappropriate environmental regulations.

United Nations Code of Conduct of Transnational Corporations: the main international initiative to establish (nonbinding) guidelines for the behavior of MULTINATIONAL CORPORATIONS. Since the 1970s an intergovernmental group of experts has been trying to formulate a comprehensive set of rights and obligations of multinational businesses and states. In 1983 this work was transferred to the Special Session of the UN Commission on Transnational Corporations. The evolving proposal was rather ambitious: it covered such different and contradictory issues as international cooperation in investments, standards of corporate behavior, technology transfer, national antitrust legislation, and foreign trade legislation. In 1992, due to unified opposition of the United States, the United Kingdom, and Japan, the UN had to abandon sixteen years of work on this code. The agency in charge of the code, the UN Centre on Transnational Corporations, was closed down. Many environmentalists and developing country theorists consider the failure of the code a major setback. They argue that it is most unlikely that businesses would voluntarily curb their own practices so as to be in line with SUSTAINABLE DEVELOPMENT.

4.2 Capital Shortages and Foreign Debt

4.2.1 PRACTICES, TRENDS, AND PROBLEMS

Capital Market Liberalization: policy reforms that relax regulations and restrictions on DEBT and equity transactions, opening up opportunities for a greater variety of institutions, often including foreign institutions, in banking, stock markets, and other financial arenas. The rationale of capital market liberalization is to provide greater opportunity and incentives for raising capital, and for increasing the chances that capital will be used efficiently. The critics of capital market liberalization point to the risks of speculative bubbles and the ease of rapid capital flight.

Debt: obligations to pay for borrowed capital or prior purchase of goods and services. Debt agreements typically specify the schedule of paying off debt and the cost of the debt (the interest rate). If a debt is not paid according to the schedule, the lender usually has the option of either declaring the debtor in default, which has legal implications that vary according to the context, or to "reschedule" the debt by reloaning the money that the debtor has failed to pay back. For both domestic and international debts, but particularly for international debts, the relationship between lender and debtor is often complicated by the limited recourse that the lender has to collect an unpaid debt; this is even more complicated when the debtor is a sovereign government or a government organ such as a state enterprise. Therefore the interest rate on loans (both domestic and international) typically reflects the lender's assessment of the likelihood of debt default. Governments and international organizations can reduce the interest rates by guaranteeing debt through their own commitment to pay back the debt if the borrower defaults. The pay-back (or "service") on international debts usually requires debtors to use internationally accepted currency, such as dollars, yen, or pounds, which must be obtained from the nation's foreign currency reserves. High levels of international debt thereby pose a drain on a nation's foreign currency reserves, sometimes requiring the government to seek debt rescheduling and loans from international institutions such as the INTERNATIONAL MONETARY FUND (IMF). Because the interest rates of most international loans vary according to the changing world interest rates, the DEBT SERVICE on a loan that was taken under favorable conditions may increase dramatically if world interest rates rise steeply, as occurred in the early 1980s, triggering a world debt crisis.

Debt Service: the capital outflow represented by amortization and interest payments on loans made to developing countries by creditors (see also DEBT). In many developing countries, debt service payments on loans exceed financial inflows. Consequently, these countries end up making a net transfer of financial resource to the creditor nations or companies. *T. S.*

Hard Loan: a loan contracted with an interest rate at the prevailing market level, in contrast to SOFT LOANS contracted at lower, concessional interest rates. Hard loans were once considered as onerous, particularly with respect to low-income borrowers, but more recent views hold that even low-income borrowers can be productive enough to benefit from and repay hard loans. Hard-loan interest rates permit the lenders to maintain the lending fund, which would be rapidly depleted with soft loans.

Investment Gap: the amount of external capital that an economy would require to augment domestic capital in order to have a high enough level of investment to grow at a targeted rate. The concept of investment gap was popular in the 1950s and 1960s as a way of gauging the foreign assistance requirements for developing nations that had explicit targets for economic growth rates, these targets often appearing in medium-term (i.e., three–six-year) economic plans. Foreign-assistance agencies could then calculate the net external loans, grants, and direct foreign investment needed to reach these targets. The investment gap concept has been criticized on several grounds: the target may not be feasible or even desirable from the perspective of long-term sustainability; the inflow of capital may be too great to be used efficiently; and the DEBT burden implications are not necessarily addressed by an investment gap framework.

Liability: probable future sacrifices of economic benefits arising from present obligations to transfer assets or provide services to other entities in the future as a result of past transactions or events. See ENVIRONMENTAL LIABILITY.

4.2.2 DEBT RELIEF

Austerity Measures: measures to reduce government spending, limit imports, and curtail demand in order to stabilize economies suffering from drains on foreign reserves, inflation, and other macroeconomic problems. Austerity (or "stabilization") measures are typically part of STRUCTURAL ADJUSTMENT LOANS that also include policy liberalization measures that truly entail adjustments in economic structures. Austerity measures are commonly among the conditions of loans from the INTERNATIONAL MONETARY FUND (IMF), the WORLD BANK, and other multilateral financial institutions. Because of the concern that austerity measures, particularly government spending cutbacks and wage limits, would place a disproportionate burden on the poor, SOCIAL SAFETY NETS are often proposed and sometimes adopted to reduce the negative impacts on low-income families.

Debt Conversion: the exchange of one form of claim on a debtor country for another form of claim, essentially to a form of DEBT that is easier for the debtor to SERVICE. For example, rather than pay a foreign debt with scarce foreign currency reserves, governments often find it easier to pay in long-term bonds, in local currency destined to purchase equity stakes in domestic companies ("debt-for-equity"), or to purchase land for conservation ("DEBT FOR NATURE SWAPS"). The original creditors are attracted to debt-for-bonds or debt-for-equity conversions when the borrowing government is unable or unwilling to meet its debt service obligations according to the original loan agreement. Other institutions, both for-profit and nonprofit, have purchased debt obligations on the secondary market (established in the early 1980s when Mexico and several other governments declared their inability to service their debts). Often buying these debt obligations at heavy discounts of the full value of the debt, these institutions negotiate with the borrowing government on debt conversions. Conservation organizations have been able to buy or restrict the uses of land, or establish conservation training programs, through the "swap" of government debt, often purchased on the secondary market at a fraction of the face value.

Debt-for-Nature Swaps (DFNs): the purchase of part of a country's foreign debt intended to convert the debt into local currency to be devoted to conservation or environmental programs in the debtor country. Debt-for-nature swaps are usually undertaken by a NONGOVERNMENTAL ORGANIZATION but occasionally by a government. Swaps can involve such projects as acquisition, maintenance or expansion of national parks and reserves, environmental training, and research programs. The idea was first proposed by Thomas Lovejoy (WWF-US) in 1984 and was soon put into practice in over twenty debt-for-nature swaps, mainly in Latin America. The debt-for-nature swap strategy, though at first sight a win-win situation, is nevertheless criticized for several reasons. Some argue that it does too little both for the solution of the developing countries' debt problem and for nature conservation. Others consider swapping a typical example of

ENVIRONMENTAL COLONIALISM, insofar as it imposes an external agenda on land use within the indebted country.

Sector Loan: an international loan, generally from a multilateral or bilateral lending agency (such as a multilateral development bank or a bilateral foreign assistance agency), earmarked for use within a sector rather than designated for particular projects. Sector loans give the recipient governments more discretion to devise projects within that sector, as opposed to having to pursue the specific projects proposed or approved by the lending or assistance agency. Nevertheless, sector loans often involve CONDITIONALITY (conditions) that binds the government to reforms of the policies for that sector.

Social Safety Nets: policies designed to protect low-income families from the negative impacts of AUSTERITY MEASURES and STRUCTURAL ADJUSTMENT LOANS, insofar as these measures reduce family incomes and government services, or create higher unemployment. Social safety nets typically involve safeguarding or increasing the government spending programs that target the poor, particularly in health and nutrition, education, and housing. They may also involve employment programs and income supplements. International organizations, such as the INTERNATIONAL MONETARY FUND (IMF) and the WORLD BANK, have been increasingly sensitive to the need to combine social safety net policies along with the loan conditions that have adverse effects on the poor.

Soft Loans: credit extended at interest rates considerably below the market costs of capital. Soft loans are often made by bilateral aid agencies and multilateral development banks and development funds, especially to low-income countries. Soft loans constitute a form of concessional lending or foreign assistance, insofar as the borrower saves on interest payments. The funding for soft loans cannot be sustained by loan repayments, and therefore must be replenished through donations. The INTERNATIONAL DEVELOPMENT ASSOCIATION (IDA), administered by the WORLD BANK, is the largest soft-loan facility in the world.

Special Drawing Rights (SDRs): the privilege of member governments to draw currencies from the general fund of the INTERNATIONAL MONETARY FUND (IMF). The volume of SDRs that a particular government can call upon depend on the contributions that that government has made to the IMF's fund, which in turn depend on the size of the nation's economy. SDRs in one sense constitute a generic currency. Beyond their SDRs, governments may borrow more funds in various currencies if they accede to conditions (CONDITIONALITY) agreed upon with the IMF.

Structural Adjustment Loans (SALs): loans made by official agencies, whether bilateral or multilateral, that are linked to policies (CONDITIONALITIES) that the recipient government is obligated to enact in order to change the structure of the economy. In particular, the INTERNATIONAL MONETARY FUND (IMF) and the WORLD BANK GROUP extend structural adjustment loans to induce governments to liberalize their economies in order to increase efficiency, reduce inflation, and redirect investment. The rationale for medium-term (i.e., three–five-year) structural adjustment loans is that economies require additional capital to ride out difficult periods of structural economic change. The elimination of the special privileges that constitute many of the efficiency-limiting economic distortions may actually enhance the equity of the economy. However, structural adjustment loans may include policy requirements to reduce spending, wages, and imports ("stabilization programs"), which have given structural adjustment programs the reputation of being harsh on low-income people dependent on government spending. One important issue concerning structural adjustment programs is how stringently the lender should impose the conditionalities; another issue is whether the lender and other entities ought to provide a SOCIAL SAFETY NET for vulnerable people adversely affected by the conditionalities.

4.2.3 MULTILATERAL LENDING INSTITUTIONS

African Development Bank: an official multilateral financial organization established in 1963 with the goal of assisting the economic development of African nations. Its current membership includes fifty African states and twenty-five non-African countries. The headquarters are in Abidjan, Côte d'Ivoire. The bank aims to spur both economic development and social progress among its members on an individual and a joint basis. It has six associated institutions through which public and private capital is channeled: the African Development Fund, the Nigeria Trust Fund, the Africa Reinsurance Corporation (Africare), the Société Internationale Financière pour les Investissements et le Développement en Afrique (SIFIDA), the Association of African Development Finance Institutions (AADFI), and Shelter-Afrique. The African Development Bank transfers funds largely through project lending, as do the other MULTILATERAL DEVELOPMENT BANKS.

Asian Development Bank (ADB): an official multilateral financial organization set up in 1966 with headquarters in Manila, Philippines. Its current objectives are to foster economic growth of member nations, reduce poverty, improve the status of women, support human development, and protect the environment. As of 1998, ADB had forty regional members and sixteen members from outside the region. Both groups of members subscribe to the bank's capital but only regional members are eligible for loans and technical assistance grants. As in most international financial institutions, voting is weighted according to a country's contribution, therefore Japan has a preponderance of power within the ADB. The bank gives special attention to the needs of smaller or less developed nations, and it gives priority to projects that contribute to the economic growth of the entire region and promote regional cooperation. In the 1980s the bank sponsored studies in several regional countries to determine how SUSTAINABLE DEVELOPMENT might be implemented in practical programs. Most of the current projects are oriented toward support of the private sector, strengthening of public sector management capacity, human resource development, and natural resource management.

See Asian Development Bank, *Economic Policies for Sustainable Development* (Manila: Asian Development Bank, 1990).

Bretton Woods System: the international monetary and financial system, named after the 1944 conference held in Bretton Woods, New Hampshire, from which the INTERNATIONAL BANK FOR RECONSTRUCTION AND DEVELOPMENT (IBRD), or WORLD BANK, and the INTERNATIONAL MONETARY FUND (IMF) were founded. In a broader sense, the Bretton Woods System refers to the international economic regime established after

World War II under the leadership of the United States and the United Kingdom, parts of which remain intact today; other aspects, such as a world monetary system pegged to the value of gold, were abandoned in the 1970s. An initiative to create an ambitious third international body to deal with trade issues, the International Trade Organization, failed in the early postwar period, leaving instead the painstaking progress toward free trade within the framework of the GENERAL AGREEMENT OF TARIFFS AND TRADE (GATT). Only in 1995 was the WORLD TRADE ORGANIZATION (WTO) founded to supervise and liberalize international trade and to settle international trade conflicts.

Compensatory Financing Facility (CFF): a loan window of the INTERNATIONAL MONETARY FUND (IMF) intended to cover balance-of-payments deficits that occur because of temporary declines in export earnings. Under the CFF, governments can borrow to finance their payments deficits when export revenues drop below a five-year period average. This is an attempt to stabilize world commodity prices, and to provide relatively easy and nonstigmatizing financing for countries that experience balance-of-payments difficulties because of world market conditions rather than faulty financial management by their governments.

Declaration of Environmental Policies and Procedures Relating to Economic Development: a statement by nine development agencies, in association with the WORLD BANK, the UNITED NATIONS DEVELOPMENT PROGRAMME (UNDP), the UNITED NATIONS ENVIRONMENT PROGRAMME (UNEP), the Organization of American States, and the Commission of the European Communities, calling for procedures to ensure that projects supported by these agencies are ecologically sound and that UN members take into account environmental considerations while formulating their economic development policies. The COMMITTEE OF INTERNATIONAL DEVELOPMENT INSTITUTIONS ON THE ENVIRONMENT (CIDIE) was established to put this declaration into effect. The declaration was also adopted by the European Investment Bank (1983) and the UNITED NATIONS FOOD AND AGRICULTURE ORGANIZATION (FAO) (1990). The effectiveness of the declaration and CIDIE as its operation mechanism are uncertain, in light of continued (though contested) criticisms that the World Bank, several development banks, and other development institutions have continued to subsidize tropical deforestation, ecologically and socially destructive water projects, and monoculture export agriculture. Under the heavy criticism, the sixteen CIDIE member organizations agreed to abide by the declaration, with special emphasis on SUSTAINABLE DEVELOPMENT, while its secretariat has been moved to the UNEP headquarters. Results of these developments are yet to be seen.

See Lynton Caldwell, *International Environmental Policy* (Durham: Duke University Press, 1996): 138–39.

European Bank for Reconstruction and Development (EBRD): a regional MULTILATERAL DEVELOPMENT BANK established in 1991 to assist Central and Eastern European countries in their transitions to market-oriented economies. The EBRD has sixty members (fifty-eight countries, the EUROPEAN UNION [EU], and the European Investment Bank). Headquartered in London, the bank aims to support economic reform, promote development of the private sector, and strengthen financial institutions and legal systems in the region. The charter recognizes an obligation to "promote in the full range of its activities environmentally sound and SUSTAINABLE DEVELOPMENT" as the bank's priority (article 2.1.vii). This is the first international financial institution to have such a proactive environmental mandate in its founding charter.

Inter-American Development Bank (IDB): the regional MULTILATERAL DEVELOPMENT BANK created in 1959 to provide financing for economic and social development in Latin America and the Caribbean. Beginning with the United States and nineteen Latin American and Caribbean member nations, the IDB has expanded its membership to include Canada, seven more Latin American and Caribbean members, and eighteen nonregional nations. Like the WORLD BANK, the IDB obtains most of its lendable capital through its bonds sold on private capital markets, and, because of the low default rate, can offer loans at interest rates modestly below the rates that Latin American and Caribbean governments can borrow from international commercial banks. However, some of the IDB's funding for grants and very low interest loans requires periodic donations from the wealthier member countries.

COMMENTARY ON THE INTER-AMERICAN DEVELOPMENT BANK

The United States has been the most influential country in shaping IDB policies and procedures, which, in addition to its headquarters in Washington, D.C., has long given rise to criticism that the IDB is a creature of U.S. policy. However, relations between the U.S. government and the IDB have often been contentious, and the Japanese government has played an increasingly important role in funding the IDB and setting its direction. In terms of its approaches to development and lending, the IDB has, like the WORLD BANK and the other regional MULTILATERAL DEVELOPMENT BANKS, expanded its scope from a focus on the "productive sectors" (i.e., industry and agriculture) and physical infrastructure to a greater emphasis on social development, community participation, and the environment. Compared to the other multilateral development banks, the IDB has often been more oriented toward sectoral programs (as opposed to single projects). *William Ascher*

International Bank for Reconstruction and Development (IBRD). See WORLD BANK.

International Development Association (IDA): the soft-loan facility of the WORLD BANK GROUP, established in 1960 to provide capital for governments unable to borrow from the WORLD BANK's normal lending facility at market rates. Rather than establishing a separate organization with its own staff, the World Bank Group member states created a separate IDA fund that is administered by the World Bank staff. Eligibility for International Development Association loans requires membership in the World Bank Group and low per capita income; in 2000, seventy-eight countries were eligible. IDA loans bear only a modest service charge rather than interest charges; the loans are for thirty-five to forty years, with a ten-year grace period prior to repayment of any principal. The existence of the IDA fund (roughly $5–6 billion is lent each year) allows the World Bank Group to extend credit to the poorest countries, and also, by blending World Bank and IDA loans for the same project or program, permits interest rates for some IDA-eligible borrowers at levels below the World Bank's prevailing normal interest rates. Because IDA funds must be replenished periodically by the wealthy nations, generally for three-year periods, the World Bank Group's management of IDA funds has exposed it to additional pressures by donor countries.

International Financial Institutions: a broad range of official international organizations that lend capital or perform other financial functions. The INTERNATIONAL MONETARY FUND (IMF) and the WORLD BANK are the most visible international financial institutions and have the largest pools of financial resources, but international financial institutions also include the regional development banks, regional monetary clearinghouses, and many other organizations. The increase in the number and importance of international financial institutions is a notable development of the post–World War II international economic system.

International Monetary Fund (IMF): the multilateral organization charged with the primary responsibility for lending to member states to avoid the exhaustion of foreign reserves and the disruption of international monetary and trade systems. The IMF was developed at the same 1944 BRETTON WOODS CONFERENCE that established the WORLD BANK, and began operations in 1947. Although member states (currently numbering more than 180 nations) can draw on a certain volume of funds from the IMF without having to meet any conditions, progressively larger loans require borrower governments to observe particular conditions, known as CONDITIONALITY (mandatory conditions that the borrower must observe), many of which have been highly controversial. Although the IMF was initially designed to address short-term balance-of-payments difficulties, it has evolved programs of lending and economic reform conditionality to promote structural adjustments of economies that chronically run trade and budget deficits. Critics of the IMF argue that its conditionalities impose hardship on indebted countries for the sake of international monetary stability, that governments lose their sovereignty when they have to submit to conditionality, that low-income people suffer under IMF AUSTERITY MEASURES, and that the IMF is a creature of the wealthy countries, especially the United States. Defenders of the IMF counter that its policy conditionalities are best for the long-term prospects of the indebted nations, even if their governments often try to place blame on the IMF in order to reduce their own political costs in adopting painful but necessary economic reforms.

See John Williamson, ed., *IMF Conditionality* (Washington, DC: Institute for International Economics, 1986).

Multilateral Development Banks: official international organizations devoted to providing loans for development projects or programs, usually in developing countries or countries recovering from war or other disasters. Member governments "own" the multilateral development banks; voting is typically on the basis of shares of the bank's capital fund rather than one-country-one-vote. Some multilateral development banks include recipient countries as members, while others do not (e.g., the Nordic Development Fund). The loan funds of multilateral development banks come most commonly from bonds floated on the international capital markets and from donations by member governments. The former source of funding is usually devoted to making loans at slightly below the prevailing commercial interest rates (owing to the lower default rates than commercial banks experience); the latter to loans with low or zero interest rates. There are over thirty multilateral development banks, ranging in geographic scope and size of capital fund from the INTERNATIONAL BANK FOR RECONSTRUCTION AND

DEVELOPMENT (IBRD, or WORLD BANK) to the smaller regional banks (e.g., the Caribbean Development Bank and the East African Development Bank). The largest multilateral development banks are the WORLD BANK GROUP, the ASIAN DEVELOPMENT BANK, the INTER-AMERICAN DEVELOPMENT BANK, the AFRICAN DEVELOPMENT BANK, and the EUROPEAN BANK FOR RECONSTRUCTION AND DEVELOPMENT (EBRD). Concerning the largest development banks, the perennial controversy is over the questions of whether the developed countries dominate the organizations' agendas, and whether the CONDITIONALITY sometimes imposed along with the loans is an infringement on the recipient country's sovereignty.

World Bank (formally International Bank for Reconstruction and Development [IBRD]): the largest lending facility of the WORLD BANK GROUP, an autonomous MULTILATERAL DEVELOPMENT BANK that provides loans and technical assistance for economic development projects in developing countries and encourages cofinancing for projects from other public and private sources. It began operations in 1946 and by 1999 had 180 member nations. The World Bank is the oldest of five affiliates of the World Bank Group, and shares its staff with the SOFT-LOAN affiliate known as the INTERNATIONAL DEVELOPMENT ASSOCIATION (IDA). The World Bank is the world's largest funder of development projects and programs (roughly US$20 billion a year), raising most of its funds from bonds sold to private investors through the international money markets. The World Bank and its "sister" institution the INTERNATIONAL MONETARY FUND (IMF) pioneered the use of CONDITIONALITY to bind borrowing governments to particular economic policy reforms and sometimes to observe environmental protection restrictions. Over the past three decades the World Bank has undergone a dramatic shift from a focus on industry and physical infrastructure to a greater emphasis on poverty alleviation, agriculture, human resource development, macropolicy reform (often through conditionality of STRUCTURAL ADJUSTMENT LOANS), governance, and environmental protection. Nevertheless, over the same period the World Bank's activities have become a focus of struggle between North and South (see NORTH VS. SOUTH) and an object of harsh criticism from environmentalists.

World Bank Group: the umbrella organization encompassing: a) the WORLD BANK, which lends to developing country governments at nearly market interest rates, both for specific development projects and programs and for STRUCTURAL ADJUSTMENT LOANS; b) the INTERNATIONAL DEVELOPMENT ASSOCIATION (IDA), which lends interest-free development assistance to the poorest World Bank Group member countries; c) the International Finance Corporation, which lends to promote the growth of the private sector in developing member countries, encourage the development of local capital markets, and stimulate the international flow of private capital; d) the Multilateral Investment Guarantee Agency (MIGA), which promotes private investment in developing countries, guarantees investments to protect investors from noncommercial risks (such as war or nationalization), and advises governments on attracting private investment; and e) the International Centre for Settlement of Investment Disputes, which promotes international investment by providing arbitration and mediation for disputes between foreign investors and host countries.

Initially comprised of the World Bank, as that institution emerged out of the BRETTON WOODS CONFERENCE along with the INTERNATIONAL MONETARY FUND (IMF), the World Bank Group expanded with the addition of other affiliates, but maintains its organizational unity through the executive leadership of a single president and over-sight by the same board of executive directors.

COMMENTARY ON MULTILATERAL LENDING INSTITUTIONS

In practice, these institutions cannot make up their minds. Are they banks, or are they development institutions? The bureaucratic incentives prevalent today show that in order to be promoted within the organization, bank project officers must give loans on a large scale. The design of the project is impor-tant for loan approval, but the evaluation and monitoring of performance are dismal. Moreover, if a project fails, the country must pay the loan back, and a new loan will be formulated to fix things up. The result of this approach is a mini-conspiracy in which bank bureaucrats push large loans because they are rewarded to do so, and bureaucrats from borrowing countries are willing to receive a loan because they too are rewarded for doing so. If the loan fails, no problem, another one will be forthcoming. This can be very frustrating for taxpayers in both countries. *Gustavo J. Arcia*

4.2.4 FOREIGN ASSISTANCE

Additionality, Principle of. See SECTION 5.5.5.

Aid Agencies. See FOREIGN ASSISTANCE.

Conditionality: with reference to international borrowing or foreign assistance, a re-quirement to achieve economic targets or to adopt regulations, laws, or policies as con-ditions to receive grants or loans. In the most common form of conditionality, "policy-based lending," donors provide loans or grants to recipients who pledge to implement an agreed-upon set of policy changes. With "allocative conditionality," donors distrib-ute assistance to countries based on their adherence to specific standards, but do not bargain with them upon reforms (at least in theory). Some scholars and policymakers criticize conditionality as placing unjustified restrictions on the rights of borrowing countries to order their own affairs. International financial institutions justify condi-tionality of their loans by the argument that there is no point in using limited avail-able resources in cases when a country's trade, fiscal, or monetary policies make it un-likely that their balance-of-payments problems can be cured. See also INTERNATIONAL MONETARY FUND (IMF).

See David Fairman and Michael Ross, "Old Fads, New Lessons," in *Institutions for Envi-ronmental Aid: Pitfalls and Promise,* eds. Robert Keohane and Marc Levy (Cambridge, MA: MIT Press, 1996): 29–52.

Development Activity Index (DAI): an inventory of all the development programs funded by the multilateral development agencies and the bilateral assistance programs of twenty-three donor nations. This computer database, available on CD-ROM, is ex-tremely helpful for identifying financing sources and for developing strategic action plans.

Development Aid (Development Assistance). See FOREIGN ASSISTANCE.

Development Agencies (Development Assistance Agencies). See FOREIGN ASSISTANCE.

Foreign Assistance: the transfer of capital and technical assistance from more developed to less developed countries. Foreign assistance often takes the form of grants and concessional lending (i.e., loans with lower interest rates than those prevailing on the private market), provided either by government ("bilateral aid") or by official international organizations ("multilateral aid"). Often bilateral foreign assistance is "tied," in that purchases by the recipient country and the technical assistance must be from the donor country. See also OFFICIAL DEVELOPMENT ASSISTANCE (ODA).

Global Environment Facility (GEF): originally a pilot program set up in 1991 to mobilize funding for four basic issues of environmental concern: OZONE LAYER DEPLETION, global warming (see GREENHOUSE EFFECT), BIODIVERSITY, and international waters. Initially administered by the WORLD BANK with the UNITED NATIONS DEVELOPMENT PROGRAMME (UNDP) and the UNITED NATIONS ENVIRONMENT PROGRAMME (UNEP) providing technical assistance, in 1994 the GEF was restructured on a permanent basis as an agency governed by a council (thirty-two members) and controlled by an assembly with universal membership. The GEF provides grants and concessional funding to countries that are eligible for a) financial assistance either within the FRAMEWORK CONVENTION ON CLIMATE CHANGE (FCCC) or the CONVENTION ON BIOLOGICAL DIVERSITY; or b) borrowing from the World Bank or technical assistance through a UNDP country program. GEF projects must be country-driven, incorporate consultations with local communities, and, where appropriate, involve NONGOVERNMENTAL ORGANIZATIONS in project implementation. Despite relatively modest funding (US$2 billion for 1994–97 in grants and concessional lending), the GEF has been a focus of major North-South contention over global environmental governance (see also NORTH VS. SOUTH). At present, the GEF seems to be the most promising mechanism for the external financing of global environmental regimes.

International Aid. See FOREIGN ASSISTANCE.

Official Development Assistance (ODA): FOREIGN ASSISTANCE provided by governments (bilateral assistance) and official multilateral institutions set up by international agreement. The overall level of ODA provides one measure of the financial effort of developed nations to support developing nations, although ODA has been criticized as being less effective than the expansion of trade opportunities because of the dependency that foreign assistance allegedly imposes on recipient nations. Looking at ODA holistically, rather than its constituent bilateral or specific multilateral efforts, focuses attention on the need for the coordination of development assistance.

United Nations Development Programme (UNDP): the world's largest grant-making development assistance organization, formed in November 1965 through the merger of the Expanded Programme of Technical Assistance and the UN Special Fund. Its governing council consists of representatives from forty-eight nations—twenty-seven developing nations and twenty-one developed nations. UNDP headquarters are in New York City while a network of offices is maintained in over one hundred countries. Since its creation, UNDP has been central to the development activities of the UN system. It has 195 member nations and provides technical assistance and preinvestment cooperation to some 170 developing and transitional countries and territories, with most of its resources allocated for the LEAST DEVELOPED COUNTRIES. UNDP programs (constituting an annual budget of over $1.7 billion in 1996) are financed through voluntary contributions from UN member states and other UN agencies. In 1996 the mission of the agency was redefined: "To help countries in their efforts to achieve SUSTAINABLE HUMAN DEVELOPMENT by assisting them to build their capacity to design and carry out development programmes in poverty eradication, employment creation and sustainable livelihoods, the empowerment of women, and the protection and regeneration of environment, giving first priority to poverty eradication." Reorientation from technical cooperation toward sustainable human development has given the agency higher visibility and a unique niche on the international scene. Through its annual *Human Development Report*, the UNDP has become instrumental in developing new concepts, standards, and policy instruments for human SUSTAINABLE DEVELOPMENT. It assists developing nations in integrating environmental concerns into development plans through programs designed to build and strengthen their capacities for sustainable development and natural resource management. However, despite its comparative advantages in assisting developing countries in capacity building for sustainable development, many believe that so far UNDP has failed to translate its visions of sustainability into operational guidelines. Whether the organization's efforts to play a leading role in implementing sustainable development will succeed depends on its ability to focus institutional restructuring and persistence of its administration, as well as on the support of individual governments that supply most of its funding.

See Poul Endberg-Pedersen et al., *Assessment of UNDP: Developing Capacity for Sustainable Human Development* (Copenhagen: Centre for Development Research, 1996).

United Nations Industrial Development Organization (UNIDO): the specialized agency of the UN established to promote industrial development, founded initially as an autonomous organization within the UN Secretariat in 1966. UNIDO's constitution stipulates its primary objective as the promotion of industrial development in developing countries and of cooperation on global, regional, national, and sectoral levels. UNIDO's mandate is also to coordinate all activities of the UN system relating to industrial development. At the request of governments, UNIDO assists developing countries in obtaining external financing for specific industrial projects. In this connection, UNIDO seeks to identify new means of cooperation and actively solicits support from, and greater participation of, both the public and private sectors of industry. Much of the leading work on project evaluation and project planning in the 1980s was conducted under UNIDO auspices.

United States Agency for International Development (USAID): an autonomous U.S. government agency that provides economic development and humanitarian assistance to developing countries. Established in 1961 by President John F. Kennedy, USAID has been the dominant channel of U.S. FOREIGN ASSISTANCE. Unlike most other large development assistance agencies, such as the WORLD BANK and counterpart bilateral development agencies, USAID has maintained sizable missions in many developing countries, although economy moves to close some of these missions were undertaken in the 1990s. As the primary vehicle for U.S. foreign assistance, USAID has been at the center of all foreign assistance controversies, including challenges to its efficiency record, and debates on how to balance the goals of improving the economic and social situations in the recipient countries versus advancing U.S. economic and political interests overseas. Like other development agencies, USAID has experimented with many development approaches, including PARTICIPATORY DEVELOPMENT.

COMMENTARY ON USAID

USAID as an institution is at present in transition. Because of congressional politics, it is affected by changing priorities in the use of American funds, and has been forced to adopt an incredible amount of bureaucratization. The net result is that many of the best and brightest staff have left. U.S. firms working on USAID contracts have been forced to absorb USAID's bureaucratic procedures, which in turn makes these firms more expensive and less competitive in the multilateral aid market. In a way, USAID is doing the complete opposite of what it preaches in the countries it assists. *Gustavo J. Arcia*

4.3 International Trade

4.3.1 PRACTICES, TRENDS, AND PROBLEMS

Corporate Average Fuel Economy (CAFÉ) **Dispute:** one of the first international disputes over trade and environmental concerns. The CORPORATE AVERAGE FUEL ECONOMY (CAFÉ) STANDARDS legislation was passed in the U.S. in 1975 and was intended to double the average fuel efficiency of the cars sold in the U.S. market within ten years. It required all vehicle manufacturers, whether domestic or foreign, to meet specific standards of fuel economy. CAFÉ standards were gradually increased during the 1980s and reached a plateau in the 1990s because of growing concern over the safety of lighter cars. When European manufacturers of luxury cars were unable to meet the standards, the EUROPEAN UNION (EU) argued that CAFÉ legislation favored domestic production. In 1994, the third GENERAL AGREEMENT ON TARIFFS AND TRADE (GATT) Panel on Environmental Trade Measures examined the debate and found that CAFÉ was intended to promote fuel efficiency, but the provision that foreign cars should be evaluated separately was found to give foreign makers less favorable conditions of competition.

Danish Bottle Case: one of the first test cases concerning contradictions between domestic environmental considerations and free trade (or common market) principles. In the 1980s, Denmark, like a few other European Community (now EUROPEAN UNION [EU]) member states, unilaterally banned the sale of nonrefillable beverage containers and required that all beer and soft drinks be marketed in reusable containers. A special DEPOSIT-REFUND SYSTEM had to be established by producers. The case of Danish bottling rules was brought to the European Court of Justice by the European Community as an example of the "risk that Member states would in future take refuge behind ecological arguments to avoid opening their markets." By the court's decision, Denmark was allowed to restrict the use of nonrefillable bottles and cans for environmental reasons. The case highlighted the need for the European Commission to coordinate its waste management and disposal policies in order to reduce the friction between two of its main objectives, namely free trade and environmental protection.

See I. J. Koppen, "The Role of the European Court of Justice," in *European Integration and Environmental Policy*, eds. J. D. Liefferink et al. (New York: Belhaven Press, 1993).

Duties: taxes or other charges on imports or exports. Import duties have been the classic mechanism for protecting domestic industry, and have come under tremendous criticism for their role in allowing inefficient industries to survive. Protective duties have sometimes been defended as necessary for nurturing "infant industries" to create future comparative advantages, but critics also question whether the duties can ever be eliminated. Duties may be used for environmental protection if they are levied against exports that are produced with high levels of pollution or resource depletion. Export duties are sometimes used to capture the wealth of the export.

Export Processing Zones (EPZs; Free Trade Zones): areas that are given special fiscal and legal status in order to attract foreign investment or stimulate domestically financed export industry. The purpose of the special status is to provide competitive advantage over neighboring regions and countries. Special status may include removal of customs duties and controls, tax holidays, financial incentives, government-financed infrastructure, and relaxation of social, environmental, and employment regulations. The advocates of EPZs argue that they promote export-led development, which contributes greatly to the incomes of people with otherwise low prospects of meaningful employment. Opponents of EPZs argue that if the area is truly competitive in export-oriented production, the subsidies are unnecessary, and that environmental damage and health impacts can be significant. See also MAQUILADORAS.

Free Trade: the international economic system of exchange unhindered by tariffs (customs DUTIES) or other "NONTARIFF BARRIERS" to trade. By the early nineteenth century, David Ricardo had demonstrated that free trade would benefit all trading partners in terms of aggregate economic welfare. However, the distribution of benefits from free trade within countries depends on specific circumstances. Movement toward a free trade international economic regime has advanced dramatically in the post–World War II period. International institutions such as the GENERAL AGREEMENT ON TARIFFS AND TRADE (GATT) and the WORLD TRADE ORGANIZATION (WTO) are specifically mandated to promote free trade; opening to trade is also often a condition for loans and grants from international financial institutions such as the WORLD BANK GROUP and the INTERNATIONAL MONETARY FUND (IMF). Current efforts are largely focused on the elimination of nontariff barriers, which, because they are embedded in marketing structures, are more difficult to eliminate and monitor than are straightforward tariffs. Critics of free trade have argued that developing countries suffer from unfair or deteriorating terms of trade, and that international trade creates inequalities of wealth and power within particular countries. Regional groupings of nations have often reduced the tariffs within the group, although sometimes such trading blocs discourage free trade with countries outside of the region. Trade liberalization is usually accompanied by substantial environmental consequences, though scholars and practitioners debate the short- and long-term linkages between the dynamics of trade and the state of the environment.

Garbage Imperialism: the export of refuse and toxic wastes from the First World to poor and less developed countries eager for foreign currency, i.e., black South Africa, China, Romania, Poland, Thailand, Ukraine, and other developing nations and societies in transition. The export of hazardous wastes from highly industrialized countries to poorer nations has grown dramatically in the last twenty-five years as the result of stricter environmental protection in the West and a search for cost advantages (the cost of dumping in China may be as little as $3 a ton in contrast to $1,500 in the U.S.). Most of the importing nations have neither the technical expertise nor adequate facilities for safely recycling or disposing of wastes and often end up with serious health and environmental risks. The issue of garbage imperialism was somewhat addressed by the CONVENTION ON THE CONTROL OF TRANSBOUNDARY MOVEMENTS OF HAZARDOUS WASTES AND THEIR DISPOSAL (BASEL CONVENTION), but even that convention was attacked by its critics for its weak stance against the toxic waste trade.

Maquiladoras: foreign-owned assembly plants, often in border areas between high- and low-wage countries, that import and assemble duty-free components for export. The term is most closely associated with plants in Mexico near the U.S. border. The official promotion of Mexican maquiladoras was initiated by the 1965 Border Industrialization Program, which allowed U.S. companies to take advantage of low-cost Mexican labor as well as proximity to the United States and to pay duty only on the "value added" (the difference between the value of the finished product and the value of the foreign-made components). Today, over two thousand Mexican maquiladoras produce electronic components, chemicals, machinery, auto parts, and other products while employing nearly 700,000 people nationwide. While this experience of economic integration provided employment and significant foreign-exchange earnings for Mexico's troubled economy and helped U.S. manufacturers compete with the low prices of East Asian goods, it also generated a host of acute environmental and social problems.

COMMENTARIES ON EXPORT PROCESSING ZONES AND MAQUILADORAS

This type of industry was also promoted in Central America and the Caribbean in the 1980s under the Caribbean Basin Initiative as a strategy for industrial development, with mixed results and much controversy over human rights and labor issues. The industry has been notorious worldwide for exploiting children, abusing women, and providing low compensation and working conditions to its employees. *Elsa Chang*

Export Processing Zones (EPZs) offer financial incentives and weak or nonexistent regulatory standards that multinational capital uses to exploit indigenous labor in "developing" nations. These workers are frequently women paid "pin money" for long hours laboring in appalling conditions. EPZs contribute little to economic development as they result in no significant technology or skill transfers.

See Cynthia Enloe, *Bananas, Beaches and Bases* (London: Pandora, 1989); Maria Mies and Vandana Shiva, *Ecofeminism* (New Delhi: Kali for Women, 1993). *Ian Welsh*

Let's not lose sight of the fact that, as vulnerable as the EPZ or maquila-dora strategy may be to abuse, some governments that have relied heavily on export-promotion strategies have improved the quality of life of their people quite dramatically over the course of several decades. Taiwan, South Korea, and Hong Kong are obvious examples. The maquiladora-heavy northern cities in Mexico have prospered more than the rest of the country. *William Ascher*

Nontariff Barriers: obstacles to trade, most generally to imports, caused by policies and economic structures other than tariffs (i.e., other than import or export DUTIES). Tariff barriers are rather easily controlled through trade agreements because they comprise highly visible taxes, but many nontariff barriers, such as the closed wholesaler-retailer merchandising systems found in some countries, are more difficult to monitor or change, just as it is difficult to hold the government responsible. Japan, in particular, has been accused of maintaining nontariff barriers through closed merchandizing systems, lengthy new product approval processes, etc. Much of the focus of international trade reform in the 1990s was on the issue of reducing nontariff barriers. However, some nontariff barriers involve environmental restrictions, on either goods produced with high levels of pollution (or consuming scarce resources) or goods whose consumption and disposal poses environmental risks. Therefore the free-trade initiative to eliminate nontariff barriers can come into conflict with environmental protection.

Protectionism: the use of tariffs and NONTARIFF BARRIERS to regulate prices of foreign goods on domestic markets in order to protect domestic industries. In terms of economic development, protectionism has been widely viewed as a major cause of stagnation because it encourages the inefficiency of domestic operations by shielding them from competition. However, in some contexts the potential COMPARATIVE ADVANTAGES of certain industries may warrant temporary protection. Yet this "infant industry" argument has often been abused. In terms of the environmental and North-South debates (see NORTH VS. SOUTH), the issue of protectionism has recently been cast in terms of so-called green protectionism. Governments of many developing countries fear that environmental standards adopted by advanced countries will be serious barriers to trade, either because they are designed and applied as protectionist measures or simply because they are too strict for developing countries' producers with limited technology to attain.

COMMENTARY ON PROTECTIONISM

In practice, the economic costs of so-called green protectionism to the developing countries are trivial compared to the costs of barriers erected in the North against labor-intensive manufactures such as textiles and apparel, and against competing agricultural commodities such as sugar or bananas. Northern protectionism in these areas forces developing countries to exploit their resources more heavily and also worsens their indebtedness. Concern over green protectionism is probably excessive, and deflects attention from much more critical trade issues. *Natalia Mirovitskaya*

Recompensing Duties. See DUTIES.

Tradable Commodity: goods that can be feasibly transported internationally and are therefore capable of being exported and imported. The significance of tradability is that a good's value depends on the international market; its "free market" value can be estimated by the so-called border price of imports. In contrast to "nontradables," tradable goods are directly affected by currency exchange rates. Overvalued exchange rates (i.e., local currency is kept artificially valuable by government limits on how many local currency units can be purchased with a unit of another currency) will decrease the domestic producer's local-currency export revenues, or will force the domestic producer to increase the international price. Therefore overvalued currency makes tradable commodities less competitive on the world market. Monetary policies therefore have a major impact on the relative costs of tradables and nontradables, and on the income distribution between those who are involved in producing each.

Trade-Related Intellectual Property Rights (TRIPS): a set of intellectual property rights subject to an agreement reached through the GENERAL AGREEMENT ON TARIFFS AND TRADE (GATT). Initiated at the URUGUAY ROUND and in effect since January 1, 1995, the TRIPS agreement is the most comprehensive international agreement on intellectual property protection, covering trademarks, copyrights, and patents, including the protection of new varieties of plants. The TRIPS agreement requires countries to treat foreigners the same as they treat their own nationals when granting intellectual property rights. It also sanctions patents on natural substances. Though governments of developing countries have a ten-year grace period before they are subject to the TRIPS, they are particularly cautious about it. From their perspective, the TRIPS agreement might have detrimental effects on TECHNOLOGY TRANSFER (including environmental technology) or local development of technology. There is also a danger that TRIPS regime will jeopardize the interests and rights of communities that developed INDIGENOUS TRADITIONAL KNOWLEDGE by enabling the patenting of this knowledge by commercial companies. See also PLANT BREEDER RIGHTS.

COMMENTARIES ON TRADE-RELATED
INTELLECTUAL PROPERTY RIGHTS

A central means of enforcing trade compliance is through the WORLD TRADE ORGANIZATION (WTO). TRIPS grant rights of commercial exploitation to those performing certain kinds of labor—typically the identification and isolation of commercially exploitable genes—irrespective of whether an indigenous means of using the same substance already exists. The best known case is the Neem tree used in India for generations until it was patented. TRIPS thus subordinate local people's knowledge of beneficial plants to the science of Western multinationals. TRIPS regimes seek to replace patent regimes, such as India's, seen as too weak to defend the rights of MULTINATIONAL CORPORATIONS to profit from their biotechnology investments. By winning patent rights over long chains of genetic materials, multinationals effectively lock cures within their laboratories. *Ian Welsh*

In assessing whether TRIPS are good or bad, one has to ask, "Compared to what?" Even if TRIPS allow companies to buy rights to exploit intellectual property, the typical alternative is for companies to develop products from local biota without paying anything to local people. In terms of the sharing of profits between the local economy and multinational corporations, increasingly the locally based corporations in developing countries such as Brazil are developing their own capacity to extract and identify pharmacologically promising molecules and ensure that they can share in the profits from emerging products. *William Ascher*

Trading Blocs: groups of nations that grant one another favorable conditions in trade, usually by lowering import duties. The formation of trading blocs can be an intermediate step toward global trade liberalization, but trading blocs, particularly in the 1950s and 1960s, have often served as barriers to trade with other countries. Although the proponents of trading blocs often say that favorable status toward other nations within the bloc does not preclude trade with other nations, it is likely that "trade diversion" will result because lower import duties will make the imports from the trading bloc partners less expensive.

Tuna-Dolphin Controversy: one of the most publicized examples of conflict between free trade and environmental protection. In 1991, a dispute panel of GENERAL AGREEMENT ON TARIFFS AND TRADE (GATT) arbitrators ruled that embargo provisions of the U.S. Marine Mammal Protection Act (MMPA) were an unfair trade barrier to Mexican fishermen seeking to sell tuna in the United States. The MMPA had banned the sale of tuna from vessels, whose fishing practices kill more dolphins than U.S. standards allow. While previous GATT panels had already dealt with four cases in which countries justified trade restrictions on the basis of environmental concerns, its approach to the U.S.-Mexico controversy has been notable for addressing two highly controversial legal issues. First, it ruled that a country could not use trade sanctions to protect the natural resources of another country or the GLOBAL COMMONS. Second, it said that no nation can ban, tax, or otherwise interfere with the imports of any product on the basis of how it was produced. The GATT panel decision on the tuna-dolphin controversy has exposed a limitation in international law that remains controversial.

4.3.2 RESPONSES AND STRATEGIES

Alternative Trading Organization (ATO): a business entity, generally nonprofit, that engages in international trade of products from developing countries with the explicit goal of promoting social goals through trade. Products imported into the markets of developed countries, mainly the European Union, the United States, and other high-income nations, include handicrafts, as well as food products such as coffee. Social goals may include raising the income of rural producers, promoting organic production, and providing markets for sustainably harvested food and forest products. Germany is the largest single market for ATOs, which may be wholesalers for products sold in ordinary retail stores or in "Third World" shops run by nonprofit organizations or which may operate their own retail outlets. *R. G. H.*

Countervailing Duty (CVD): import DUTIES imposed by the importing country on goods from the exporting country that are subsidized, and thus have an unfair advantage in trade. With reference to the environment, CVDs have been proposed to correct for the implicit subsidy of lax environmental regulation in the exporting country. The reasoning is that, whether or not another country chooses to protect its environment from production externalities, other producers that are more environmentally benign should not be held at a competitive disadvantage. *R. P.*

Eco-Cooperation Agreements: a Dutch initiative that links national foreign and trade policy to sustainability strategies domestically and overseas. The government of the Netherlands signed eco-cooperation agreements, for instance, with Bhutan and Costa Rica, both of which are at the forefront of sustainability initiatives.

COMMENTARIES ON ECO-COOPERATION AGREEMENTS

Eco-cooperation agreements, like the ones signed in 1994 by the Dutch government with Costa Rica, Bhutan, and Benin, reward current sustainability initiatives rather than past environmental performance. This approach of providing international economic rewards for sustainable policies is fundamentally different from traditional forms of environmental or development cooperation between North and South. Joint projects undertaken under the 1994 eco-cooperation agreements are beneficial for the SUSTAINABLE DEVELOPMENT of all four partners. For instance, Costa Rica and Bhutan, which harbor immense biodiversity and currently pursue sound conservation policies (approximately one quarter of their national territories has protected status), are involved in the development of a Dutch government policy on biodiversity, cooperate in a unique research project on sustainable methods of exploiting natural resources, and take common positions at the international environmental negotiations. Through the 1994 eco-cooperation agreements, roughly one hundred organizations became involved in joint projects, including the Dutch-based multinationals Phillips and Ahold, environmental investment funds, banks and other financial institutions, universities, and NONGOVERNMENTAL ORGANIZATIONS. Such initiatives, based on principles of reciprocity and public-private partnerships, are very promising. *Natalia Mirovitskaya*

It is curious and sobering that the government of the Netherlands chose Costa Rica as one of the beneficiaries of an eco-cooperation agreement. While Costa Rica has indeed been a leader in some environmental initiatives, it also suffered one of Latin America's highest rates of deforestation and biodiversity loss in the 1980s and 1990s. The implication is that it is indeed very difficult to decide which nations and governments ought to be recognized and rewarded for efforts at sustainability and environmental protection. *William Ascher*

Environmental Trade Measures (ETMs): 1) the use of taxes, regulations, or direct bans applied to internationally traded goods in order to facilitate multilateral environmental cooperation. Trade measures are used to provide incentives for participation in environmental agreements (MONTREAL PROTOCOL ON SUBSTANCES THAT DEPLETE THE OZONE LAYER, CONVENTION ON THE BAN OF THE IMPORT INTO AFRICA AND THE CONTROL OF TRANSBOUNDARY MOVEMENTS AND MANAGEMENT OF HAZARDOUS WASTES WITHIN AFRICA [BAMAKO CONVENTION]), to encourage compliance with the treaties (CONVENTION ON THE CONTROL OF TRANSBOUNDARY MOVEMENTS OF HAZARDOUS WASTES AND THEIR DISPOSAL [BASEL CONVENTION], CONVENTION ON INTERNATIONAL TRADE IN ENDANGERED SPECIES [CITES]), to prevent FREE-RIDING by nonmembers, among other reasons. There are at least twenty-four international environmental agreements that contain trade provisions. However, the practical impact of these measures to environmental regimes' effectiveness is not yet well understood. WORLD TRADE ORGANIZATION (WTO) officials object to using trade measures in international environmental agreements and maintain that these agreements should require a waiver from the WTO.

2) unilateral regulations that authorize restrictions on international trade by invoking considerations of environmental protection. They include import prohibitions, product standards, standards governing production of natural resource exports, and mandatory ecolabeling schemes. The United States, as the world's largest single market, has been pressured by the U.S. environmental movement to use ETMS to protect the environment beyond the U.S. jurisdiction. In particular, the United States used trade restrictions to end commercial whaling and driftnet operations that endangered marine mammals in the North Pacific. Exporters affected by these measures often claim that ETMS are intended to protect the domestic economy from foreign competition and therefore are in violation of basic world trade principles that allow countries to discriminate against an imported good only on the basis of the product's characteristics, not on the basis of the process by which it was produced (see TUNA-DOLPHIN CONTROVERSY). The dispute resolution panel of the GENERAL AGREEMENT ON TARIFFS AND TRADE (GATT), and more recently the WORLD TRADE ORGANIZATION (WTO), has addressed the controversy over the legitimacy of specific ETMS. See also INTERNATIONAL ORGANIZATION FOR STANDARDIZATION (ISO).

See Gareth Porter and Edith Brown, *Global Environmental Politics* (Boulder: Westview Press, 1996), 129–36.

International Organization for Standardization (ISO): a worldwide, nongovernmental federation of national standards bodies from some 130 countries, one from each country. It was established in 1947 to promote the development of standardization and related activities in the world with a view to facilitating international exchange of goods and services, and developing cooperation in the sphere of intellectual, scientific, technological, and economic activity. ISO brings together the interests of producers, users (including consumers), governments, and the scientific community in its work, which is carried out through some 2,400 technical bodies. ISO's work results in international agreements that are published as International Standards, and cover such topics as air quality, building and construction, fertilizers and pesticides, mining, nuclear energy, natural gas, and water quality. The two best-known ISO series are ISO 9000, which provides a framework for quality management and quality assurance, and ISO 14000, which provides a similar framework for environmental management. International standardization is market-driven and therefore based on voluntary involvement of all interests in the marketplace.

ISO **14000:** a set of global environmental management standards developed by an international technical delegation of the INTERNATIONAL ORGANIZATION FOR STANDARDIZATION (ISO) and adopted by the ISO in October of 1996. It is a performance-based system for industry to undertake environmentally oriented continual improvement. It does not enforce environmental compliance, but provides a framework for companies to develop voluntary environmental management systems designed to achieve increased levels of environmental accountability and improved performance. Adding to the credibility of the standards is the requirement that for a company to receive a certification it must be scrutinized by an independent auditor. Governments and trading blocs have the option to require certification in accepting imports, and consumer-conscious merchandisers may also shun export-producing companies that do not qualify for ISO 14000 certification. The ISO 14000 series of standards and guidance documents fall into two broad categories. Environmental Management Systems (EMSs) specify how to set up the system as well as how to evaluate and manage its performance. The second category of standards addresses product stewardship, environmental design throughout product life cycle, and labeling principles. Although ISO 14000 is a voluntary system, the standards are framed by an ambitious set of guiding principles. The environmental aspects of performance cover air emissions, releases to water, waste management, contamination or degradation of land, use of materials, and environmental community issues.

See Perry Johnson, *ISO 14000: The Business Manager's Guide to Environmental Management* (New York: Wiley and Sons, 1997).

COMMENTARY ON ISO 14000

There is a crucial assumption that we must keep in mind at all times when using ISO 14000: we assume that an adequate management system will lead to emission reductions. ISO 14000 deployment often does not contain direct measurement of pollution reduction, but instead depends on the assumption that certain practices will lead to emission reductions. This is an untested, even though plausible, assumption. Process does not guarantee outcome but must be verified to validate ISO 14000 effectiveness in any application. *Gerald Emison*

Principle of Equity in Sharing Environmental Space: an approach used by developing-country theorists in debates over global environmental management. It is based on the notions of ENVIRONMENTAL SPACE and ECOLOGICAL FOOTPRINT. According to Canadian ecologist William Rees, who pioneered work on ecological footprints, four to six hectares of land are needed to maintain the lifestyle of the average Westerner. However, in 1990 the total available productive land in the world was an estimated 1.7 hectares per person. Most Northern countries and urban regions in the South consume much more than their fair share, depending on trade or natural capital depletion—therefore, populations of these regions appropriate additional environmental space either from elsewhere, or from future generations. This phenomenon of unequal and unfair appropriation of global natural resources and environmental space has been a point of a broad and basic disagreement between developing and developed nations regarding the equity of the existing international economic and environmental system, the values that ought to underlie this system, and the legal principles consistent with these values.

See William E. Rees, *Sustainable Development and the Biosphere: Concepts and Principles* (Chambersburg, PA: ANIMA Books, 1990); Mathias Wackernagel and William E. Rees, *Our Ecological Footprint: Reducing Human Impact on the Earth* (Philadelphia: New Society Publishers, 1996).

Production and Process Method (PPM): a label for the characteristics of a product's origins and processing. The question of whether a given product's PPM ought to be taken into account in regulating its trade is one of the core issues in the current trade and environment debate. According to international trade rules, a country may block imports of some products when their physical characteristics differ from national standards. However, countries may not ban imports of the products based on the process and production methods (as was reaffirmed in the TUNA-DOLPHIN case). The arguments against the use of PPM-based restrictions are economic, political, and environmental. First, the principle of COMPARATIVE ADVANTAGE upon which international trade is based posits that nations should be able to procure benefits from their natural endowment, which entails different methods of production. Second, allowing some economically powerful nations, like the United States, to dictate environmental policies unilaterally might destabilize world trade and environment. Third, it is not economically wise or even feasible to create universal standards. On the other side, proponents of the PPM-based restrictions argue that it could be an effective mechanism in protecting domestic, foreign, and shared environments. Most experts agree that there should be principled exceptions to the general rule prohibiting discrimination on the PPM basis; however these exceptions should be negotiated at international forums rather than adopted by unilateral state legislation.

Winnipeg Principles on Trade and Sustainable Development: a document prepared by Canada's International Institute for Sustainable Development (IISD) that proposes a set of sustainability yardsticks against which international policies could be measured. These principles link environmental, trade, and development policies to the issue of sustainability through the criteria of efficiency and cost internalization, equity, environmental integrity, SUBSIDIARITY (i.e., allocation of responsibilities for local and non-localized environmental issues), international cooperation, science and precaution, and openness. The Winnipeg Principles establish a common basis for evaluation for those concerned with managing all three prime systems: ecological, economic, and social.

See Ted Schrecker and Jean Dalgleish, eds., *Growth, Trade and Environmental Values* (London, Ontario: IISD, 1994).

4.3.3 GLOBAL TRADE INTEGRATION INITIATIVES

General Agreement on Tariffs and Trade (GATT): a multilateral treaty that set forth rules of conduct for international trade. Trade restrictions had contributed to global economic depression during the 1930s, bringing about a need for regulating trade barriers (tariffs as well as quantitative, administrative, and technical restrictions). The agreement was signed in 1947 in Geneva by the representatives of twenty-three countries. The GATT Articles defined the basic rules of the world trade system and eventually governed trade relations between 123 governments that accounted for approximately 90 percent of world merchandise trade. First and foremost, GATT served as the framework for negotiating broad trade agreements, through multiyear "rounds" of negotiations (the Kennedy Round, the Tokyo Round, the URUGUAY ROUND) usually designed to reduce specific tariffs and other barriers to trade, and to enforce these agreements. GATT implementation had many implications on international environmental policy. Its rules covered border adjustments of domestic policies, the application of trade standards, the use of subsidies, and the use of trade measures with public policy goals. GATT convened dispute panels to arbitrate disagreements among GATT members. The decisions by GATT dispute panels on trade and environmental regulations (such as the U.S.-Mexico TUNA-DOLPHIN CONTROVERSY) were presented by GATT critics as free trade threats to the global environment. In 1994, by the final act of the Uruguay Round of trade negotiations under GATT, the "temporary" organization was replaced by the new permanent body, the WORLD TRADE ORGANIZATION (WTO).

United Nations Conference on Trade and Development (UNCTAD): a permanent intergovernmental body established in 1964. Headquartered in Geneva and comprised of 188 member states, UNCTAD is an organ of the UN General Assembly in the field of trade and development. Its principal goals are to maximize the trading, investment, and development opportunities of DEVELOPING COUNTRIES in order to ease their integration into the world economy, and to cope with challenges resulting from globalization. UNCTAD essentially operates as a forum for multilateral discussion and negotiation, most visibly through a series of major conferences that take place at four-year intervals. Yet UNCTAD also convenes many commissions, working groups, and negotiating sessions on topics ranging from trade regulations and commodity agreements to DEBT RELIEF. Several international trade agreements have emerged from UNCTAD negotiations, including the 1980 Multilaterally Agreed Equitable Principles and Rules for the Control of Restrictive Business Practices and the 1989 Agreement on a Global System of Trade Preferences (GSTP) among developing countries. Developing countries have long looked at UNCTAD, a one-member, one-vote organization, as a more useful vehicle for pressing their interests than the weighted-voting international organizations or consensus organizations (e.g., the WORLD BANK GROUP and the GENERAL AGREEMENT ON TARIFFS AND TRADE [GATT]). Governments of the developing countries have therefore focused on UNCTAD to push for major transformations of the international economy to a NEW INTERNATIONAL ECONOMIC ORDER, and UNCTAD conferences have often served as forums for such proposals. The governments of DEVELOPED COUNTRIES have often been skeptical about UNCTAD as a vehicle for reaching practical international agreements. It is not surprising that many UNCTAD accomplishments focus on cooperation among developing countries. The International Trade Center (ITC) is the arm of UNCTAD and the WORLD TRADE ORGANIZATION (WTO) for operational and enterprise-oriented aspects of international trade development. UNCTAD IX (Johannesburg, 1996) embraced the concept of SUSTAINABLE DEVELOPMENT as a central theme for the organization's work program.

Uruguay Round: negotiations on the liberalization of world trade which took place in the GENERAL AGREEMENT ON TARIFFS AND TRADE (GATT) framework in 1985–94.

The Uruguay Round resulted in a substantial degree of tariff liberalization, detailed rules on the application of trade policy measures, new multilateral regulations to cover intellectual property and trade disputes, and the reduction of "discriminatory" aspects of regional trade agreements. The Uruguay Round also linked the various agreements within a formal institutional framework with an integrated dispute settlement mechanism, and created the WORLD TRADE ORGANIZATION (WTO). The agreements emerging from the Uruguay Round were signed by 111 countries. The Uruguay Round provided the first opportunity to infuse global trade negotiations with environmental considerations. It adopted a Decision on Trade and Environment, as well as a Decision on Trade in Services and Environment.

See United Nations Conference on Trade and Development (UNCTAD), *Trade and Environment Report* (New York: United Nations Publications, 1994).

World Trade Organization (WTO): an international organization that began operations in 1995 to promote international trade through the reduction of trade barriers, resolution of trade disputes, and the provision of trade information and training. The WTO was formulated in the URUGUAY ROUND (1985–94) of the GENERAL AGREEMENT ON TARIFFS AND TRADE (GATT), which the WTO succeeded. Whereas GATT was an international agreement that called for a small and ostensibly temporary secretariat to oversee extended rounds of trade negotiations, the WTO is a full-blown international organization with a secretariat of roughly 500 staff and a mandate to promote international trade of not only goods but also services and intellectual property. Previous attempts to form an "International Trade Organization," parallel to the INTERNATIONAL MONETARY FUND (IMF) and the WORLD BANK GROUP, foundered in the early post–World War II era because of the difficulties of developed and developing countries to reach accord on the principles and structure of the international trade system. The WTO therefore reflects the convergence of views on international trade that emerged slowly throughout the postwar period.

4.4 Regional Economic Cooperation and Integration

Amazonian Cooperation Treaty: an agreement signed in 1978 between Bolivia, Brazil, Colombia, Ecuador, Guyana, Peru, Surinam, and Venezuela with the purpose of "carrying out joint efforts and actions to promote the harmonic development of . . . Amazonian territories." It refers to the Amazon Basin, an area larger than Australia that stretches from the Atlantic coast of Brazil to the Andes in Peru. Still mostly covered with tropical rainforests, Amazonia is considered by some ecologists to be the "lungs" of the world because it absorbs great quantities of carbon dioxide and gives out large amounts of oxygen. However, development in the form of mining, ranching, roads, and dams is threatening vast tracts of forest. The role of Amazonia in global atmospheric circulation as well as in the global pool of BIODIVERSITY has evoked substantial international concern over the trends of its development. The objective of the Amazonian Cooperation Treaty is "to bring about fair and reciprocally profitable results that benefit preservation of the environment, as well as conservation and rational use of the natural resources of those territories." The treaty applies to the Amazonian watershed territories of the contracting parties and promotes rational utilization of shared hydrological and biotic resources and the creation of conditions favorable for social and economic development of the region. Though initially extremely weak and unable to prevent resource devastation in Amazonia, the treaty received new impetus with the start of international negotiations on global environmental threats, which increasingly emphasize the value of the Amazonian environment. Since the late 1980s, the signatories established the Amazonian Special Environmental Commission and the Special Commission on Indigenous Affairs. In 1989 the presidents of the Amazonian countries adopted a strongly pro-environmental Amazonian Declaration. From that point, an ambitious set of concrete programs of regional cooperation was formulated, though most of the programs are still in the process of implementation. Despite this delay, some observers are encouraged by the initial results.

See Rodolfo Rendon, "Regimes Dealing with Biological Diversity from the Perspective of the South," in *Global Environmental Change and International Governance,* eds. Oran Young, George Demko, and Kilaparti Ramakrishna (Hanover, NH: University Press of New England, 1996), 173–77.

Asia-Pacific Economic Cooperation Forum (APEC): the main intergovernmental organization for economic cooperation in the Asia Pacific region, established in 1989. The APEC forum focuses on promoting trade liberalization and investment facilitation, as well as economic and technical cooperation. Its eighteen member economies—Australia, Brunei Darussalam, Canada, Chile, China, Hong Kong, Indonesia, Japan, the Republic of Korea, Malaysia, Mexico, New Zealand, Papua New Guinea, the Philippines, Singapore, Taiwan, Thailand, and the United States—had a combined gross domestic product of more than US$13 trillion in 1995, approximately 55 percent of total world income and 46 percent of global trade. APEC member economies also account for over half of the world's emissions of pollutants, energy use, and food production and consumption. In recognition that the Asia Pacific region's rapidly expanding population and economic growth would sharply increase pressures on the environment, APEC leaders agreed on the need for joint action to ensure sustainable economic prosperity of the region. In 1996 the APEC members adopted a Declaration and Action Program for cooperation on SUSTAINABLE DEVELOPMENT, including three key areas (sustainable cities, cleaner production and technologies, and sustainability of the marine environment), and later endorsed concrete, action-oriented strategies in these areas. A fundamental concern is that while APEC is operating on the basis of consensus and the premise of equality of members, real differences in members' economic levels could jeopardize its long-term efforts to address sustainable development. For instance, during deliberations on sustainable cities some countries focused on cleaner production as a "fix-it" strategy, whereas others emphasized that poverty eradication must be the basis for any solution. Given rapid economic and population growth in the region, the question of whether APEC members can genuinely cooperate on the environment is of global importance.

Association of South East Asian Nations (ASEAN): an intergovernmental organization founded in 1967 by the Bangkok Declaration. Its original membership consisted of Indonesia, Malaysia, the Philippines, Singapore, and Thailand, with Brunei joining in 1984 and Papua New Guinea having observer status. ASEAN's main objectives include the acceleration of economic growth, social progress, and cultural development through joint endeavors; the promotion of regional peace and stability; and the encouragement of collaboration and mutual assistance on matters of common interest. One of the main ASEAN initiatives is the Regional Seas Programme that is supposed to include regional oil-spill contingency planning, activities related to land-based pollution sources, research on mangrove ecosystems, and the study of environmental problems of offshore exploitation and exploration. The ASEAN nations usually take a common standing at UN forums dealing with environment and development.

European Community. See EUROPEAN UNION (EU).

European Union (EU): the major regional organization of most Western European nations, which oversees cooperation on economic, political, social, and environmental issues. Current members are Austria, Belgium, Denmark, Finland, France, Germany, Greece, Ireland, Italy, Luxembourg, the Netherlands, Portugal, Spain, Sweden, and the United Kingdom. Many more European nations, including Eastern European countries and Eurasian states like Turkey, have applied for EU membership. The EU has taken a lead role in international environmental diplomacy and policy making. The European Union came into existence in 1994, replacing the European Economic Community (EEC) in accordance with the MAASTRICHT TREATY. From an international development perspective, the EU represents the most ambitious example of regional economic integration in the world, with far more extensive integration of labor, social legislation, and policy coordination than other regional integration arrangements such as the NORTH AMERICAN FREE TRADE AGREEMENT (NAFTA). From an environmental perspective, the Directorate General for Environment, Consumer Protection, and Nuclear Safety, within the Commission of the European Union, initiates and implements environmental projects and recommends policy initiatives for the European Council of Ministers, which has the final decision-making authority. Integrated EU policy, if adopted by the Council of Ministers in the form of regulations, directives, and decisions, becomes mandatory for member states. EU members also have an information agreement requiring members to notify the commission on proposed national laws, regulations, and procedures relating to the environment. The EU commission determines whether proposed regulations are in line with EU legislation and policy. Both the EU commission and member states can bring noncompliance lawsuits before the European Court of Justice. Since 1973, coordinated efforts of the EU and its predecessors have been undertaken through its "Action Programmes." The Fifth Action Programme, in force from 1993 to 2000, focused on environment and SUSTAINABLE DEVELOPMENT.

See Lynton Caldwell, *International Environmental Policy* (Durham: Duke University Press, 1996), 164–71.

Maastricht Treaty: the treaty of the EUROPEAN UNION (EU) signed in 1992 at Maastricht, the Netherlands, as an amendment to the Rome Treaty that established the European Economic Community (EEC) in 1957. Through the Maastricht Treaty, the contracting parties (Belgium, Denmark, France, Germany, Greece, Ireland, Italy, Luxembourg, the Netherlands, Portugal, Spain, and the United Kingdom) greatly strengthened the economic, political, environmental, and monetary integration that had existed under the EEC framework. The Maastricht Treaty opened the way for the single European currency, uniform social programs for member countries, and greater integration for raw materials management. It also included specific references to environmental policy and SUSTAINABLE DEVELOPMENT.

North American Free Trade Agreement (NAFTA): a treaty signed in December 1992 by the governments of Canada, Mexico, and the United States to establish greater regional economic integration. In effect since January 1, 1994, the NAFTA arrangements are intended to eliminate import duties and other barriers of trade among the three signatory countries and to create a continent-wide FREE TRADE ZONE, to come into complete effect over a ten-year period. The agreement also strives to promote fair competition, increase cross-investment, protect INTELLECTUAL PROPERTY RIGHTS, and establish a framework for further cooperation among the countries. Supplemental agreements to strengthen NAFTA in the areas of the environment, labor, and import surges are under way. Environmentalists are concerned that under the current NAFTA text, existing U.S. environmental standards could be challenged as a "trade barrier" and that companies located in the United States may be encouraged to move to Mexico to avoid U.S. environmental regulations. *T. S.*

North American Agreement on Environmental Cooperation (NAECC): an agreement concluded by the governments of Canada, Mexico, and the United States in September 1993 in association with the NORTH AMERICAN FREE TRADE AGREEMENT (NAFTA). The objectives of the NAECC include promoting SUSTAINABLE DEVELOPMENT through cooperative economic and environmental policies; cooperation on development and improvement of environmental laws, procedures, and practices; enhanced enforcement of environmental laws; promotion of transparency and public participation; and the promotion of economically efficient and effective environmental measures. The NAECC established the Commission for Environmental Cooperation to monitor the environmental conditions affected by increased trade. It also set up groundbreaking rules to ensure public participation in the work of this commission. Environmentalists in all three countries use these provisions to illuminate policies and issues of environmental concern and to press for action. Evidence on the success of the NAECC is still inconclusive.

Organization for Economic Cooperation and Development (OECD): an intergovernmental organization established in 1961 to coordinate the economic policies of its members. Headquartered in Paris, the OECD's original membership consisted of countries participating in the Marshall Plan's post–World War II reconstruction in Europe: Austria, Belgium, Denmark, France, Germany, Greece, Iceland, Ireland, Italy, Luxembourg, the Netherlands, Norway, Portugal, Spain, Sweden, Switzerland, Turkey, the United Kingdom, as well as Canada and the United States. The following countries later joined the OECD: Japan (1964), Finland (1969), Australia (1971), New Zealand (1973), Mexico (1994), the Czech Republic (1995), Hungary, Poland, and the Republic of Korea (1996). The OECD objectives are to achieve the highest sustainable economic growth, employment, living standards, and financial stability among its member nations by fostering their policies to this end. Such policies are supposed to contribute to the development of the world economy and expansion of world trade as well. One of the OECD bodies, the Development Assistance Committee, coordinates financial aid and technical assistance to developing countries. Some use the term "OECD countries" to refer to the industrialized, developed nations; the membership of developing countries such as Turkey and Mexico make this quite misleading.

Though primarily oriented toward the promotion of economic growth, the OECD has been active for the last two decades in coordinating the environmental and natural resource policies of its members. Its Environment Committee, established in 1970 at the ministerial level, studies various aspects of human-nature interaction, reviews environmental performance and policies of its members, and makes concrete recommendations on pollution abatement with an emphasis on its prevention rather than cure. The majority of work undertaken by OECD in the 1990s was designed to evaluate and recommend the most effective policy options.

See Organization for Economic Cooperation and Development (OECD), *Managing the Environment* (Paris: OECD, 1994); Organization for Economic Cooperation and Development (OECD), *Evaluating Economic Instruments for Environmental Policy* (Paris: OECD, 1997).

Organization of Petroleum Exporting Countries (OPEC). See under SECTION 6.7.2.

South Pacific Commission (SPC): an international organization established in 1947 by the then colonial powers to ensure the economic and social stability of the Pacific Islands. Currently, the SPC has twenty-seven members (thirteen island developing countries, nine island territories, and five developed countries, including the United States, the United Kingdom, France, Australia, and New Zealand). The SPC provides technical advice, assistance, and training in such diverse areas as agriculture and plant protection, rural development and technology, environmental management, and marine resources. Though environmental issues were not on the organization's initial agenda, they have become one of the major targets of its work. A special project on nature conservation initiated by the SPC resulted in the establishment of another regional organization, the SOUTH PACIFIC REGIONAL ENVIRONMENT PROGRAM (SPREP).

South Pacific Forum: an organization founded in 1971 by seven South Pacific countries (Australia, the Cook Islands, Fiji, Nauru, New Zealand, Tonga, and Samoa) to develop a collective response to regional and international issues. Later, the Federated States of Micronesia, Kiribati, the Marshall Islands, Niue, Palau, Papua New Guinea, the Solomon Islands, Tuvalu, and Vanuatu joined the forum, increasing its membership to sixteen. Members of the forum meet annually to discuss various concerns, including security, economy, and environment. In recent years, issues of SUSTAINABLE DEVELOPMENT of island states were included into the agenda. The forum secretariat, headquartered in Suva (Fiji), addresses problems of regional development with a strong emphasis on trade, shipping, civil aviation, telecommunications, energy, and economic issues. The environment unit within the secretariat seeks to coordinate sustainable development policies at the regional level, in cooperation with the SOUTH PACIFIC REGIONAL ENVIRONMENT PROGRAM (SPREP).

South Pacific Regional Environment Program (SPREP): a comprehensive program for environmental management in the South Pacific initiated by regional organizations with UN support in 1978. The program is responsible to its member governments for the technical implementation of the Action Plan for Managing the Natural Resources and Environment of the South Pacific Region. The action plan addresses such issues as coastal area management, protected areas and species conservation, management of natural resources, education and training, waste management, and pollution control. Recently, the SPREP mandate was updated to incorporate recommendations of AGENDA 21 and to take into account the region's responsibilities under the FRAMEWORK CONVENTION ON CLIMATE CHANGE (FCCC) and various biodiversity conventions.

4.5 Security and Sustainable Development

Atmospheric Test Ban Treaty: an international agreement that committed the parties to refrain from testing nuclear explosives in the atmosphere, space, and the oceans. It permitted the exploding of such devices in other environments, such as underground, if radioactive debris is not deposited beyond the borders of the country doing the testing. The treaty was negotiated in 1963 by the United States, the Soviet Union, and the United Kingdom, following the extensive series of atmospheric tests in 1961-62 by the two superpowers, which distributed measurable amounts of radioactive fallout globally. The three original parties complied with the treaty by conducting all subsequent nuclear tests underground, while France and China rejected the agreement and continued aboveground testing until 1974 and 1980, respectively. The prohibition of aboveground tests is now widely viewed as a principle of international customary law, given that no known nuclear tests have been conducted in the atmosphere since 1980 and that there are now 122 states party to the treaty. The Comprehensive Nuclear Test Ban Treaty, concluded in 1996, prohibits all nuclear testing, including those conducted underground.

See Marvin S. Soroos, *The Endangered Atmosphere: Preserving a Global Commons* (Columbia: University of South Carolina Press, 1997), 83–109. M. S. S.

COMMENTARY ON ATMOSPHERIC TEST BAN THEORY

> While usually categorized as an arms control agreement, the test ban treaty of 1963 did little to restrain the nuclear arms race because the primary competitors—the United States and the Soviet Union—conducted a large number of underground tests to continue the development of nuclear arsenals. The significance of the treaty is rather as an environmental agreement intended to curb the injection of radioactive substances into the atmosphere. *Marvin S. Soroos*

Convention on the Prohibition of the Development, Production, and Stockpiling of Bacteriological (Biological) and Toxic Weapons and Their Destruction (Biological and Toxic Weapons Convention): an international agreement that was initiated by the United Nations General Assembly in December 1971 and came into force on 26 March 1975. The convention had 156 signatories by 1997. It requires parties to undertake the complete destruction or diversion to peaceful uses of such weapons within nine months of assuming their obligations. The Biological Weapons Convention has been widely recognized as a precursor to a possible future agreement on chemical weapons and affirms the undertaking by its parties to continue negotiations to that end. The convention is often singled out as the first genuine example of a multilateral negotiated disarmament agreement achieved under the general auspices of the UN. However, in the 1990s at least ten countries were proved or suspected to have developed or used biological weapons.

Convention on the Prohibition of the Development, Production, Stockpiling, and Use of Chemical Weapons and on Their Destruction (Chemical Weapons Convention [CWC]): an international agreement concluded in 1993 that came into force in 1997. By 2000, the convention had 127 members. The CWC is designed to curb the production and use of chemical warfare agents. Parties to the CWC are subject to specific restrictions on trade in chemical products with non-parties as well as certain production limitations. To ensure that certain precursor chemicals are not illegally diverted to the production of chemical weapons, both producers and users of specified quantities of the chemicals may be required to submit annual declarations of their activities to national authorities, who in turn pass information to a new international agency, the Organization for the Prohibition of Chemical Weapons. The CWC potentially could result in new domestic and international regulations on a broad range of commercial chemical products and therefore has been of extreme concern to private industry.

Convention on the Prohibition of Military or Any Other Hostile Uses of Environmental Modification Techniques (Environmental Modification Convention (ENMOD, 1977): an international agreement adopted by the United Nations in 1977 with the goal of restricting environmental manipulations for hostile interstate purposes. The main provisions of the convention are rather limited: restrictions are placed only on environmental modification techniques that have "widespread, long-lasting or severe effects." In general, ENMOD is weak and not widely accepted.

Ecocide: intentional destruction of the natural environment through bombing, setting fire, oil spills, use of defoliants, scorched-earth tactics, and other means. As a wartime strategy conducted by governments or other warring parties, ecocide is an important source of environmental change, often resulting in a flow of ENVIRONMENTAL REFUGEES. However, until now ecocide has not been specifically addressed by international law, though legal scholars have proposed a "Convention on the Crime of Ecocide."

See Richard Falk, "Proposed Convention on the Crime of Ecocide," in *Environmental Warfare: A Technical, Legal and Policy Appraisal*, ed. Arthur Westing (London: Taylor and Francis, 1984), 45–49.

Environmental Refugees: persons who are forced to migrate due to natural or human-induced environmental disruption that endangers their lives or critically affects their quality of life. Currently, some one billion people may be displaced for environmental reasons and many of them will cross international borders as a result. Researchers and policy makers define refugee-generating environmental changes differently, ranging from natural calamities to gradual natural disasters (desertification, degradation of land, deforestation, and global sea-level rise), technological accidents, and development or politically induced changes of the habitat. Environmental refugees do not fall under the standard definition of *refugee* provided by the 1951 United Nations Convention Relating to the Status of Refugees and its 1967 Protocol and therefore have no legal protection and receive only limited international assistance.

See Jodi Jacobson, *Environmental Refugees: a Yardstick of Habitability* (Washington, DC: Worldwatch, 1988); Shin-wha Lee, "In Limbo: Environmental Refugees in the Developing Countries," in *Conflict and the Environment*, ed. Nils Peter Gledisch (Dordrecht: Kluwer, 1997), 273–92.

COMMENTARY ON ENVIRONMENTAL REFUGEES

"Environmental migrant" would probably be a more accurate term than refugee since such population movements tend to be overwhelmingly incremental and reflect the search for improved conditions rather than a life-or-death necessity for immediate flight. Population movements in the face of natural disasters such as hurricanes are usually over relatively short distances and rarely result in permanent relocation. However, the term *refugee* sticks because of its rhetorical potency. It is also interesting to note that since the Second World War the major environmental migration in the United States has been from the relatively secure conditions of the Northeast to the largely desert areas of Southern California and the Southwest or to low-lying and hurricane-prone Florida. *Steve Rayner*

International Environmental Security: a term for security concerns arising from environmental degradation, or environmental degradation arising from political and military conflict. The term has come into fashion since the end of the Cold War. Traditional conceptions of security focusing upon national military security have given way to a more expansive view of the important threats to human well-being in the realms of economics, energy, food, and the environment. The term *environmental security* has been used in two ways. The first refers to international environmental issues, such as disputes over water resources, which may become a cause of violent conflict or warfare, and thus a threat to military security. The other conception of environmental security calls attention to the ways in which environmental changes, such as depletion of the ozone layer, directly jeopardize human welfare even without heightening the threat of military confrontations.

See Richard Ullman, "Redefining Security," *International Security* 8 (1983): 129–53. *M. S. S.*

COMMENTARY ON ENVIRONMENTAL SECURITY

While the term "environmental security" has been embraced in many circles over the past decade, it has not been without controversy. Traditional security specialists contend that efforts to broaden definitions of security may render the term meaningless in view of basic differences in the types of threats addressed in the military and environmental realms. Some environmentalists express the concern that the "greening of security" would open the door to an unwelcome involvement of military establishments in the implementation of environmental policies. *Marvin S. Soroos*

North Atlantic Treaty Organization (NATO): a military and political alliance established in 1947 through the North Atlantic Treaty. The treaty tied Western European countries and their North American allies with a pledge that a military attack on one of the parties would be regarded as an attack against all, and that any necessary action would be taken in response. NATO military cooperation also places some national troops under a NATO united command. The rationale for NATO's creation and initial activities was the containment of the Soviet Union's power and influence in Europe; in the wake of the breakup of the Soviet Union, many Russians now fear that the essential rationale for NATO has become the containment and restriction of Russia. NATO's current nineteen members are Belgium, Canada, the Czech Republic (since 1999), France, Germany (since 1955), Greece (since 1952), Hungary (since 1999), Iceland, Italy, Luxembourg, the Netherlands, Norway, Poland (since 1999), Portugal, Spain (since 1982), Turkey (since 1952), the United Kingdom, and United States. Following the end of the Cold War, NATO structures and policies have been adapted to the new European security environment, culminating in the full membership of the Czech Republic, Hungary, and Poland in 1999. Meantime, cooperation with nonmember countries and collective crisis management has risen prominently on the NATO agenda. Through its so-called third dimension activities, NATO deals with the issues of ENVIRONMENTAL SECURITY. The NATO Science Committee funds advanced research workshops and advanced study institutes and also provides collaborative research grants for a wide range of environment-related projects (i.e., Ecological Restoration and Management of Military Training Areas). The environment has been defined as one of the priority areas of the current Workplan for Dialogue, Partnership and Cooperation, which comprises ministers of the NATO countries and former socialist countries.

Nuclear Winter: a theorized combination of dramatic changes in the world climate that would be caused by smoke and fallout from the use of nuclear weapons. The concept was brought to public attention by Carl Sagan in 1983. Dust from a nuclear explosion and soot from subsequent fires would inevitably block out sunlight for months on end; mean temperatures would drop to subfreezing levels and all remaining animal and plant life on earth would be threatened with extinction. Therefore the survivors of a nuclear war would face extreme cold and famine in addition to other effects of the nuclear explosion. In the famous *Science* paper, "Nuclear Winter," Sagan and his colleagues presented the results of their calculations of the nuclear war scenario, based on a computer model of the earth's atmosphere. The development of "nuclear winter" models by Soviet scientists in cooperation with their Western colleagues in the mid-1980s basically confirmed the conclusions of Sagan's team, encouraging better understanding of global issues and greater pressure on the superpowers to end the arms race. However, the concept of nuclear winter has not escaped criticism; the accuracy of its projections was questioned, as well as its assumption that the enormous loss of life of any nuclear exchange is not a sufficient reason in itself to condemn nuclear weapons.

See R. P. Turco, O. B. Toon, T. P. Ackerman, J. B. Pollack, and C. Sagan, "Nuclear Winter: Global Consequences of Multiple Nuclear Weapons Explosions," *Science* 222 (December 23, 1983): 1283.

5

Decisionmaking

5.1 Analytical Tools for Environmental Decisionmaking

5.1.1 ADJUSTING NATIONAL ACCOUNTS TO REFLECT ENVIRONMENTAL AND RESOURCE CONSIDERATIONS

Avoidance Costs Approach (Prevention Cost Approach): a concept of cost accounting pioneered by Herman Daly, R. Hueting, and Paul Ekins, and later elaborated in the 1993 *UN Handbook of National Accounting*. It suggests that from a comprehensive economic point of view, national accounts must include the costs of strategies necessary to avoid negative environmental impacts of economic activities. However, the calculation of avoidance costs is possible only on the basis of economic modeling, which does not fit into the framework of traditional accounts and therefore requires more research.

See United Nations Statistical Office, *Integrated Environmental and Economic Accounting. Handbook of National Accounting, Studies in Methods* (New York: United Nations, 1993). *R. P.*

Eco-Domestic Product (EDP; Green or Environmentally Adjusted Net Domestic Product): an aggregate measure used in ENVIRONMENTAL ACCOUNTING. It is obtained by subtracting the costs of natural resource depletion and environmental degradation from the net domestic product. If economic activities do not cause depletion or degradation of the natural environment, the net domestic product and the eco-domestic product should be identical.

See Kumio Uno and Peter Bartelmus, eds., *Environmental Accounting in Theory and Practice* (Dordrecht, Netherlands: Kluwer, 1998); Wouter Van Dieren, ed., *Taking Nature into Account* (New York: Springer-Verlag, 1996).

Ecological Capital: a notion used in ENVIRONMENTAL ACCOUNTING to define the contribution of the environment to economic activity. Ecological capital performs three distinct functions: the provision of resources for production; the absorption of wastes; and provision of basic conditions within which production is possible at all (i.e., basic survival services such as those maintaining climate and ecosystem stability).

See Wouter Van Dieren, ed., *Taking Nature into Account* (New York: Springer-Verlag, 1996).

Environmental Accounting (Green Accounting): 1) a system of national economic accounting that includes adjustments to reflect pollution costs and the depletion of natural resources. The basic idea of environmental accounting is that the depletion of nature's capital—natural resources and environmental services—has a real cost to society and should be reflected in national accounts the same way as the depletion of physical business assets. There is no single model for environmental accounting; approaches vary according to the specific purpose and requirements by individual countries. Three main approaches are: a) the direct adjustment of the System of National Accounts (SNA) to incorporate environmental effects; b) the development of satellite accounts outside and complementary to the core SNA; and c) independent natural resource and environmental accounts linked to the national accounts. The WORLD RESOURCES INSTITUTE and the WORLD BANK pioneered environmental accounting, which was endorsed by AGENDA 21. In 1993 the United Nations Statistical Office proposed the System of Integrated Environmental and Economic Accounting (SEEA). Despite its potential usefulness, the SEEA has not yet been incorporated formally into any nation's national accounting system.

See United Nations Statistical Office, *Integrated Environmental and Economic Accounting* (New York: United Nations, 1993).

2) in corporate accounting, environmental accounting usually refers to ENVIRONMENTAL AUDITING, but may also include the costing of environmental impacts caused by the corporation.

Environmentally Adjusted National Income (ENI): an aggregate measure used in ENVIRONMENTAL ACCOUNTING obtained by adding the net income received from abroad to ECO-DOMESTIC PRODUCT. Some scholars also suggest the additional deduction of the net cost of cross-boundary pollution. See also ENVIRONMENTAL RESOURCE ACCOUNTS.

Environmentally Adjusted Net Domestic Product. See ECO-DOMESTIC PRODUCT.

Environmentally Defensive Expenditures (EDEs): the expenditures associated with the mitigation of environmental damages through economic activities. EDEs are understood as expenditures aimed to compensate for, repair, and restore ecosystems subjected to environmental losses and damages originating in the past; to avoid the emission of additional pollutants to the environment; and to anticipate burdens on and damage to the environment caused by economic growth. Examples of EDEs are the costs of cleaning polluted air, water, and soils, treating damaged forests, and treating pollution-related illnesses.

See Wouter Van Dieren, ed., *Taking Nature into Account* (New York: Springer-Verlag, 1996).

Environmental Resource Accounts: a framework for summarizing the supplies, uses, and improvement or deterioration of ENVIRONMENTAL RESOURCES, generally in quantitative terms. The stocks and flows of environmental resources are recorded in terms of mass, quality, area, population, and diversity. Examples are air quality accounts, forest accounts, water resource accounts, wildlife resource accounts, land accounts, etc.

Therefore, in principle, environmental resource accounts can be extended to incorporate the quality aspects of natural resources. These accounts aim at measuring pollution flows from economic activities to natural resources or serve to draw a picture of the quality of a particular natural resource. See also ENVIRONMENTAL ACCOUNTING.

See Organization for Economic Cooperation and Development (OECD), *Natural Resource Accounts: Taking Stock in OECD Countries* (Paris: OECD, 1994), 6–7.

Framework for Indicators of Sustainable Development (FISD): a conceptual framework for environmental, social, and economic indicators developed by the United Nations Statistics Division in 1994 in accordance with the recommendations of the UNITED NATIONS CONFERENCE ON ENVIRONMENT AND DEVELOPMENT (UNCED).

Genuine Savings. See under SECTION 1.1.6.

Green Accounting. See ENVIRONMENTAL ACCOUNTING.

Green Gross Domestic Product (GGDP). See ENVIRONMENTALLY ADJUSTED GROSS DOMESTIC PRODUCT.

Gross World Product Resource Accounting. See NATURAL RESOURCE AND ENVIRONMENTAL ACCOUNTS (NREA).

Natural Patrimony Accounting: a French system of ENVIRONMENTAL ACCOUNTING that includes all components of nature that can be changed by human activity. It comprises NONRENEWABLE RESOURCES, environmental media and living organisms of ecosystems, the agents that may affect natural assets and systems, and the impacts of human beings on nature. All resources and impacts are estimated both in monetary and in physical terms.

See Jacques Theys, "Environmental Accounting in Development Policy: The French Experience," in *Environmental Accounting for Sustainable Development*, eds. Yusuf J. Ahmad, Salah El Serafy, and Ernst Lutz (Washington, DC: World Bank, 1989).

Natural Resource and Environmental Accounts (NREA): a system of ENVIRONMENTAL ACCOUNTING that aims at collecting, within a consistent framework, quantitative and qualitative information on both the state of natural resources and their evolution. Primarily in physical terms, they describe the stocks and flows of resources, the flow of resources between the environment and the economy, and the flow of resources within the economy. The purpose of NREA is twofold: to provide policymakers with an information base on natural resources and to contribute to the general awareness of environmental issues at each level of decision making and by the general public. Two main categories of NREA can be distinguished: ENVIRONMENTAL RESOURCE ACCOUNTS and MATERIAL RESOURCE accounts.

See Organization for Economic Cooperation and Development (OECD), *Natural Resource Accounts: Taking Stock in OECD Countries* (Paris: OECD, 1994), 8. R. P.

Resource Accounting. See NATURAL RESOURCE AND ENVIRONMENTAL ACCOUNTS (NREA).

Resource Adjusted Net Domestic Product (RANDP): the domestic product adjusted for both the depreciation of humanmade capital and the depletion or degradation of natural resources. The RANDP is thus a central concept in NATURAL RESOURCE AND ENVI-RONMENTAL ACCOUNTS (NREA). While resource depletion and degradation are very important for gauging the risks to sustainability, some have criticized the adequacy of this measure because it cannot signal or capture future risks of ecosystem collapse. R. P.

Resource and Pollutant Flow Accounts (RPFA): one of the types of SATELLITE ACCOUNT-ING techniques. They are usually conceived as physical extensions to the input-output accounts, when data on the physical flow of natural resources (such as input of energy to production processes) and a physical flow of wastes and emissions are provided for each production and final demand sector. When directly linked to input-output accounts, RPFAS can be applied to a wide variety of policy issues. For instance, they can be used to measure the impact of existing and prospective environmental regulations and taxes, and provide the link between international trade policies and the pollution burden associated with a particular structure of trade. Set up in physical terms, RPFAS differ from other types of green accounting techniques that contain a mixture of physical and monetary data. Various kinds of RPFA have been used in Canada, Germany, and the Netherlands. See also GREEN ACCOUNTING and NATURAL RESOURCE AND ENVIRONMENTAL ACCOUNTS (NREA).

See Timothy O'Riordan, ed., *Ecotaxation* (New York: St. Martin's, 1997).

Satellite Accounts: separate accounts of environmental, resource, and social trends that are not directly incorporated into the SYSTEM OF NATIONAL ACCOUNTS (SNA). A satellite accounting approach complements the economic information drawn from standard national accounts rather than modifying it. The rationale of satellite accounts is that changes in the long-established standard national accounting system are unlikely to be accepted or well understood, but this system can be augmented by auxiliary accounts that combine physical information from environmental statistics and natural resource accounts with economic information from national accounts. Satellite accounts may entail the disaggregation of conventional national accounts with regard to environmental aspects, valuation of natural resources, and of nonmarket services of the environment, and valuation of environmental damage due to economic activity.

See Organization for Economic Cooperation and Development (OECD), *Natural Resource Accounts: Taking Stock in OECD Countries* (Paris: OECD, 1994), 7. R. P.

Stress-Response Environmental Statistical System: system of ENVIRONMENTAL AC-COUNTING developed in Canada. It specifies measures that exert stress on the environment (stress and stressor statistics), measures of the effects on the environment (environmental response), and measures of policy response (both collective and individual).

System of National Accounts (SNA): the standardized set of definitions of national-level economic statistics used worldwide to develop comparable and consistent information on GROSS DOMESTIC PRODUCT (GDP), GROSS NATIONAL PRODUCT (GNP), NATIONAL INCOME, and a host of other economic quantities and trends. The SNA is maintained and periodically revised by the United Nations Statistics Division, relying on a series of expert workshops. The efforts to introduce adjustments for environmental degradation and natural resource depletion into the main SNA have been blocked by concerns over the lack of uniformity of the many GREEN ACCOUNTING approaches, the lack of quantitative cost information, and the resulting variation in estimates using different methods or assumptions. Thus far, the UN Statistical Division has called for SATELLITE ACCOUNTS to reflect environmental and depletion impacts, rather than incorporating them directly into the SNA.

User Cost Approach: one of the approaches of accounting for depletable resources proposed by Salah El Serafy first in 1981. The method is intended to adjust the national income of countries that rely on depletable resources in order to gauge their true or sustainable income. It is based on the concept of the user cost of a depletable resource — the argument that extraction has an inherent opportunity cost deriving from the depletability of the resource.

See Salah El Serafy, "The Proper Calculation of Income from Depletable Natural Resources," in *Environmental Accounting for Sustainable Development*, eds. Yusef Achmad, Salah El Serafy, and Ernst Lutz (Washington, DC: World Bank, 1989).

COMMENTARIES ON NATURAL RESOURCE AND ENVIRONMENTAL ACCOUNTS

Efforts to account for environmental matters in traditional accounting methods are numerous and mostly welcome. At the level of the firm many of these fall under the general label of "Full Cost Accounting," or FCA. At the regional and national or macro levels these efforts are labeled variously GREEN NATIONAL ACCOUNTS, GREEN GROSS DOMESTIC PRODUCT (GGDP), or the greening of the gross domestic product. Yet the quest for an all-encompassing, single measure or system has so far been unsuccessful. It may even be misguided since a more reasonable goal might well be to produce appropriate and suitable, that is, contextually relevant and understandable, measures of the impact of different kinds of economic activity on the environment and resources (and the reverse). Thus, in places where ECOTOURISM exists and figures prominently in a nation's economic well-being, including environmental assessments related to this activity makes sense. For other nations, in different touristic circumstances, it probably does not. One needs to consider natural resource accounting for at least three general reasons: a) to focus better on the relationships between the environment and the economy; b) to include environmental considerations in decision making, including decisions reached by popular as well as narrower political and business interests; and c) to contribute to a more robust national economic accounting system.

Three common means exist for valuing environmental costs — considered as the reduction in the number and/or quality of goods and services provided by nature as a consequence of (usually negative) environmental impacts. Damage cost refers to consumers' willingness to pay for what has already been lost or damaged as a consequence of human activity. This is often referred to as "demand side" costing. So-called HEDONIC PRICING studies, TRAVEL COST METHODS, hypothetical market studies, and guesses about the political willingness to pay all fall within the damage cost methodology. None has proven particularly interesting or useful. Restoration cost, as noted under RESTORATION ECOLOGY, refers to what it will cost to fix damage already done. Its basic conceptual weaknesses are also mentioned in the commentary on restoration ecology. Avoidance cost is the price to be paid to prevent damage in the first place. In general, avoidance cost methods underlie much "end of tail pipe" and simple clean-up rules and regulations. *Garry D. Brewer*

Such attempts to broaden the SYSTEM OF NATIONAL ACCOUNTS to take account of environmental costs and benefits and natural resource depletion have become increasingly popular in recent years. However, the results of efforts to operationalize various methodologies lag behind the conceptual work in this area. One reason is the complexity of quantifying the range of environmental impacts in an economy so that these can be integrated into the national accounts. The main problem, though, is that the exercise fails the "so what" test. Proponents of this approach have been unable to articulate clearly the policy rationale for going through this elaborate and complex exercise other than that it provides a clearer picture of the productive capacity of economies. Given the difficulties and cost involved in such exercises and the residual uncertainties concerning many of the estimates, this rationale is not compelling especially in a world in which simpler and potentially more useful cost-benefit analyses must often be forgone due to resource limitations. *Sudhir Shetty*

The virtue of environmental accounting, even if incomplete, is that it unmasks resource-depletion aspects of some economic growth patterns that otherwise would be seen as spectacular. Thus, Indonesia's official annual gross domestic product (GDP) growth rate from 1971 to 1984 was a seemingly impressive 7.1 percent; but adjusting for *only* the depreciation of petroleum, timber, and Javanese soils brings the adjusted annual rate down to only 4 percent. This does not even count the depletion or deterioration of natural gas, coal, hard minerals, water, etc.

See Robert Repetto, William Magrath, Michael Wells, Christine Beer, and Fabrizio Rossini, *Wasting Assets: Natural Resources in the National Income Accounts* (Washington, DC: World Resources Institute, 1989). *William Ascher*

5.1.2 BENEFIT-COST ANALYSIS

Benefit-Cost Analysis (Cost-Benefit Analysis): a framework and set of methods for the systematic evaluation of the benefits and costs of project investments or broader policies. Although benefit-cost analysis can have the modest objective of merely identifying and assessing each benefit or cost, the more common aspiration is to measure each benefit and cost according to a common metric that permits the calculation of an overall value for society, typically expressed as a rate of return or net present value. The complications of benefit-cost analysis therefore entail the issues of: a) evaluating (usually "monetizing") outcomes that are difficult to measure in quantitative terms (e.g., the value of biological diversity); b) valuing benefits and costs according to their societal values rather than the prevailing market or administered prices; c) selecting a time discount rate to evaluate the present value of benefits or costs expected in the future; and d) agreeing on whose benefits and costs ought to be included in the assessment (e.g., should the impacts on noncitizens be included?) A variety of methods, such as HEDONIC PRICING, CONTINGENT VALUATION, TRAVEL-COST, and MITIGATION COSTS, have been developed to try to measure environmental impacts so that they can be incorporated into benefit-cost analysis.

COMMENTARIES ON BENEFIT-COST ANALYSIS

The challenges to cost-benefit calculation are far wider-ranging than technical problems of calculating the relative value of nonmarket goods. They include the challenge of applying a utilitarian calculus in situations where significant constituencies reject the moral premises of utilitarianism. See EFFICIENCY, ECONOMIC. *Steve Rayner*

The major pitfall of cost-benefit analysis is not in the exercise of systematically assessing costs and benefits; it is in asserting the cost-benefit results as the only relevant information. The underlying premise that benefit-cost analysis is an assessment of utility also raises the issues of whether benefit-cost analysis can be compatible with either according status to the intrinsic merit of ecosystem conservation or the assertion of rights. The seemingly unobjectionable idea that the policy, program, or project with the greatest net benefit ought to be chosen simply has no room for the alternative conceptions of rights (something ought to be done, or ought not to be done, *regardless* of the costs or benefits) or the legitimacy of voting outcomes. Yet this problem can be remedied by allowing multiple types of claims to be put on the table. *William Ascher*

Cost-Effectiveness Analysis (CEA): economic and engineering analysis used to determine how to minimize the cost of achieving a specified set of developmental or environmental objectives. For example, in the acid deposition field the objective might be to meet target loading of sulfur at minimum cost over a large region, taking into account that control costs vary from industry to industry, and that the cost of control increases with the severity of control. Cost-effectiveness analysis presumes that the objectives are appropriate, as distinct from BENEFIT-COST ANALYSIS, which evaluates whether achieving an objective is worth the cost. *R. P.*

COMMENTARY ON COST-EFFECTIVENESS ANALYSIS

A major challenge in conducting cost-effectiveness analysis is identifying all the costs. Costs to particular individuals or firms are often easier to identify than those that fall on the public. Since cost-effectiveness analysis depends for its legitimacy not on selecting objectives but on selecting courses of action, the consequences of the courses of action must be thoroughly considered. Cost-effectiveness is often used when dealing with public interventions, which necessarily have some incidence on the public. A bias may be introduced in cost-effectiveness analysis if only those costs that are easily identified, such as costs to individuals or firms, are included and costs that are more difficult to identify, such as public costs, are omitted. *Gerald Emison*

Costs of Environmental Damage: the economic and social costs of environmental damage, usually divided into three broad categories: a) human capital costs (health consequences of environmental damage—sickness, premature death, etc.); b) productivity costs (reduced productivity of natural resources and physical capital, disruption of environmental services such as the natural cleansing of water, or the yield from fisheries, spending more time on cleaning and maintaining houses and other buildings); and c) loss of environmental quality, or amenity costs (a loss of a clear view, a pristine lake, a mature forest, and clean and quiet neighborhoods. *R. P.*

Discount Rate: the factor by which future benefits and costs are adjusted to determine their present value. The calculations of discount rates typically reflect the interest rates that would permit a given present value to appreciate in the future, and the preference of people for current spending over future gains (time preference). The present value of future benefits less costs in the typical BENEFIT-COST ANALYSIS are thus discounted by the annual discount rate r according to the formula *Present Value = Sum of Future Values/$(1 + r)^i$*, where i is the number of years into the future for each value. Discount rates have been interpreted as a measure of the myopia or selfishness of current generations, although its relationship to the interest rate also permits an interpretation that high discount rates simply reflect the high returns on investment.

Opportunity Cost. See under SECTION 1.1.4.

Shadow Price: the price attributed to a good or service to reflect its true societal value, generally conceived as the price that would prevail in a competitive, free-market economy. Shadow prices are widely used in project evaluation to determine economic returns; for assessing financial returns, actual prices cannot be ignored.

Shadow Projects: projects undertaken to offset environmental or resource degradation caused by other projects. Governments sometimes mandate shadow projects to allow firms to compensate for the damage or depletion of these other projects, thus in principle permitting economically rewarding activities without sacrificing environmental quality or the resource endowment. *R. P.*

Social Cost: the OPPORTUNITY COST to society of a given output or outcome, measured as the amount of compensation that would restore the utility levels of those who incur costs as a result of that output's production or other actions. Social cost is a crucial element of BENEFIT-COST ANALYSIS and PROJECT EVALUATION. *R. P.*

Social Cost-Benefit Analysis. See BENEFIT-COST ANALYSIS.

5.1.3 ECOSYSTEM MANAGEMENT AND LAND-USE PLANNING

Adaptive Management: management practices that facilitate continual adjustments to changing conditions and to lessons of success and failure. Adaptive management is accomplished through performance feedback and built-in mechanisms for reassessing and changing policies and practices. Adaptive environmental and natural resource management is particularly important when there are uncertainties about ecosystem behavior and risks of irreversibility. *R. P.*

COMMENTARIES ON ADAPTIVE MANAGEMENT

Perhaps the most important feature of adaptive management is that it emphasizes learning-by-doing, and takes the view that resource management policies can be treated as experiments from which managers can learn. Adaptive management rejects the view that managers can control and predict yields. Instead, it makes the assumption that ecosystems are inherently unpredictable, uncontrollable, and nonlinear. As an ecosystem management approach that rejects the expert-knows-best positivism of modernist science, adaptive management opens the door to epistemological pluralism. See also RESILIENCE and INDIGENOUS KNOWLEDGE.

See C. S. Holling, *Adaptive Environmental Assessment and Management* (New York: Wiley, 1978); Fikret Berkes and Carl Folke, eds., *Linking Social and Ecological Systems* (Cambridge: Cambridge University Press, 1998). *Fikret Berkes*

This is what we do anyway—create policies to address perceived problems, see how the policies work out over time, modify them in response to changing natural and social conditions. The trick is to shorten the amplitude (i.e., the scale of the policies) and frequency (i.e., the time intervals between policy modifications). Accomplishing this trick requires a change in mindset about environmental policies and science. Policies are not solutions to problems, they are experiments that will always yield surprises. Science is not a crystal ball for dictating policy by predicting the future of nature, but a tool for assessing the state of nature and the real-time impacts of policy decisions. Adopting this mindset requires that policymakers and scientists embrace uncertainty rather than try to overcome it. But the cultures of politics and science are not conducive to the requisite humility. *Daniel Sarewitz*

Ecosystem Management: a comprehensive approach to the conservation and management of various ecosystems (forests, wetlands, marine ecosystems, etc.) aimed at addressing the ecosystem as a whole and to promote multiple resource values. Ecosystem management combines natural science knowledge about the resource in question, human social factors, and consideration of the political, cultural, institutional, and organization capacities to forge a new approach for the conservation of resources. It should be applied in space and time at a variety of levels. At the macro level, it involves maintenance of a regional balance between resource protection and resource utilization to maintain biodiversity, productivity, and ecological functionality. At the micro level, it involves detailed management planning and control to maintain structural and biotic diversity through time and site productivity, ecological functionality, and essential environmental protection service. *T. S.*

Ecosystem management is terrific in theory, but in practice is often a pretext for top-down, technocratic resource and environmental management. The implication that good ecosystem management requires both an overarching appreciation of the entire ecosystem and the capacity to manage the ecosystem as a unit may greatly exaggerate the demand for system-wide information and technical analysis, elevating the power of technical analysts over the STAKEHOLDERS and often delaying immediate coping actions in favor of further information gathering and technical analysis. The idea that the whole ecosystem must be managed as a unit can also strengthen the hand of central authorities insofar as they can claim that the entire ecosystem is their responsibility, undermining the prerogatives of local stakeholders and authorities. Top-down decisionmaking may increase the likelihood that policymakers will choose single, "one approach for all areas" approaches, thus reducing the opportunities for local experimentation and responsiveness to local preferences. *William Ascher*

Habitat Management Plan (Habitat Conservation Plans [HCPs]): a formal plan for protecting terrestrial or marine habitat essential to the survival of an ENDANGERED SPECIES. The controversies over habitat management plans typically center on how stringent they should be. For example, the U.S. Endangered Species Act strictly regulates activities on private, as well as public, land that would negatively impact listed species or their habitats. This has led in some areas to great difficulties for private landowners, who find many economic uses foreclosed. Therefore, some of the resulting habitat management plans have been developed as a compromise solution, allowing some development (e.g., urban or agricultural uses) on a portion of a species' range, provided a suitable core area is set aside for permanent habitat protection. *R. G. H.*

COMMENTARY ON HABITAT MANAGEMENT PLANS
(HABITAT CONSERVATION PLANS)

More appropriately called Habitat Conservation Plans (HCPs), these plans have come under criticism for compromising the integrity of habitat protection under the Endangered Species Act of 1972. In 1982 HCPs were added to help balance property rights and land protection for rare and endangered species. HCPs allow private landowners to use their land and reduce the stock of an endangered species and habitat in exchange for conservation commitments. The landowner receives immunity from the ESA for a negotiated time period that can be up to one hundred years. Between 1982 and 1992 only fourteen HCPs had been approved between private landowners and the U.S. Fish and Wildlife Service. When the Clinton administration entered office, HCPs began to be used more widely. Today there are 211 in operation and another 200 being planned. Critics say that HCPs are often a better deal for developers or landowners than the species and give landowners a license to kill endangered species and destroy habitat. Proponents of HCPs state that without the plans, landowners have no incentive to conserve the species or the habitat. *Toddi Steelman*

Key Area Approach (Enclave Strategy): the strategy of protecting distinctive habitats and biological diversity by protecting a limited, "key" area where the habitat is most distinctive or the BIODIVERSITY densest. This is the principle of habitat protection and biodiversity conservation that has been traditionally applied by most developed nations. It is based on the assumption that conservation can be achieved by protecting certain areas, rather than on a country-wide or production-related basis. Many international and national conservation organizations have criticized this approach as being inadequate in addressing broader interactions.

Land Use Planning: the management and allocation of land uses, based on systematic assessment of land use values and options, often conducted through formal land use plans and regulations. Land use planning spans a range of actions that may be undertaken by private and public owners of land, in both rural and urban contexts. Individual private property owners may wish to plan how best to allocate their land for mixed uses, such as agriculture and forestry. Local governments may plan to develop a vision for the growth of their town or city. National agencies can use planning activities to allocate the many competing land uses on publicly owned lands. Regulations can be enacted to help implement land use planning. ZONING is the most commonly used form of land use regulation. *T. S.*

COMMENTARY ON LAND USE PLANNING

The manner and pattern in which land is used can substantially impact environmental quality. There is a growing recognition of the air- and water-quality impacts of the use of land. The pattern of land use in an area can induce polluting behavior. Traditional suburban single-family detached dwelling units have been shown to produce substantial degradation in water quality through nutrient pollution. Highly dispersed residential and commercial activities have been associated with increased use of cars, leading to ozone and carbon monoxide air pollution. *Gerald Emison*

Multiple Resource Use Management: resource management that takes into account all relevant resources; in U.S. environmental legislation, the management of public lands and their various resources to meet the demands for multiple uses. Therefore, multiple resource use management requires the balancing of competing land uses, including recreation, range, timber, minerals, watershed, wildlife and fish, and natural scenic, scientific, and historical values. U.S. environmental legislation calls for multiple resource use management to find the combination that will meet the present and future needs of the American people, but in practice the determination of a workable (and politically acceptable) balance depends on frequent and often heated interactions between U.S. resource agencies and the various public and interest groups.

See U.S. Government, Federal Land Policy and Management Act (103–43 USC1702).

5.1.4 ENVIRONMENTAL IMPACT ASSESSMENT AND TOOLS

Comparative Risk: the notion that risk regulation decisions should explicitly consider the risk implications of all decision alternatives. For example, when government considers the question of whether to require separate seats and the use of seatbelts (and therefore the purchase of a ticket) for small children in airplanes, it must consider not only the risk benefits for children riding on planes, but also the costs associated with those children whose families will choose alternative, and less safe, forms of transportation in the face of such a regulation. *D. S.*

Convention on Environmental Impact Assessment in a Transboundary Context (Espoo Convention): an international agreement adopted in 1991 and in force since 10 September 1997. The convention is of regional scope (open to European and North American countries) and as of mid-1999 had twenty-seven member states and the EUROPEAN UNION. The agreement seeks to encourage international cooperation in evaluating environmental impacts and preventing significant adverse impacts, especially those that cross national boundaries. It requires notification and consultation among states and encourages public participation for major projects under consideration that are likely to cause significant adverse transboundary impacts. It also encourages the development of consistent national policies and practices for preventing, mitigating, and monitoring significant adverse environmental impacts. The effectiveness of the Espoo Convention still remains to be seen.

Cumulative Environmental Impact Assessment. See ENVIRONMENTAL IMPACT ASSESSMENT.

Environmental Audit: a systematic analysis of the measurable impacts of a project, program, or other action, in comparison with preproject conditions and earlier predicted impacts. The environmental audit measures the relative accuracy of the prediction of impacts and tries to assess the causes of departure of actual from predicted results. By identifying unanticipated impacts, the environmental audit contributes to their management through mitigation and compensation. Environmental audits are being developed to monitor the environmental performance of corporations, alongside the customary audit of financial performance. Such audits are being introduced as a response to the demand by the public and shareholders for corporate responsibility toward the environment.

See Mostafa K. Tolba and Osama A. El-Kholy, eds., *World Environment, 1972–1992: Two Decades of Challenge* (London: Chapman and Hall,1992), 640. *R. P.*

Environmental Impact Assessment (EIA): an ex ante evaluation of the likely environmental impacts of a project, program, procedure, or regulation on the biological and geophysical environment and human health and well-being. Environmental impact assessments are particularly prevalent and important for large-scale construction projects such as roads and dams. In many jurisdictions, environmental impact statements are required before the action can obtain government approval. Since the 1970s, environmental impact assessments have increasingly included socioeconomic impacts. The challenge of environmental impact assessment is to identify and forecast relevant outcomes, some being very indirect consequences of the proposed action. To be fully effective, environmental impact assessment must be integrated into project or program evaluation that determines whether to proceed and how to achieve the best project or program design. *R. P.*

COMMENTARIES ON ENVIRONMENTAL IMPACT ASSESSMENT

Underlying the use of EIAs is the belief that full disclosure of consequences will lead to improved decisions. Inherent in the use of EIAs is the assumption that a community has a convergence of values concerning environmental protection. If these impacts are clearly articulated, this reasoning assumes the proper decision will obtain. At its heart EIA depends on procedural rationality. *Gerald Emison*

Environmental impact assessments are used as tools to predict the consequences of proposed action in over half the nations of the world. The use of EIAs began formally in the United States with the passage of the National Environmental Policy Act in 1969. In most cases, EIAs have had less influence than intended. EIAs are often criticized on the grounds that they are crafted to ratify preexisting decisions or that the information gathered is provided too late in the decision cycle to make a difference. Consequently, the existence of an EIA does not mean that it has had a substantive impact on the decision to move forward with a proposed project. *Toddi Steelman*

There has been a marked improvement in the seriousness and sophistication of environmental impact assessments over the years, not only because of methodological improvements, but also because both governments and infrastructure companies have grown to appreciate the risks of political backlash, lawsuits, and tarnished reputations that can arise out of the failure to anticipate environmental consequences. *William Ascher*

Environmental Impact Statement: a document prepared by an agency or firm on the environmental consequences of its projects and programs. While an environmental impact statement could report on the environmental impact assessments for projects and programs in development, more typically it is confined to assessments of impacts that have already occurred. Many environmental impact statements have been criticized for not adequately tracing the secondary consequences of impacts, and thus underestimating environmental damage.

Environmental Risk: a measure of the probability that damage to life, health, property, and/or the environment will occur as a result of a particular hazard. The process by which environmental, health, and safety risks are measured or quantified is called risk assessment. In practice, risk assessment involves a series of measurements, judgments, and decisions resulting in a quantitative expression of the magnitude of the risk posed by a substance or action. Risk assessments are controversial because of disputes over the metric in which they are expressed, the quantifiability of risk, and the assumptions that underlie the analysis. Nevertheless, risk assessments are commonly used by environmental agencies and other policymakers in reaching policy decisions. Related concepts include: a) risk-tradeoff analysis (RTA) and risk-risk analysis (RRA) evaluate environmental decisions by highlighting the risks that can be created by an activity intended to reduce risk—for instance, disinfection of water with chlorine might create a cancer risk while failure to disinfect creates risk of microbial contamination; b) risk aversion is the tendency of some people and organizations to avoid risk even if the possible gains have the same utility as the possible losses, adjusting for the likelihood of each outcome. That is, someone who is risk averse may not be willing to risk losing $100 on an even bet with which he or she could win $200. In addition, many people express a greater willingness to pay to avoid new risks than to remove otherwise identical existing risks; and c) risk management is the process by which organizations balance, avoid, and minimize environmental risk. For government, it refers to how the government balances (and otherwise uses) risks in its policy decisions. In the context of business decision making, it refers to how businesses seek to minimize risk or to balance risk with other goals.

See John D. Graham and Jonathan B. Wiener, eds., *Risk versus Risk: Tradeoffs in Protecting Health and Environment* (Cambridge, MA: Harvard University Press, 1995); W. Kip Viscusi, ed., "Risk-Risk Analysis Symposium," *Journal of Risk and Uncertainty* 8 (1994): 5–122. *D. S.*

Full Cost Accounting (FCA): a management tool used to identify, quantify, and allocate the direct and indirect environmental costs of ongoing operations. FCA identifies and quantifies environmental performance costs for a product, process, or project. FCA considers four levels of costs: a) direct costs such as labor, capital, and raw materials; b) hidden costs such as monitoring and reporting; c) contingent liability costs such as for fines and remedial action; and d) less tangible costs such as public relations and good will. The use of FCA goes beyond the earlier practice of focusing solely on costs that directly affect a company's bottom line.

See Michael A. Berry and Dennis A. Rondinelli, "Proactive Environmental Management: A New Industrial Revolution," *Academy of Management Executives* 12 (2) (1998): 38–50. *D. A. R.*

COMMENTARY ON FULL COST ACCOUNTING

The emergence of full cost accounting is beginning to reshape the concept of environmental accounting and making it essential to business success. Firms find sound business reasons to account for the full costs of environmental performance. First, many environmental costs can be eliminated by simply changing operational and housekeeping practices. Second, environmental costs in the form of wasted raw material add no value to a process or product and under most circumstances constitute potential cost savings. Third, understanding the environmental costs and performance of processes and products leads to more accurate pricing and value of goods and services. Under traditional accounting systems, these potential savings and business opportunities may be obscured in overhead accounts and otherwise overlooked. *Dennis A. Rondinelli*

Natural Resource Damage Assessments: in the U.S. context, damage assessments undertaken to fulfill the legal right of the trustees of natural resources (local governments, state governments, and Native American nations) to collect damages from firms or individuals who release hazardous substances that damage or destroy environmental resources. Under the Comprehensive Environmental Responses, Compensation, and Liability Act of 1980 (CERCLA), the U.S. federal government established guidelines for the methodologies that have legal standing in court through which trustees may make claims for damages. Both the TRAVEL COST METHOD and the CONTINGENT VALUATION METHOD have attained this status, and the use of these methodologies may not be challenged by the parties to the dispute. The case of the *Exxon Valdez* oil spill is one of the most notable examples in which these techniques were used, although the related court case was eventually settled out of court. Nevertheless, the National Oceanic and Atmospheric Administration (NOAA) commissioned a "blue ribbon" panel of experts to investigate contingent valuation and make recommendations as to the best ways it could be employed in assessing the monetary value of damage from oil spills and other compensable environmental accidents. These findings have been published in the *Federal Register.*

See James R. Kahn, *The Economic Approach to Environmental and Natural Resources* (New York: Dryden Press, 1995). *R. A. K.*

5.1.5 INFORMATION AND MONITORING

Earthwatch: a program of global environmental assessment performed on a continuing basis by different United Nations agencies and coordinated by the UNITED NATIONS ENVIRONMENT PROGRAMME (UNEP). Its main objectives are to: a) review and evaluate environmental conditions to identify gaps in knowledge and need for action; b) research environmental problems of potential global and transboundary importance; c) monitor certain environmental variables; and d) facilitate exchange of information among governments and scientists.

The monitoring function of Earthwatch is provided through the GLOBAL ENVIRONMENTAL MONITORING SYSTEM (GEMS) while the INTERNATIONAL REFERRAL SYSTEM FOR SOURCES OF ENVIRONMENTAL INFORMATION (INFOTERRA) and the INTERNATIONAL REGISTER OF POTENTIALLY TOXIC CHEMICALS (IRPTC) serve as clearinghouses for environmental information.

Global Environmental Monitoring System (GEMS): a part of the EARTHWATCH program created in 1975 by the UNITED NATIONS ENVIRONMENT PROGRAMME (UNEP) as a collaborative effort of the United Nations system to monitor and assess the world environment. Its objectives are to strengthen monitoring and assessment capabilities in participating countries, to increase the validity of environmental data and information, to produce global and regional assessments in selected fields, and to compile environmental information at the global level.

Global Monitoring System: a set of environmental observation systems providing monitoring of different elements of the global environment. It includes the World Weather Watch (in operation since 1963), Integrated Global Observing System (founded jointly by the WORLD METEOROLOGICAL ORGANIZATION [WMO] and the INTERGOVERNMENTAL OCEANOGRAPHIC COMMISSION in 1967), the GLOBAL ENVIRONMENTAL MONITOR-

ING SYSTEM (GEMS) (1975), and the Global Atmospheric Watch (established by WMO in 1989). In the 1980s it became obvious that these systems, each designed for a specific purpose, were not adequate to monitor the earth system as a whole for the purposes of global change research. Therefore the INTERNATIONAL COUNCIL OF SCIENTIFIC UNIONS (ICSU) and some UN organizations designed three additional interconnected global observation systems (the Global Climate Observing System, the Global Ocean Observing System, and the Global Terrestrial Observing System). These systems have been in the planning stage since 1992; however, because of inadequate funding they have not yet been implemented.

International Referral System for Sources of Environmental Information (INFOTERRA): one of the most comprehensive international mechanisms for the exchange of environmental information within and among nations. It was established in 1975 by the UNITED NATIONS ENVIRONMENT PROGRAMME (UNEP) and together with the GLOBAL ENVIRONMENT MONITORING SYSTEM (GEMS), GLOBAL RESOURCE INFORMATION DATABASE (GRID), and the INTERNATIONAL REGISTER OF POTENTIALLY TOXIC CHEMICALS (IRPTC) assists UNEP's mandate to monitor, assess, and disseminate information on the environment. INFOTERRA is a decentralized system operating through a worldwide network of national environmental institutions designated and supported by their governments, coordinated at UNEP headquarters. INFOTERRA consists of 173 National Focal Points (NFPs), located in nearly every United Nations member country, linking approximately 8,000 national and international institutions and experts from various organizations in the UN system, NONGOVERNMENTAL ORGANIZATIONS (NGOS), businesses, governments, and academia.

National Reports on Sustainable Development: documents produced by a few countries in the aftermath of the Rio Conference (UNITED NATIONS CONFERENCE ON ENVIRONMENT AND DEVELOPMENT [UNCED]; Earth Summit) that reflected a national vision of this concept and recommendations for the implementation of sustainability. For instance, the U.S. President's Council on Sustainable Development issued a report entitled *Sustainable America: A New Consensus for Prosperity, Opportunity, and Healthy Environment for the Future.* The report includes a vision statement and fundamental beliefs on sustainability, indicates changes at all levels of decision making (government, business, and community) that should occur to achieve SUSTAINABLE DEVELOPMENT, and provides a wide range of recommendations and actions with which to implement them. The basic theme of the report is that economic, environmental, and social equity issues are intertwined and should be addressed together.

See U.S. President's Council on Sustainable Development, *Sustainable America: A New Consensus for Prosperity, Opportunity, and Healthy Environment for the Future* (Washington, DC: PCSD, 1996).

State-of-the-Environment Reports (SERS): government reports published by most countries on an annual basis to provide up-to-date information on the quality of the environment and natural resources. Typically, these reports are organized around a few main topics: an inventory of natural resources and environmental services, an assessment of their state and quality, a baseline against which to compare dynamics, and monitoring over time. Some SERS outline action programs for addressing environmental problems. The earliest SERS were produced in the 1970s and first provided rather limited information on environmental quality. Moreover, in the former Soviet Union and other former socialist countries, SERS were considered classified information. Currently, the most elaborate SERS are produced in Canada, Germany, the Netherlands, Japan, and the United States.

5.1.6 PLANNING

Capacity 21: an initiative launched in 1993 by the UNITED NATIONS DEVELOPMENT PROGRAMME (UNEP) to assist developing countries in formulating SUSTAINABLE DEVELOPMENT strategies and to enhance national and local capacities to implement related programs. The initiative complements capacity-building programs of the GLOBAL ENVIRONMENT FACILITY (GEF) and the MONTREAL PROTOCOL ON SUBSTANCES THAT DEPLETE THE OZONE LAYER's Multilateral Fund, all aimed at building local capacities of developing countries to implement AGENDA 21. Capacity 21 assists recipient countries to design sustainable development policies, adapt appropriate technologies, and develop mechanisms for public participation in decision making. To become a recipient of Capacity 21 assistance, a government must demonstrate a strong commitment to Agenda 21 principles, develop a national policy framework, and provide assurances that program beneficiaries and STAKEHOLDERS would be involved at appropriate stages.

Environmental Action Programme for Central and Eastern Europe (EAP): a set of recommendations developed following the first "Environment for Europe" Conference (Dobris, Czechoslovakia, 1991) with the main objective of addressing the most urgent environmental problems in Central and Eastern European countries in a cost-effective and efficient manner. The EAP recommended that reducing threats to human health be one of the most important criteria for setting priorities for action at the international, national, regional, and local levels. The program's recommendations included a mix of policy tools (market-based instruments along with realistic regulations and enforcement mechanisms), institutional strengthening, and carefully selected and well-prepared investments. The EAP's distinct feature was its new (at that time) approach to environmental management—emphasizing the need to build on the positive linkages between economic development and the environment; policies to tackle the underlying causes rather than the symptoms of environmental problems; and the use of cost-effectiveness as a criterion in allocating financial resources. The EAP argued that urgent actions should be taken in the "window of opportunity" following the collapse of the old regimes, when environmental considerations could be factored into the process of economic reconstruction at least cost. As of 2000, none of the Central and Eastern European governments were reporting success in seizing this window of opportunity.

Green Planning: a term originally applied to plans produced (mainly in developed countries) to address escalating environmental problems. Formally this term was first introduced in 1989 when the Canadian Environment Ministry prepared, under strong public and political pressure, a Green Plan for Canada. Recently, the green planning terminology has been used in a broader context to embrace a series of SUSTAINABLE DEVELOPMENT initiatives. In practice, most green plans developed in the world still remain centered on environmental issues.

International Network of Green Planners (INGP): an organization founded in 1992 by a group of experts from governments and international agencies engaged in strategic environmental management. It now has over 200 members and is coordinated by a secretariat headquartered in the Dutch Ministry of Environment. The main objective of INGP is to foster information exchange and learning among strategy practitioners and green planners from developed and developing countries.

National Environmental Action Plans (NEAPs): mechanisms of environmental policy that were elaborated in many countries over the last fifteen years. An NEAP usually describes a country's major environmental concerns, identifies the principal causes of problems, and formulates policies and actions to deal with the problems. In addition, when environmental information is lacking, the NEAP should identify priority environmental information needs and indicate how essential data and related information systems will be developed. It provides the preparation work for integrating environmental considerations into a country's overall economic and social development strategy. Recently, the elaboration of NEAPs has become a condition of the WORLD BANK's lending. See also STATE-OF-THE-ENVIRONMENT REPORTS.

National Environmental Policy Act of 1969 (NEPA): a statute, Title 42 of the U.S. Code, Ch. 55, that requires U.S. federal agencies to consider, on a routine basis, the environmental implications of the actions they propose to undertake or permit. For certain categories of major actions, that review is required to take the form of an ENVIRONMENTAL IMPACT STATEMENT. *D. S.*

COMMENTARY ON NATIONAL ENVIRONMENTAL
POLICY ACT OF 1969

While standards of monitoring and implementation vary, NEPA has directly inspired similar legislation in at least forty-one countries around the world from Australia to Venezuela.

See Nicholas Robinson, "EIA Abroad: The Comparative and Transnational Experience," in *Environmental Analysis: The NEPA Experience*, eds. Stephen G. Hildebrand and Johnnie B. Cannon (Boca Raton, FL: Lewis Publishers, 1993). *Steve Rayner*

National Sustainable Development Strategies (NSDSs): strategies developed (or under consideration) in a number of developed and developing nations after the 1992 UNITED NATIONS CONFERENCE ON ENVIRONMENT AND DEVELOPMENT (UNCED), to guide the implementation of AGENDA 21 at the national level. These strategies ideally "should build upon and harmonize the various sectoral economic, social and environmental policies and plans that are operating in the country" (chapter 8, Agenda 21, UNCED, 1992). Comparisons of existing strategies demonstrate different approaches applied by developing and developed countries. For instance, in developing countries most of the approaches followed a basic framework previously developed for NATIONAL CONSERVATION STRATEGIES, were promoted by donors who provided financial support and technical assistance, and were formulated by expatriate technical experts and advisers. NSDSs in developed countries are internally generated and funded, use local expertise, and are therefore extremely diverse.

See Barry Dalal-Clayton, *Getting to Grips with Green Plans* (London: Earthscan, 1996).

Policy Dialogue: structured but nonbinding "informal" consultation with STAKEHOLD-ERS. The policy dialogue can be seen as the ongoing process of clarifying objectives, exploring the possible outcomes of policy options, assessing the degree of consensus, and determining which groups hold which positions. Some technical organizations and advocacy groups view their role as pursuing a policy dialogue with the government, with the end of convincing the government to reform its policies. Organizers of policy dialogues often use decision-support systems (backcasting, simulations, and policy exercises) to assess the consequences of suggested policy options. One common approach to international environmental policy dialogue is to convene short (e.g., two–three-day) retreats for different stakeholders (government officials, academics, labor and business leaders, and their technical advisers), to provide them the opportunity to establish personal rapport, develop networks, and informally exchange ideas of different ways to address specific environmental issues.

5.1.7 VALUATION OF ENVIRONMENTAL SERVICES AND NATURAL RESOURCES

5.1.7.1 Value Concepts

Ability to Pay. See WILLINGNESS TO PAY.

Aesthetic Value of Nature: 1) the intrinsic value that people attribute to nature for its own sake; essentially equivalent to EXISTENCE VALUE. As such, the notion of the inherent value of the natural world was pivotal to the development of North American conservation theory and environmental ethics. It is reflected in the 1940 Convention on Nature Protection and Wildlife Preservation in the Western Hemisphere (WESTERN HEMISPHERE CONVENTION) and the 1972 Convention Concerning the Protection of the World Cultural and Natural Heritage (WORLD HERITAGE CONVENTION), which defined the world heritage to include "physical and biological formations . . . of outstanding universal value from the aesthetic . . . point of view." The aesthetic value of the natural world has been also recognized by the CONVENTION ON INTERNATIONAL TRADE IN ENDANGERED SPECIES (CITES), the CONVENTION ON BIOLOGICAL DIVERSITY (CBD), and a number of other international environmental agreements along with other mostly instrumental values, such as "economic" and "scientific."

2) the value of nature derived from the gratification of enjoying beauty, grandeur, or other aesthetic qualities of ecosystem components. As such, the aesthetic value is an amenity value to be enjoyed (and valued) by people according to their aesthetic preferences. These aesthetic values may or may not correspond to the "naturalness" of the ecosytems. For example, many people find human-shaped landscapes (such as the English countryside of hedgerows and small stands of woods) to be of greater aesthetic value than less manipulated landscapes.

Bequest Value: the value attributed to providing or preserving environmental benefits for future generations. Bequest values may be narrowly focused on the descendants of the particular individual, or more broadly on future generations. Bequest values may or may not accurately represent the interests and values of the future generations themselves, because of the uncertainty of knowing the preferences of these generations.

Consumptive Use Value: the value of resources derived directly from the products and services that use up the resources. Consumptive use value is distinguished from both EXISTENCE VALUE and other nondepleting enjoyment of resources (e.g., appreciation of beauty).

Costs of Environmental Damage. See under SECTION 5.1.2.

Environmental Costs: costs connected with the actual or potential deterioration of natural assets and the quality of environmental services due to economic activities or human habitation. The actors who cause environmental costs are not necessarily those who bear the costs. When environmental costs are not borne by those who create them, the costs are considered to be negative EXTERNALITIES.

Environmental Values: the values that people attribute to various aspects of the ecosystem and environmental services. Some environmental values, termed "direct use values," reflect tangible benefits, such as timber, fish, and tourism revenues. These direct use values can be further subdivided into CONSUMPTIVE VALUES, reflecting uses that reduce the availability of benefits (such as a factory using clean water in ways that reduce the water's purity), and nonconsumptive uses, such as the tourism revenues based on the enjoyment of beautiful landscapes. Yet there are also nonuse values, such as the BEQUEST VALUE that reflects the desire to leave ECOSYSTEMS intact or improved for future generations and EXISTENCE VALUES derived from the gratification of knowing that an environmental service or a component of the ecosystem is intact (for example, the willingness to pay so that Central American forests can have jaguars, even if one has no expectation to directly "experience" or benefit from the jaguar). Environmental aspects may also be valued for leaving open the option of benefiting in the future. Determining environmental values is greatly complicated by the fact that many such values do not involve tangible objects whose sale prices can directly indicate how much they are valued. Indirect pricing methods, such as HEDONIC PRICING, and survey methods such as the CONTINGENT VALUATION approach of asking people about their WILLINGNESS TO PAY to preserve the environment for the benefit of others, have been developed. However, these approaches remain controversial. There is also an important debate over whether environmental values, as expressions of human preference, can fully capture the standing of the ecosystem (or "nature") itself, inasmuch as the ecosystem may be accorded rights beyond the value that humans attribute to it. See also VALUATION. *R. P.*

COMMENTARY ON ENVIRONMENTAL VALUES

In the United States, environmental values have shifted over time. Up until the 1960s, environmental values focused primarily on the instrumental or direct use value of natural resources. Timber was meant to build homes. Water was intended to irrigate crops and turn turbines. Land was intended for cultivation. After the end of World War II, rising incomes, increased leisure, and a growing scarcity of resources slowly led to a shift in environmental values. Forests, water, and land tended to be viewed and valued more for their intrinsic worth than they were prior to the 1960s. Forests were valued for their habitat, water was used to restore ecosystems, and land was preserved as open

Existence Value: the value of a natural asset stemming exclusively from the knowledge that it exists rather than from any uses or benefits derived from it. It is therefore the label that is often attached to an individual's WILLINGNESS TO PAY to ensure that flora, fauna, and their habitats will continue to exist even though the individual may believe that he or she will never personally use those resources. The existence value would not include the interest in having the element available for one's descendants, friends, relatives, or humanity—these are benefits that are captured instead by the concept of BEQUEST VALUE. Because the existence value is not typically engaged in economic transactions, the CONTINGENT VALUATION METHOD is often applied for measuring it.

Intrinsic Value: the value, or "worth," of nature itself, regardless of whether it serves as an instrument for satisfying individuals' needs and preferences. For instance, many arguments for preserving BIODIVERSITY are based on the moral premise that all living entities have a fundamental intrinsic worth and therefore should be "saved" from extinction. There are some problems with this approach: intrinsic value cannot exist without a conscious valuer; attempts by humans to take intrinsic values into account are constrained by human preferences rather than the absolute merit of nature. The value that individuals derive from just knowing that ecosystems are preserved "in their own right" is usually described as EXISTENCE VALUE. See also ENVIRONMENTAL VALUES. *R. P.*

Instrumental Value: value that is exclusively attributed to an object of environmental protection on the basis of its significance to human beings. Instrumental values may be further subdivided into economic and noneconomic value categories. See also ENVIRONMENTAL VALUES.

Nonconsumptive Use Value. See ENVIRONMENTAL VALUES.

Option Value: the additional value of an asset, particularly a natural-resource endowment, that comes from the possibility that the asset has currently unknown benefits. The option value is therefore an individual's willingness to pay for the privilege of preserving assets (such as flora, fauna, and habitats), above and beyond their known benefits, to keep open the option of using or otherwise enjoying these assets in the future. *R. P.*

Resource Rent. See under SECTION 3.4.3.

Use Value: in economics, the total utility of a good as opposed to its sale value, the latter being set by the value of the marginal utility of exchanged units of the good rather than its total utility. In resource and environmental economics, the use value reflects the utility of the goods and services that a natural asset provides, as distinct from other components of value such as EXISTENCE VALUE and INTRINSIC VALUE. One problem in valuation is that the use value, commonly associated with the extraction of the resource, is often more apparent and more easily measured than other components of value, making it more dominant in the analysis.

Vicarious Benefit: the value that comes from knowing that others derive a use value from an asset. The concept of vicarious benefit is one aspect of VALUATION that goes beyond the selfish USE VALUE that an individual might enjoy. The BEQUEST VALUE is one component of vicarious benefit. *R. P.*

5.1.7.2 Valuation Techniques

Contingent Valuation Method (CVM): an approach to valuing public goods that relies on stated preferences; usually elicited by survey questions. Contingent valuation asks how much people would be willing to pay to get specified improvements or to avoid decrements in them (WILLINGNESS TO PAY), or how much they would require as compensation for doing without a particular good (WILLINGNESS TO ACCEPT). The aim of contingent valuation is to elicit valuations—or "bids"—that are close to those that would be revealed if an actual market existed. The process works in reverse if the aim is to elicit a willingness to accept (WTA): bids are systematically lowered until the respondent's minimum WTA is reached. Due to the hypothetical nature of contingent valuation questions, the method is susceptible to biases. Willingness to pay estimates are sometimes biased when respondents distort their responses because they believe that they are actually engaged in bargaining, or are eager to appear to be generous. *R. P.*

Direct Valuation. See VALUATION OF ENVIRONMENTAL SERVICES AND NATURAL RESOURCES.

Dose-Response Technique: 1) a method for estimating the impact of pollutant or other harmful substances by establishing the impacts (or "responses") of a given level (or "dosage").

2) a market approach to valuation of environmental resources that uses market prices for the environmental damage that the reduction of an environmental resource causes. The dose-response technique aims to establish a relationship between some cause, such as a given dose of pollution, and the ensuing environmental damage. When individuals are unaware of the impact of a change in environmental quality, then their economic behaviors or subjective valuations are inappropriate measures; therefore dose-response procedures that do not rely on individuals' preferences can be used. This technique is used extensively where dose-response relationships between some causes and output/impacts are known. For example, it is used to look at the effect of pollution on health, physical depreciation of material assets such as metal and buildings, aquatic ecosystems, vegetation, and soil erosion. However, the approach is mainly applicable to environmental changes that have impacts on marketable goods and therefore is not suitable for valuing nonuse benefits. *R. P.*

Energy Valuation: an approach to determining the costs of a program by determining the volume of energy involved. It attempts to replace monetary valuation, for instance, in accounting or project costing by energy values. The underlying hypothesis is that, in the final analysis, all goods and services are generated by solar energy. The motivation for this theory is that direct substitution between capital and energy inputs in production does not adequately indicate how much energy is saved. This is because energy and other inputs are not independent, inasmuch as additional capital or labor in turn requires additional energy. See also ECOLOGICAL ENERGETICS PARADIGM; ECOLOGICAL ECONOMICS.

See Robert Costanza, "Embodied Energy, Energy Analysis and Economics," in *Energy, Economics and Environment*, eds. Herman E. Daly and Alvaro F. Umaña (Boulder, CO: Westview Press, 1981).

COMMENTARY ON ENERGY VALUATION

One can see interesting analogies between energy valuation and the Marxist labor theory of value, as well as with Bogdanov's *Ecological Economics* and Stanchinsky's *Ecological Energetics*. Energy valuation tries to free the analysis from the strong dependence on financial, as opposed to physical, accounting. There is no indication, however, that the U.S.-based proponents of energy theory of valuation were aware of the efforts of their Russian predecessors. *Natalia Mirovitskaya*

Hedonic Pricing: a form of "revealed preference" valuation that uses market-based prices to infer the values of nonpriced goods and services. For example, the selling prices of comparable homes with and without a scenic view can be compared to determine the value of the scenic view; wages of workers exposed to an occupational risk can be compared to wages of those who are not exposed to that risk. Hedonic pricing estimates an implicit price for environmental attributes by looking at real markets in which those characteristics are effectively traded. Thus, "clean air" and "peace and quiet" are effectively traded in the property market since purchasers of houses and land do consider these environmental dimensions as characteristics of property. The attribute "risk" is traded in the labor market. High-risk jobs may well have a "risk premium" in the wages to compensate for the risk. Great care must be exercised to determine that the cases compared are truly comparable. *R. P.*

Household Production Function: a component of an indirect approach to VALUATION OF ENVIRONMENTAL SERVICES AND NATURAL RESOURCES that places value on environmental resources by examining expenditures on commodities or services that are substitutes or complements for the environmental characteristic. Thus, noise insulation is a substitute for a reduction in noise at source; travel is a complement to the recreational experience at the recreation site (it is necessary to travel to experience the recreational benefit). Household production functions are used in the TRAVEL COST APPROACH and the HEDONIC PRICING APPROACH. *R. P.*

Maintenance Valuation (in Environmental Accounting): a method of measuring imputed environmental costs by determining the costs of maintaining the system at the same level. The value of maintenance cost depends on the choice of avoidance, restoration, and replacement or prevention activities. It can be directly measured only when the maintenance actually keeps the system at the same level of quality or integrity; otherwise the analyst must estimate the maintenance costs.

Replacement Cost Technique: a method for measuring the value of an asset by determining the cost of replacing or restoring it to its original state. The replacement cost technique is widely used for valuation because it is often easy to make such estimates; information on replacement costs can be obtained from direct observation of actual spending on restoring damaged assets or from professional estimates of what it costs to restore the asset. However, the approach is frequently problematic as an element in a cost-benefit analysis because it does not truly measure the benefits from the asset. In some circumstances, the costs of replacing or restoring an asset exceed the benefits of the asset; in others, the benefits may be greater than the replacement or restoration costs. The approach is useful for cases in which remedial work must take place because of some other constraint such as a water quality standard or an overall constraint not to let environmental quality decline (sometimes called a "sustainability constraint"). In these circumstances replacement costs might be allowable as a first approximation of benefits or damage. *R. P.*

Surrogate Market: a market for private goods and services that reflects nonpriced environmental services. The goods or services bought and sold in these surrogate markets permit inferences regarding how much environmental services are valued. For example, the housing market, insofar as it permits comparisons of prices of dwellings that are equivalent except for differences in location, can reveal the higher prices that people are willing to spend for dwellings in areas with better air quality or greater tree cover. These techniques are sometimes preferred by policymakers because they rely on actual choices rather than the hypothetical choices involved in the DIRECT VALUATION approaches such as CONTINGENT VALUATION. Surrogate market approaches include HEDONIC PRICING and HOUSEHOLD PRODUCTION FUNCTION techniques. The latter includes the TRAVEL COST METHOD.

Travel Cost Method (TCM): an indirect valuation method that uses the visitors' travel costs to recreational areas or other attractions as a proxy for the value of the recreational activity or attraction. The visitation rate expresses the amount of the recreational experience "bought" by these expenditures. Variations in travel costs and visitation rates can be used to estimate a demand curve, which is used to compute the consumer surplus of the activity, i.e., the recreational value. In contrast to the methods that rest on responses to hypothetical questions, such as the CONTINGENT VALUATION APPROACH, the travel cost method is based on revealed preferences. *R. P.*

Valuation of Environmental Services and Natural Resources: the calculation of the utility of resources in their current state or under alternative scenarios of use. Valuation permits the costs and benefits of environmental and resource aspects to be incorporated into benefit-cost analysis and project evaluation. Valuation is typically quantitative, and assessed in monetary terms in order to permit the comparison of costs and benefits across different categories. Valuation is unavoidably tied to human values (i.e., people attribute value to the environment and natural resources, whether through the amounts they pay or by expressing their willingness to pay). Therefore, valuation cannot address the merit or standing of nature apart from the utility that people express, whether for their own benefit (USE VALUE), the gratification of knowing that various elements of the ecosystem persist (EXISTENCE VALUE), or out of an ethical sense of stewardship. Methods of valuation range from the direct measurement of the prices of resources or environmental services, to indirect inferences of the value of nonpriced resources and services. The most contentious debate over valuation is whether the less tangible, less "economic" elements of ecosystems can be captured to the same degree as the more obvious economic elements. See CONTINGENT VALUATION; HEDONIC PRICING; SURROGATE MARKETS; TRAVEL COST METHOD.

If one believes that the environment is worthy of protection for its own sake, then valuation theory and practice is potentially both inappropriate and absurd. Consider, for comparison, the application of valuation methodology not just to the environment, but to a broad range of other moral dilemmas facing society, such as gender equity, abortion, affirmative action, ethnic tension, eugenics, capital punishment, euthanasia, and limits of freedom of expression. Do we want our most fundamental moral dilemmas to be redefined as methodological problems? Should we, as a further step, replace Supreme Court justices with economists? *Daniel Sarewitz*

Willingness to Accept (WTA): the amount of compensation that an individual would accept in exchange for forgoing a good or service or suffering from a particular cost. As a measure of value of non-priced goods and services, the willingness to accept is often used in conjunction with the WILLINGNESS TO PAY. Typically, the willingness to accept is significantly greater than the willingness to pay, because the willingness to pay is constrained by the individual's ability to pay. The explanation may be that people are more averse to losses than to gaining benefits of equal utility, and people see only limited substitutes for forgone goods and services such as environmental quality.

See David W. Pearce and Dominic Moran, *The Economic Value of Biodiversity* (London: Earthscan, 1994). *R. P.*

Willingness to Pay: the amount that an individual would pay for a good or service *if* that individual's payment could ensure that the good or service would be provided. The willingness to pay could be revealed by a host of economic decisions, including the actual payment for a good or service or indirectly through the payment for another asset that is associated with that good or service. The willingness to pay concept permits the valuation of nonmarket entities (such as a national park or a reduction in pollution) by determining the price that people report they would pay or forgo. The method of CONTINGENT VALUATION depends crucially on whether the expressions of willingness to pay are accurate reflections of the value that respondents attribute to particular goods and services. One virtue of the willingness to pay approach is that it provides a monetary value of the good or service; limitations include the possibility that people asked to express such willingness may deliberately or inadvertently distort their true willingness.

See David W. Pearce and Kerry Turner, *Economics of Natural Resources and the Environment* (Baltimore: Johns Hopkins University Press, 1990), 125. *R. P.*

COMMENTARY ON WILLINGNESS TO PAY

"Willingness to pay" is not a "concept"; it is a symptom of the psychosis induced by parsing the world into internalities and externalities. In the developing world, for example, willingness to pay must surely be buffered by ability to pay; one can reasonably imagine that citizens of Laos are willing to pay less to preserve wilderness than citizens of the United States, yet who would like to argue that wilderness is therefore more valuable for rich people than for poor people? Willingness to pay techniques "commodify" nature and thus undermine the idea of intrinsic value. They also imply that nature is in fact worth nothing to those who would pay nothing for it. One's valuation of nature, however, may be unrelated to one's actual dependence upon it. *Daniel Sarewitz*

5.2 Environmental Policy Instruments and Principles

Environmental Policy: government policy that explicitly intends to promote environmental protection, conservation, and rational use of natural resources. However, many policies that are not specifically directed at these objectives still have major impacts on the environment, conservation, and resources. Environmental policy includes regulations to prohibit or limit pollution and resource depletion; incentive policies (including tax measures) to encourage environmental improvements to discourage pollution and depletion; and direct government efforts to clean up, protect, or restore ecosystems.

COMMENTARY ON ENVIRONMENTAL POLICY

Environmental policy is fundamentally about two questions: 1) How are we going to use the environment, including consumptive (see CONSUMPTIVE USE VALUE) (e.g., oil) and nonconsumptive (e.g., endangered species) uses (see ENVIRONMENTAL VALUES), and 2) who gets to decide, or who has authority and control? These two questions may not be addressed explicitly or systematically by policy participants. There are many cases where the second question is of more interest to policy actors than the first. Environmental policy can be conceived as a process with a beginning and an end, even though initiation or termination may not be clear or sharp. Throughout this process, information must be gathered for planning. It must be openly debated for its adequacy and meaning. This is often followed by the setting of rules or guidelines about how the environment will be used. These are then implemented, followed by resolution of disputes that may arise as the rules are fully implemented. Policy may be appraised or monitored for its effectiveness. And finally the policy will be ended because the policy problem has been resolved or another approach is taken. This sequence of activities or decision process characterizes all environmental policy processes.

See Tim W. Clark and Ronald D. Brunner, "Making Partnerships Work in Endangered Species Conservation: An Introduction to the Decision Process," *Endangered Species Update* 13, 9 (1996): 1–4. *Tim W. Clark*

Environmental Policy Instruments: the specific government measures that constitute an environmental policy often addressed to both pollution problems and conservation. ENVIRONMENTAL REGULATIONS that impose negative sanctions on pollution or resource depletion have been challenged in the past two decades by the development of ECONOMIC (OR MARKET-BASED) INSTRUMENTS that steer economic agents' behavior in environmentally beneficial directions by reshaping the benefits and costs of alternative actions.

Environmental Protection. See under SECTION 1.2.4.

5.2.1 DIRECT ENVIRONMENTAL REGULATIONS AND POLITICS

Abatement: a decrease in the level of a nuisance, such as pollution or congestion; or the act of decreasing a nuisance. Pollution abatement in industry, for example, may occur in response to regulation, consumer demand for cleaner processes, or the diffusion of less pollution-intensive technology. Costs incurred by reducing (in intensity or absolute level) a nuisance such as pollution or congestion are called abatement costs. The cost per unit of abatement usually increases as the level of the nuisance approaches zero. Therefore, it is assumed that abatement costs are higher in developed countries. *R. P.*

Allowance: a license or permit to emit a given amount of pollution, with essentially the same meaning as "marketable emissions permit," "emissions reduction credit," and "tradable discharge permit." An allowance is a limited "use right" that in some systems is tradable. The allocation of allowances is typically done after the relevant environmental agency determines the maximum amount ("cap") for that type of pollution (e.g., sulfur dioxide). Yet often the cap reflects the existing level of pollution, inasmuch as the polluters may be judged to have had prior property rights. The trade of allowances contributes to economic efficiency by allowing firms with greater production possibilities to buy the allowances from firms with lesser expectations about production. *R. P.*

Best Available Technology. See under SECTION 5.3.1.

Command and Control Approach: the general approach to environmental policy that relies on direct government specification of both the means and the standards of environmental protection. This approach is embedded in ENVIRONMENTAL POLICIES in most countries. The basic premise behind the command and control approach is that the government has the capability both to formulate the optimal regulations and to enforce them efficiently. See also DIRECT (ENVIRONMENTAL) REGULATIONS.

COMMENTARIES ON COMMAND AND CONTROL

In the United States, the federal government has steadily increased its enforcement of environmental regulations making business executives and owners liable for environmental pollution. The U.S. Environmental Protection Agency (EPA) takes hundreds of enforcement actions against businesses every year leading to prison sentences and heavy fines. But regulatory enforcement is expensive for both government and business. The total costs of complying with environmental laws over the past twenty-five years have easily exceeded $1 trillion. About $120 billion is spent annually for pollution abatement and control. Current estimates of compliance costs under the new Clean

Air Act Amendments alone are on the order of $50 billion a year. In the United States and many other industrialized countries, however, environmental legislation was adopted piecemeal, creating a complex regulatory process. In 1970 there were about 2,000 federal, state, and local environmental rules and regulations in the United States; today there are more than 100,000. Environmental regulations are listed in over 789 parts of the Code of Federal Regulations. A command and control system for environmental management became the foundation for scores of environmental, health, and safety programs and thousands of federal, state, and local standards, regulations, and guidelines within which businesses must operate. This complex system imposes tremendous compliance burdens on corporations and especially on small- and medium-sized firms. Scientists have long known that the natural environment is a closed and limited system in which the quality of air, water, and land are interrelated. Consequently, environmental managers are discovering that the complex command and control approach to environmental protection, which often addresses only one environmental medium at a time, can merely be an expensive way of moving pollutants around. For example, water treatment often results in the collection of hazardous wastes or toxic sludge that must be deposited on land or incinerated and returned in some form into the air. Air pollution controls often depend on the treatment of liquids and gases that are disposed of in water or land. At the same time, stringent land protection regulations shift disposal of pollutants and solid waste back to water treatment facilities or incinerators.

See Dennis A. Rondinellli and Michael A. Berry, "Industry's Role in Air Quality Improvement: Environmental Management Opportunities for the 21st Century," *Environmental Quality Management* 7 (4) (1997): 31–44. *Dennis A. Rondinelli*

Such measures are the most commonly used in environmental regulation in industrial and developing countries. They work through legally imposed, and typically undifferentiated, restrictions placed on all sources of environmental damage either in terms of the quantity of pollution or resource use or the technology of production, resource extraction, or pollution control. Such policies are opposed almost universally by economists on the grounds that they are more costly (in achieving a given environmental goal) than measures that would explicitly price environmental damage and fail to provide the right long-term incentives to develop technologies that reduce environmental damage. In actual applications, however, at least in industrial countries, command and control measures to control pollution tend to be used flexibly with most involving a considerable amount of negotiation involved in their application as well as the use of "carrots" (subsidies) along with "sticks" (penalties). These features may mean that such policies may not be as costly in achieving their pollution control goals as might first appear. *Sudhir Shetty*

Command and Covenant Approach: an approach to environmental compliance under which the government determines performance goals (or minimum acceptable levels of risk of environmental degradation) while implementing agents (states, local districts, or individual businesses) design their own enforceable alternative compliance methods (covenants). The command and covenant approach is typically proposed as an alternative to the COMMAND AND CONTROL approach. The command and covenant approach provides economic actors with an incentive to achieve an equivalent level of environmental performance in the most efficient way. Compliance with established performance goals is monitored by independent auditors while limited enforcement resources are used against those who fail to meet their environmental obligations. The command and covenant approach is the basis of such theoretical concepts of next-generation environmental policy as alternative compliance, PROJECT XL, "hybrid regulation," as well as the existing "dual" system of environmental regulations in Germany and the "covenant" system in the Netherlands.

See E. D. Elliott, "Toward Ecological Law and Policy," in *Thinking Ecologically*, eds. Marian R. Chertow and Daniel C. Esty (New Haven: Yale University Press, 1997).

Direct (Environmental) Regulations ("Command and Control" Regulations): measures aimed at reducing environmental degradation by specifying pollution limits and production technologies to meet emission standards. Polluters are legally obliged to comply, and various penalties such as fines, imprisonment, or the closure of offending sites are used to enforce compliance. Direct regulations are particularly important for toxic or dangerous materials and health-related pollutants where no margin of error can be tolerated. They are more effective in controlling limited pollution problems, when individual sources have similar pollution characteristics, when impacts are understood, and when technological options are known. Direct regulations are generally designed to control END-OF-PIPE pollution emissions. However, in stipulating a BEST AVAILABLE TECHNOLOGY, direct regulations give companies no incentive to go beyond regulatory norms and can lock them into less efficient technologies when the required technologies cease to be the most efficient available. Direct regulations are also less effective in controlling numerous "nonpoint" pollution sources, comprised of individual households, small enterprises, and farms, and are expensive to administer.

Ecological Funds. See ENVIRONMENTAL FUNDS.

End-of-(the)-Pipe Regulations and Technologies: regulations that specify the use of specific equipment designed to reduce, transform, or impound emissions created in the production process. This is in contrast with production technologies that create less pollution in the first place. Often encouraged by conventional environmental protection regulations that focus exclusively on pollution outcomes or require particular end-of-pipe devices, end-of-pipe technologies (catalytic converters, recycling, sewage treatment) do not contribute to production efficiency or reduction in materials consumption. Therefore, many environmental experts and policymakers advocate supplementing or replacing end-of-pipe regulations with economic instruments that provide incentives for POLLUTION REDUCTION.

Too much is made of the distinction between these and pollution prevention measures. Intrinsically, there is nothing wrong with end-of-pipe technologies. Their use has yielded considerable and relatively low-cost reductions in pollution, as with technologies such as catalytic converters, electrostatic precipitators, and scrubbers. The problem that is often attributed erroneously to the use of end-of pipe technologies is actually due to the reliance of regulatory regimes in many countries on "technology-forcing" regulations, which require the use of specific control or production technologies so as to reduce pollution. *Sudhir Shetty*

Environmental Funds (Ecological Funds): special funds established by most central and eastern European countries, largely financed by the revenues from pollution CHARGES. However, there is increasing interest in capitalizing them by proceeds from privatization, donor grants, and loans from international financial institutions. Environmental funds disburse their revenues through a variety of mechanisms, providing financial assistance to the private or public sector for investments in achieving environmental objectives. These funds offer a mechanism for relaxing the tight budgetary constraints on government spending. They are also a way of lessening opposition to raising pollution charges if enterprises see that revenues raised from such charges are being used to finance ENVIRONMENTAL EXPENDITURES. See also NATIONAL ENVIRONMENTAL FUNDS.

See United Nations Economic and Social Council, *Draft Integrated Report on Environmental Financing*, 25 June 1995, p. 8, para. 25. *R. P.*

Unfortunately, environmental funds have serious long-term disadvantages. Off-budget environmental funds receive tied revenues, such as road funds financed by gasoline taxes, and have a very mixed record around the world. They tend to distort government spending decisions and, if they become widespread, they can narrow the scope for fiscal policy to an unacceptable degree. Spending for environmental purposes must eventually be evaluated on an equal footing with other budgetary programs and justified by reference to the benefits that are generated relative to the cost of raising revenue or of not spending it to meet other objectives. Funds tied to specific types of investment have a tendency to become self-perpetuating bureaucracies and are likely to set charges at a level that is too low in order to maximize revenues over the longer term. Environmental funds should therefore be established with a limited mandate to ensure that their performance is reviewed at regular intervals and that they cease to operate after a period of perhaps ten years. In general, the disbursement of money from such funds should be linked to implementation of the priorities and measures identified in a national (or regional) environment action plan.

See United Nations Economic and Social Council, *Environmental Action Plan for Central and Eastern Europe* (New York: UN Economic and Social Council, 1994). *Renat Perelet*

Environmental Regulation: the approach to environmental protection that relies on negative sanctions to prohibit or reduce pollution or resource depletion, or to require particular technologies to accomplish these ends. Environmental regulation often requires elaborate rules and strong enforcement capacity. See also DIRECT (ENVIRONMENTAL) REGULATIONS; COMMAND AND CONTROL APPROACH.

Environmental Standards: quality standards generally defined through a set of limits on particular pollutants or conditions. Typically, the environmental standards for a given area or facility define the threshold for government restrictions. When environmental standards are exceeded, governments usually fine or prohibit further activity. When environmental standards are set on the basis of thresholds determined by health or aesthetic concerns, they may clash with economic efficiency. Heated debates occur over the appropriate levels of environmental standards, whether they should be uniform across different areas, and how they should be enforced. See also AMBIENT STANDARDS; ECONOMIC INSTRUMENTS; EFFLUENT CHARGES.

Grandfathering: the practice of exempting existing activities from new regulations, or allowing recipients of benefits to continue to receive them even after the general benefit program has been terminated. Grandfathering is often defended on the grounds that putting previously acceptable activities under stricter regulation violates the original property rights, and that future investments will be deterred if investors fear that the rules will be changed after their investments have been made. The practice of grandfathering polluting activities slows the reduction of pollution.

Iron Triangle: the alliance among regulatory agencies, legislative bodies, and the regulated industry that prevents appropriate regulation or meaningful reform in that industry. The iron triangle alliance is attractive to the staff of the agencies and legislature because they come to rely on information, expertise, or funds from the industry. The incentive for reforming the industry or enacting regulations that run counter to the interests of industry is thereby reduced. See also CAPTURE THEORY. *T. S.*

Nonattainment Areas: geographical areas that fail to meet environmental standards. As mandated by preexisting environmental regulations, such areas are subject to various restrictions in economic activity, such as prohibiting new industry. It is typically assumed by environmental regulators that the threshold level of pollution (see THRESHOLD INSTRUMENTS) that distinguishes a nonattainment area poses significantly more risk and damage than do lower levels. However, the objections to declaring an area as a nonattainment area are often based on the argument that the standard is set arbitrarily, and therefore penalizes the economic growth of the area without adequate justification. Developing effective and fair policies toward nonattainment areas is also complicated by the fact that the factors causing nonattainment may partly or even largely originate outside of the area in question.

Offsetting: the practice of reducing emissions from one facility in order to be able to increase emissions in another (perhaps new) facility without violating rules against increasing total permissible emissions. For instance, a company may build a new facility in a NONATTAINMENT AREA only if it installs stringent controls and obtains emission reductions from other sources to offset the emissions of the new facility. See also BUBBLE POLICY.

Regulations: government measures that explicitly require or restrict specific actions on the part of other actors. Regulation has been the main instrument of environmental management, involving specification of responsibilities to redress environmental damage, and restrictions on actions, e.g., on emission of pollutants, use of substances, waste disposal, and location of factories. While the regulation of acutely harmful activity will always be a crucial part of environmental protection, a growing movement has sought alternatives to regulation because of the difficulty and expense of monitoring regulatory violations and enforcing regulations. As a result, environmental regulation and standards are gradually being supplemented by ECONOMIC POLICY INSTRUMENTS. See also COMMAND AND CONTROL APPROACH; DIRECT (ENVIRONMENTAL) REGULATION; ENVIRONMENTAL REGULATION. R. P.

Standard-Setting: the establishment of environmental standards with legal or regulatory status such that the failure to meet these standards has consequences for the parties held responsible. Typically, the failure to meet standards (being "out of compliance") subjects the responsible parties to fines or the requirement to reduce their pollution-generating activities. AMBIENT STANDARDS set maximum levels of a pollutant in the receiving medium (air, water, and soil); they are typically determined by examining the overall vulnerability of the ecosystem. EMISSION STANDARDS set maximum amounts of a pollutant that may be given off by a plant or machine, sometimes according to what can be achieved using the BEST AVAILABLE TECHNOLOGY or by trying to estimate the volume or concentration of emissions that is compatible with meeting the ambient standards. Standard-setting may focus on establishing uniform standards across areas, or setting different standards in different areas; it may focus predominantly on health and ecosystem functioning, or on a more balanced benefit-cost approach. Different government and international agencies use strikingly different standard-setting techniques. The BENEFIT-COST APPROACH, which examines the costs of higher standards as well as their benefits, is often regarded as the method that contributes most to maximizing societal welfare if done properly. However, many criticize this approach for being too lenient toward pollution and polluters, and question whether the environmental costs can be measured as fully as the economic losses. R. P.

Threshold Instruments: policy instruments designed to restrict the exploitation of species that are "endangered" (such as fish stocks, whales, terrestrial mammals) and, more recently, to protect the resilience of ecosystems, particularly where ecological functions and resilience are sensitive to the mix of species or to biodiversity losses. R. P.

Zoning: a method of land use control that specifies permissible uses for each parcel of land within a municipal jurisdiction or other spatially bounded area. It originated in the United States in the first decade of the twentieth century, in response to the chaotic mixture of uses that characterized the country's large and rapidly growing cities. For example, New York City enacted zoning in 1916 as a method of keeping the noisy and congested garment district (and its working-class denizens) from encroaching on an elite retail district. New York's system, like many others of the time, specified manufacturing, retail, and residential districts, and decreed that the uses be strictly separated. Zoning also began to regulate the physical form of development, for example by regulating building heights in each type of district, street setbacks, and minimum lot sizes. With help from the U.S. Department of Commerce, which encouraged each state to pass a model act giving municipalities the authority to adopt zoning, the new idea spread rapidly throughout cities and towns during the first three decades of the century. In 1926 the U.S. Supreme Court declared that zoning was not an unconstitutional limit on private property rights, as it simply protected properties from undesirable adjoining uses—"a pig in the parlor rather than in the barnyard," as the court described it (*Euclid v. Ambler Realty*). A zoning system consists of two elements: a zoning ordinance, which defines the zones and lists uses permitted in each, and a zoning map, which assigns each land parcel to a zone. Zoning is generally backed by a comprehensive plan and is regarded as a major tool, along with infrastructure investment programs, of plan implementation.

More recently, zoning has gained increasing popularity as a tool for managing public lands, including forests and parklands. For example, "wilderness" is a form of zone, in the sense that each parcel of land so designated must be managed in a specified way (e.g., no roads, limited suppression of wildfires). Once the nature of the wilderness zone is defined, individual land areas can be designated as wilderness and be subjected to that management regime. A form of zoning is also found in UNESCO's World Biosphere Reserves. A protected area might be divided into a core, periphery, and buffer zones, with only scientific uses allowed in the core, tourism and extractivism in the periphery, and agriculture and human settlement in the buffer.

See Richard Babcock, *The Zoning Game* (Madison: University of Wisconsin Press, 1956). R. G. H.

5.2.2. MARKET-BASED INSTRUMENTS

5.2.2.1 Principles of Charges and Taxation

Beneficiary Pays Principle: the rule of requiring the beneficiaries of high-quality environments to compensate resource users for the ongoing costs of maintaining ecological functions, environmental services, and attributes that do not bring market benefits and are not required of all people. This principle requires that any additional costs associated with the provision of positive nonmarket benefits, and the costs of desisting from resource extraction, be compensated. This is in strong contrast to the principle that natural-resource exploiters should be penalized for deciding not to maintain the system that provides ecological functions, for example by extracting timber or draining wetlands in order to capture their private economic value. The choices among these principles depend, in part, on the perspectives on PROPERTY RIGHTS. *R. P.*

Benefit Principle of Taxation: a basic principle of traditional taxation theory that holds that tax burdens should be allocated among tax payers in accordance with the benefits they receive from the provision of public goods. It is based on the premise that people purchase goods and services from the government. Whether a tax system based on the benefit principle would be regressive, progressive, or neutral in terms of the tax burden on different income groups depends on the nature of government benefits and attributions of benefits to these groups. However, such attribution is extremely difficult to accomplish, because determining how much different individuals or groups benefit from some government functions carries with it a large degree of arbitrariness. For example, spending on national defense may be an equal benefit for all residents, insofar as their lives are protected, or it can be argued that defense spending provides greater benefits for the wealthy because they would have more to lose from external attacks. *R. P.*

Cost Recovery: the practice of charging the users of resources and environmental services for the costs of providing those resources and services. FULL COST RECOVERY entails charging for the resources per se, infrastructure investment, operating expenses, maintenance, and repairs. When cost recovery is not achieved, systems must be subsidized, and often end up undercapitalized. Resource users also have an incentive to overexploit natural resources when cost recovery is not achieved.

Disincentives: any negative sanction or market-based mechanisms that discourage individuals, firms, governments, or other entities from engaging in an activity. Disincentives to engage in conservation or in sustainable economic development activities are often created inadvertently by government policies. Market disincentives have become increasingly popular for discouraging polluting or resource-depleting activities because of their frequently lower costs of administration and greater effectiveness in comparison with DIRECT REGULATIONS.

Pigovian Tax Principle: the charge imposed on economic actors who cause damage to others (negative EXTERNALITIES), especially pollution damage. If the Pigovian tax (named after the British economist Alfred Pigou) is set at the level of the damage that the action causes to others, then the actor will fully "internalize" the costs of the action, and therefore would be more likely to engage only in those actions that have positive societal benefits. There is nothing in this logic that requires that the harmed parties will be compensated for the damage they suffer.

Polluter Pays Principle (PPP): the rule that producers should be required to pay fully for the damage of pollution, and that the prices for goods and services produced through polluting activities ought to reflect these costs. The polluter pays principle was introduced on the international level by the Organization for Economic Cooperation and Development (OECD) in 1973 and recommended by the OECD as a norm for international pricing in 1982. See also VICTIM PAYS PRINCIPLE. *R. P.*

This principle is as widely misunderstood as it is invoked. Contrary to accepted wisdom, it gives little guidance in designing cost-effective environmental policies. By specifying a particular distribution of the costs of environmental protection, all it provides is a useful starting point for environmental policy making. The principle has two limitations as a guide to environmental policies. First, although it can be interpreted as either requiring polluters to pay only the costs of control and cleanup (standard PPP) or, in addition, to pay for the damages due to pollution (extended PPP), neither interpretation implies anything about the choice of cost-effective pollution control policies, such as charges or tradable permits. Second, the principle is even less helpful in designing policies to address pollution problems when polluters are numerous or hard to identify or when pollution impacts are transnational. *Sudhir Shetty*

Victim Pays Principle (VPP): in contrast to the POLLUTER PAYS PRINCIPLE, the victim pays principle maintains that the victim of pollution or environmental degradation shall pay the party responsible not to engage in the offending activity. The victim pays principle is typically only invoked if the polluter or resource depleter is deemed to have the property rights to engage in the polluting or depleting activity, often established by long-standing practice. *R. P.*

5.2.2.2 Instruments of Charges and Taxation

Administrative Charges. See CHARGES.

Assurance Bonding System. See ENVIRONMENTAL ASSURANCE BONDING SYSTEM.

Banking Scheme/Program: an arrangement for pollution charges under which enterprises can defer payment of the charges in exchange for making larger reductions in emissions or paying higher emission charges in the future. The amount "borrowed" is usually limited, so that the length of time allowed for "paying back" the deferred sum of pollution charges is short, for example four years. The banking scheme can be useful when the government is unwilling in the short run to bankrupt heavy polluters that are unable to pay their pollution charges. However, it is credible only if the authorities make clear that enterprises that fail to "repay their loans" will be closed down or otherwise severely punished. The banking program allows existing firms to reduce emissions below existing levels and to store the excess reductions as "credits." In principle, banking encourages firms to invest in new CONTROL TECHNOLOGIES and to recoup their investment upon subsequent sale or use of the emission reduction credits. A banking system can also reduce the search costs for firms seeking to buy credits: buyers can go to the bank for a list of potential sellers. In practice, very few firms have banked emission credits.

See Organization for Economic Cooperation and Development (OECD), *Climate Change: Designing a Tradable Permit System* (Paris: OECD, 1992). *R. P.*

Carbon Tax: a tax imposed on the production of fossil fuels, based on each fuel's carbon content. The main objective is typically the reduction of fossil fuel consumption in order to reduce the emission of GREENHOUSE GASES. Some environmentalists advocate carbon taxes as more efficient and more practical than direct limitations on either fossil fuel consumption or industrial activity, and suggest that carbon taxes be employed in order to adhere to the reduction in carbon emissions called for by the KYOTO PROTOCOL. Carbon taxes per se are still relatively rare (although many countries have stiff taxes on hydrocarbon fuels), but they have been adopted in the Scandinavian countries and the Netherlands.

Charges: government taxes for resource uses or environmental damage. Charges provide price signals to producers and consumers that more fully reflect the societal costs of using natural and environmental resources and services. Charges on resource uses and polluting behaviors can strengthen incentives to reduce resource depletion and pollution, and can also generate revenues to fund environmental management and provide companies the correct incentives to significantly reduce pollution. However, charges are often seen as "penalties for bad behavior," or as revenue-generating instruments, with little thought of obtaining the efficient levels for charges. Theoretically, charges should equal the societal value of the depleted resources that were not owned by the depleter, or the environmental damages to society that are not permitted by prevailing property rights. The government's expenditures for ABATEMENT must also be taken into account. Four types of charges are most often used: a) administrative charges are mainly intended to finance direct regulatory measures, such as licensing and control activities of authorities. Administrative charges may contribute to environmental efforts if the revenues add to the budgets of environmental authorities, but in practice the revenues often go to the general budget; b) emission (or effluent or pollution) charges are generally calculated on the volumes of pollutants emitted into the air, water, or soil, often varying according to the type of pollutant. EMISSION OR EFFLUENT CHARGES have often been effective in water pollution control policy, yet they have generally played a limited role in the field of air pollution control where direct regulations have always been the main instrument; c) product charges (or taxes) are introduced in order to discourage the consumption of products harmful to the environment. They may be levied on products that pose environmental problems when used in production processes, consumed, or disposed of because of their bulk or toxicity; and d) user charges are intended to recover the costs of publicly provided resources, infrastructure, or resource treatments. They are common with respect to both inputs (such as water supplies) and discharges (such as wastewater). *R. P.*

Deposit Refund Systems: ENVIRONMENTAL INSTRUMENTS designed to encourage recycling or to cover the costs of environmentally sound waste disposal by requiring a deposit to be paid upon acquiring a potentially polluting product. The deposit is refunded if the product or its residues are returned for disposal and recycling. Deposit refund systems are widely applied to disposable products such as beverage containers. Deposit refund systems are being extended to disposable batteries, lubricating oils, tires, electronic equipment, consumer durables such as refrigerators and stoves, and even to automobiles. Deposit refund systems encourage companies and consumers to take a LIFE-CYCLE APPROACH to product management. Designed appropriately, they can provide an accurate reflection of environmental costs, but their effectiveness may be limited by deposits that are too low to secure high levels of compliance. *R. P.*

Cleverly designed deposit refund systems can create employment for the poor (collecting bottles, recycling appliance or automobile parts) as well as reducing trash disposal and economizing on resources. The city of Curitiba, Brazil, has had notable success with a policy of buying recyclables from poor residents in exchange for food and bus tickets. *Robert Healy*

Earmarking (of Revenues) (Hypothecation): designation of revenue uses prior to their collection, allocated either to particular public expenditures or to particular agencies. Along with the introduction of green taxes and charges, earmarking has become an issue of heated debate between environmental and fiscal branches of government, industry, consumers, and other STAKEHOLDERS of environmental policy. Opponents of earmarking (mainly fiscal authorities and some academics) argue that it distorts the distribution and efficiency of public expenditure. The business community (in a surprising alliance with environmentalists) argues that the revenues from environmental financial incentives should be recycled back into the industry and either spent on technology subsidies to affected enterprises or used for general environmental research. Such patterns of earmarking have been introduced in Germany, France, Sweden, and the Netherlands, and are widespread in the transitional countries of Central and Eastern Europe.

Economic (or Market-Based) Instruments: measures for environmental protection that try to induce polluters to reduce pollution in order to maximize profit. Through charges and taxes on polluting activities and products, and incentives such as emission reduction credits, economic instruments replace the more traditional approach of prohibiting or reducing pollution through regulation.

Critics assert that economic instruments are unacceptably imprecise when it comes to identifying actual reductions that can be or are achieved. This imprecision, coupled with a bias to avoid imposition of costs on individual economic units, results in systematic underachievement of pollution reduction. Critics further argue that the most direct way to achieve socially desirable reductions is through direct fiat and enforcement programs. Doing otherwise would result in any social efficiency gain through economic instruments being more than offset by the shortfall in actual achieved emission reductions. *Gerald Emison*

Although their use is broadly supported by economists, such measures have been rarely used in environmental policy. Even when they have been employed, the levels of pollution charges, for instance, are set too low to affect behavior significantly. Their limited application in the real world is likely a combination of several factors—the resistance from resource users and polluters to the redistributive effects of charges or tradable permits and from citizens and environmentalists to placing a price tag on environmental quality, the complexity of introducing such measures so that there is little or no cost saving from their use instead of regulatory measures, and a lack of understanding among policymakers of the potential cost savings from the use of such measures in some circumstances. *Sudhir Shetty*

The traditional prohibition approach has the problem of banning pollution whether or not the polluting activity is, on balance, socially worthwhile. One potential advantage of economic instruments is that they can permit socially worthwhile polluting activities as long as markets operate well enough to allow adequate private returns on these activities. Another advantage is that creating business incentives to reduce pollution relieves regulators from rigidly fixing technology requirements, and encourages polluters to seek pollution-reducing innovations. In addition, economic instruments are often less expensive to administer and adjudicate than complicated regulations. Nevertheless, economic instruments may be ineffective when markets do not operate efficiently; for example, when information is poor, property rights are unclear, or enforcement and collection are inadequate. Another problem is that economic instruments may have less certainty in terms of the resulting levels of pollution; when society must be protected from pollution concentrations above particular thresholds, explicit prohibitions may be indispensable. *Renat Perelet*

Eco-Taxation. See TAXATION.

Effluent (or Emission) Charges (or Taxes): charges to be paid by polluters on their discharges. These charges are based on the volume of emissions or the concentration of pollutants in the emissions. This form of ECONOMIC INSTRUMENT neither bans the emission nor dictates the technology for reducing the emission. The polluter can choose to pay the emission charges, treat the effluent in order to reduce the charge, or cut back on production. Thus when the emissions have high societal costs, the emission charge should be high (reflecting the negative EXTERNALITY of the pollution), and polluters will be induced to invest in pollution prevention or to desist from the polluting activity. The revenues raised may be earmarked for pollution monitoring, abatement, or clean-up. Setting effluent or emission charges high to collect revenues may clash with the motive of setting the charges at the level of societal damage such that economic activity with net societal benefit is still privately profitable. *R. P.*

Emission Rights (or Credits): tradable emission credits designed to put a ceiling on particular types of pollution, reduce pollution-abatement costs, and encourage the most productive activities that create emissions. The government first establishes emission standards that set the ceilings for specific pollutants, and then assigns emissions rights. Rights holders can pollute up to the ceiling on their emission rights, or sell or lease some or all of their rights. Pollution reduction and abatement are thus rewarded by the opportunity to sell or lease emission rights; rights are also likely to be purchased by producers who can benefit from lower abatement costs. Higher overall ceilings on the emission rights granted by the government in a particular area will allow higher levels of emissions in places where purification costs are highest. There are various forms of emissions trading systems, such as BUBBLING; NETTING; OFFSET; and BANKING. *R. P.*

Emission Taxation: See EFFLUENT CHARGES.

Environmental Assurance Bonding System: a variation of the DEPOSIT REFUND SYSTEM that requires producers to post a bond equal to the worst-case damages they could inflict on the environment, in advance of any activity. Worst-case damage scenarios would be established by the regulatory authority with the best information available and with the advice of independent scientists. If resource users demonstrate that damages to the environment were less than the amount of the bond (over a predetermined length of time), this difference and a portion of earned interest would be refunded. If damages occur, the bond would be used to rehabilitate or repair the environment, or to compensate injured parties. The burden of proof is shifted from the public to the resource user. If the estimates are accurate, the bonds ensure that the funds available for environment mitigation are equal to the potential harm, and may discourage polluters more effectively than standard pollution charges that are imposed only after the pollution occurs. Assurance bonds are an extension of the POLLUTER PAYS PRINCIPLE to the "polluter pays for uncertainty as well." Critics of assurance bonding may raise the question of whether the regulatory authority can accurately gauge the severity of worst-case scenarios. *R. P.*

Environmental Impact Tax. See TAXATION.

Environmental Liability Insurance: insurance that covers the risk that an insured party may be held liable for environmental damage. This type of risk-sharing instrument reduces the danger that a firm found responsible for major environmental damage would go bankrupt.

See United Nations Environment Programme, *Economic Instruments for Environmental Protection* (New York: UNEP, 1993).

Fees: charges for services or for the right to engage in particular activities. Fees, in contrast to most taxes, are based on the amount or cost of use. For example, a school tax is often based on the value of the property owned by the taxpayer; a school fee is generally based on how many children the feepayer has in the school system. However, in some circumstances when a particular level of government lacks the authority to impose taxes, it may impose fees to serve the same purpose. Pollution or resource-use "fees" are often used interchangeably with the terms "charge" or "tax."

Green Tax. See POLLUTION TAX; CARBON TAX.

Hypothecation. See EARMARKING.

Individual Transferable Quotas (ITQ): tradable permits used in the management of natural resources, primarily fishery stocks. These are transferable property rights allocated to individual fishing operators as an entitlement to a particular level of catch over a given period. The fixed right to harvest any amount within a quota reduces the need to race to catch the fish, allowing the fisher to allocate his or her activities over the fishing season in a most efficient manner. The aggregate level of quotas restricts the total catch, while the provision for trading allows rationalization and, subsequently, economic efficiency of the industry. ITQ systems have been intensively used in New Zealand, and, though on more restricted basis, in Australia, Canada, and Iceland. The experience of these countries is testimony that the ITQ system generates higher economic rent than alternative management regimes and prevents overcapitalization of the industry. However, in view of the substantial variations in prices for quotas for individual species, ITQ systems have been questioned in terms of the efficiency of price determination in the quota market and the measurement of the aggregate economic benefit from the operation of the regulated fishery.

See Organization for Economic Cooperation and Development (OECD), *Evaluating Economic Instruments for Environmental Policy* (Paris: OECD, 1997), 70–76.

COMMENTARIES ON INDIVIDUAL TRANSFERABLE QUOTAS (ITQS)

The movement toward ITQs is part of the trend to privatize property in an open access situation. The idea behind ITQs is to provide an incentive to conserve the resource that has been overexploited. ITQs hold some promise as a way to regulate fisheries that appear on the verge of collapse. However, ITQs are probably not appropriate for all fisheries and all fishing situations. The decision to use ITQs should take into account the status and size of the fishery, the biology of the fish, and the history and culture of the communities where the technique is tried. For instance, ITQs may be more successful in situations where the fishery is overcapitalized, where fishers are using too much gear, and where too many fish are being taken. Likewise, ITQs may work best where there are a small number of fishers, a small geographic area, and small number of fish processors. Nonmigratory species, single species, and healthy fish stocks lend themselves to ITQ management. Finally, ITQs can more likely be expected to succeed in cultures where there is a tradition of individuality and property ownership. *Toddi Steelman*

ITQs hasten the privatization of fisheries, where living creatures become tradable commodities, and their relationship to other living things becomes meaningless. In most countries, ITQs have resulted in the concentration of fisheries in the hands of a few large fishing companies that accumulate the quotas transferred by individual fishers who find it costs more to fish than stay at home. In New Zealand, five large companies own 65 percent of the quota. Canada follows the same pattern. In Alaska, a Community Development Quota (CDQ) program was established in 1991, assigning a portion of the total allowable catch of pollock, halibut, and sablefish for harvest by eligible western Alaska communities. Bering Sea coastal communities set up partnerships with large corporations in industrial-scale seafood production. Up to $20 million in annual profits must be spent on local employment and fisheries development. However, the program has led to the concentration of the fisheries in the hands of a few large companies. The CDQ program is highly touted as a way to address chronic unemployment and social problems and share directly in a multimillion dollar fishery. There are calls to replicate it in the Russian Far East. Rural development is financed by fish in the sea, not by state governments. We have seen CDQ communities urge increases in quotas to maintain jobs and development funds; quash dissent or criticism of their industrial partners; and disregard indigenous systems of knowledge and prudent use of their own resources. By becoming partners in large-scale resource extraction, local people are removed from direct, meaningful control and involvement in their region's resources. They are left with infrastructure and outside investments they cannot sustain when the fish are gone. *Susanne Swibold and Helen Corbett*

International Tradable Emission Permits: MARKETABLE POLLUTION PERMITS to be traded internationally as well as within nations. International tradable emission permits have been proposed as a policy instrument for use in the global climate regime to limit the build-up of GREENHOUSE GASES. The principle is that each country is given an initial endowment of permits to emit carbon dioxide, with permit holders being able to trade permits in a competitive permit market. Paradoxically, while at the domestic level the initial distribution of permits does not affect the efficiency properties of the solution, in the case of global environmental threats, as was shown by G. Chichilnisky et al., only a limited number of initially allocated permits leads to a superior outcomes for all nations.

See Graciela Chichilnisky, Geoffrey Heal, and David A. Starrett, *International Emission Permits: Equity and Efficiency* (San Francisco: Annual Congress of ASSA, 1996).

Market Alternatives to Direct Regulation. See ECONOMIC INSTRUMENTS; TAXES; TRADABLE PERMITS, DEPOSIT REFUND SYSTEMS

Marketable Pollution Permits (Negotiable Permits): permits required by all sources in order to emit pollutants under a transferable emission permit system. The control authority issues precisely the number of permits needed to produce the desired emission level. Each permit specifies exactly how much the firm is allowed to emit, and any emission by a source in excess of those allowed by its permit would cause the source to face severe financial penalties. However, these permits are freely transferable; they can be bought and sold. In theory, this produces a cost-effective allocation by issuing the appropriate number of permits and letting the market system do the rest. A real world example is the U.S. Acid Rain Program.

See Tom Tietenberg, *Environmental and Natural Resource Economics* (New York: HarperCollins, 1996). *R. A. K.*

Negotiable Permits or Vouchers. See PERMITS.

Noncompliance Fee: fee applied when polluters violate environmental regulations. The fee may be set at a rate proportional to the benefits that the polluter achieves by not complying with the environmental regulations. The administration of noncompliance fees may be impeded by measurement problems and legal restrictions.

Output Taxes (Product Taxes or Product Charges): taxes levied on products that are environmentally harmful when produced or consumed. The rates can be based on some product characteristic (e.g., a charge on sulfur content in mineral oil; a carbon tax on fossil fuels; charges on the phosphorous content of fertilizers) or on the product itself (e.g., a charge on freon, batteries, or packaging materials). In practice, output taxes may be levied through tax differentiation leading to more favorable prices for "environmentally friendly" products. Output taxes are sometimes confused with PIGOVIAN TAXES, although they are levied on the output rather than on emissions. *R. P.*

Performance Bonds: payments by potential polluters to authorities before an operation that could be harmful to the environment begins. If the operation complies with environmental regulations, the payments are refunded; otherwise the initial payment is forfeited. Despite their apparent simplicity, performance bonds are less common than other instruments and are applied primarily in cases of clear environmental damage, such as surface mining. The limited use of performance bonds in developed counties can be attributed to problems with monitoring polluting activities and to legal restrictions on contracting. Australia, Norway, Sweden, Canada, and the United States use slightly different variations of these instruments. *R. P.*

Permits: official authorizations to engage in certain activities, particularly the authorization to operate a facility within specified limits of pollution. The innovation of MARKETABLE PERMITS (also called "negotiable permits") allows permit holders to transfer their authorization or right to pollute to others who buy or lease these permits. See also ALLOWANCE. *R. P.*

Pollution Tax: a charge for pollution, usually set near the level of damage that the pollution is estimated to impose on actors other than the polluter. Pollution taxes are also called PIGOVIAN TAXES. Pollution taxes discourage economic activities that would occur if the polluter did not have to pay for polluting. In some cases, however, pollution taxes come to be valued for their revenue, and are therefore set at levels that permit production that is not justified in light of its total costs and benefits. Pollution taxes are generally set for categories of pollution (e.g., taxes on the emission of particular pollutants such as sulfur dioxide), which precludes setting the tax at precisely the costs of the damage. Thus, taxing pollution in practice is rarely as straightforward as most economists assume in theory. *R. P.*

Sewage Charge: charges for the use of sewage services, usually charged according to the volume of effluent discharged into the sewage system. See also CHARGES.

Subsidies. See under SECTION 1.1.4.

Taxes: charges imposed by government on private and state actors to generate revenues and often to influence activities. Taxes are typically assessed on particular economic activities, including the holding of assets, in accordance to the volume and economic importance of those activities, although some taxes are imposed on each unit at the same rate. Tax systems can be characterized as regressive, neutral, or progressive depending on the incidence of the tax burden on tax payers of different income levels.

COMMENTARY ON TAXES

The importance and effectiveness of taxes as means for steering economic activity and redistributing income have recently come into question, because tax incentives frequently lead to abuses while also denying the government important revenues. Many ostensibly progressive tax systems have proven to be distributionally neutral or nearly so. On the other hand, tax revenues adequate to cover the spending determined by the government are crucial, because without such tax revenues, governments often resort to highly distorting means to spend (e.g., printing more money, resulting in inflation) or attempt to accomplish government objectives through inefficiency-generating manipulations of the economy (e.g., instead of providing income supplements through the budget, requiring employers to hire unneeded labor). Environmental taxes (technically, pollution charges) are legitimate instruments for reducing pollution, as long as they are set at the level of the pollution damage. Tax differentiation for products with differing pollution effects provides one mechanism for setting environmental taxes at appropriate levels (e.g., charging more for more polluting fuels), and are relatively easy to implement from an administrative point of view, since they are embedded in existing tax systems. *William Ascher*

Tradable Pollution Discharge Permits. See MARKETABLE POLLUTION PERMITS.

Tradable Pollution Rights. See MARKETABLE POLLUTION PERMITS.

User Charges. See CHARGES.

5.2.3 OTHER ENVIRONMENTAL PROTECTION ACTIONS

Civic Environmentalism: a bottom-up process of problem solving and decision making in relation to environmental problems in a specific location that involves nongovernmental actors. Civic environmentalism as the model of environmental governance comes from a civilian tradition identified by Alexis de Tocqueville. While not rejecting the resources of government, civic environmentalist movements rely on the energy and resources of all types of voluntary associations, citizens, and businesses in local environmental protection.

COMMENTARY ON CIVIC ENVIRONMENTALISM

Civic environmentalist movements require broad citizen interest and often the emergence of a "shadow community" that includes citizens, scientists, and frontline staff from government agencies. Second, the success of civic environmentalism depends on an ability of state and federal agencies to modify their traditional roles and to assist the grassroots initiatives in a creative but nonintrusive manner. The third element of civic environmentalism is the moral authority given to the process by a consensus reached on a particular environmental issue by the "shadow community" along with the reputations of its various members. Civic environmentalism is very different from the traditional pattern of environmental management but it is believed to be a very promising component of modern environmental governance.

See John DeWitt and Marian Mlay, "Community-Based Environmental Protection: Encouraging Civic Environmentalism," in Ken Sexton, Alfred A. Marcus,

Timothy D. Burkhardt, and William Easter, eds., *Better Environmental Decisions* (Washington, DC: Island Press, 1999): 353–76. *Natalia Mirovitskaya*

Community Advisory Panels (CAPS): a model of environmental linkage between firms and the communities in which they operate. It was first developed by chemical companies in an effort to create links between their manufacturing plants and nearby communities when public confidence in the chemical industry was undermined after two disastrous industrial accidents, the 1976 explosion in Seveso, Italy and the 1984 accident in Bhopal, India. The creation of CAPS allows the public to express their safety concerns with plant managers and work with them on the issues of pollution prevention, emergency management, and hazardous waste management. The Chemical Manufacturers' Association, which initially developed the idea of CAPS, recommends that its membership be representative of the community in age, gender, socioeconomic and ethnic profile, as well as in stakeholders' interests. Some critics of CAPS are concerned about the possibility of cooptation by the industry. Thus far, the impact of CAPS over the environmental performance of the companies has been mixed. A few panels have been dissolved, while others have been successful in securing better warning systems and corporate cleanups.

See Nevin Cohen, Caron Chess, and Frances Lynn, *Fostering Environmental Progress: A Case Study of Vulcan Chemical's Community Involvement Group* (New Brunswick, NJ: Center for Environmental Communication, Rutgers University, 1995); Nevin Cohen, Caron Chess, and Frances Lynn, *Improving Dialogue: A Case Study of the Community Advisory Panel of Shell Oil Company's Martinez Manufacturing Complex* (New Brunswick, NJ: Center for Environmental Communication, Rutgers University, 1995).

Community-Based Environmental Protection (CBEP): a set of approaches that governmental agencies use to encourage, support, and participate in solving local environmental problems through community-based effort. Like other collaborative management processes, CBEP is time-consuming, can lead to unexpected results, and can sometimes fail. However, in successful cases it can provide additional information, avoid or end litigation, enlist greater public support for environmental policy, and find new means to meet its goals. See also PROJECT XL.

See John DeWitt and Marian Mlay, "Community-Based Environmental Protection: Encouraging Civic Environmentalism," in *Better Environmental Decisions*, eds. Ken Sexton, Alfred A. Marcus, Timothy D. Burkhardt, and William Easter (Washington, DC: Island Press, 1999): 353–76.

COMMENTARY ON COMMUNITY-BASED
ENVIRONMENTAL PROTECTION

In the United States, the Environmental Protection Agency (EPA) initiated CBEP in the early 1980s with its Great Lakes and Chesapeake Bay programs. In both cases, the EPA, in cooperation with the states, established procedures for setting goals for environmental cleanup, conducting new research, consulting with stakeholders, and providing better information to the public about environmental problems and responses. It is believed that in both cases the EPA was able to initiate a civic process that greatly contributed to its goals. *Natalia Mirovitskaya*

Consumer Groups. See under SECTION 3.2.

Dirty Dozen: 1) a list of the twelve members of Congress who annually achieve the lowest scores on the League of Conservation Voters' (LCV) Scorecard, an index of environmentalism. The LCV, an American environmental lobbying organization, publishes the list to pressure members of Congress to be more pro-environment in their votes on proposed legislation. *D. S.*

2) a list of the twelve U.S. companies whose environmental performance is assessed as the worst during the year, as judged by a consortium of environmental organizations. The motivation is to turn consumers away from the companies, in order to prod them to make improvements to escape the list.

Ecological Funds. See ENVIRONMENTAL FUNDS.

Environmental Covenants. See under SECTION 5.3.2.

Environmental Labeling (Ecolabeling): indication of the environmental impact-related attributes of a product or service. Environmental labeling is done by private or public institutions and is used to help consumers select environmentally friendly goods or services, as well as to offer a market incentive to manufacturers and service providers willing to qualify for the label or "certification." Environmental labels may be used, for instance, by hotels that preserve natural habitats on their grounds and minimize pollution, or by companies that use only natural ingredients in their production or avoid animal testing. *R. P.*

National Environmental Funds (NEFs): national-level financial mechanisms set up in developing countries in the early 1990s as an alternative to traditional aid mechanisms. The NEFs receive international grants for environmental protection, manage these funds, and disburse them at the national level. Most national environmental funds are independent of national governments and are administered by boards composed of representatives of donors and various national STAKEHOLDERS. Whether this attempt at partnerships among international, state, and nongovernmental actors turns out to be an effective mechanism of SUSTAINABLE DEVELOPMENT still remains to be seen. See also ENVIRONMENTAL FUNDS.

Policy-Relevant Research. See under SECTION 1.3.2.

Project XL: a model of environmental partnership between government, company, and community, developed by the U.S. Environmental Protection Agency (EPA) as an experimental program to give companies greater flexibility in environmental management if they cooperate with government agencies and communities. Project XL offers businesses flexibility in meeting air and water quality standards as long as the company achieves superior environmental performance and the regulatory agency and a community advisory group agree to its plan of environmental management. To date, only a few Project XL pilot projects have been conducted and their success has been assessed as mixed. To be used on a broader basis it requires statutory authorization as a standard approach to environmental problem solving.

See Environmental Protection Agency, *EPA Response to Comments on the Intel XL Pilot* (Washington, DC: EPA, 1996); available online at: http://www.epa.gov/ProjectXL.

Public Environmental Awareness: the level of information and sensitivity about the state of the environment maintained by different groups and societies. Public environmental awareness is a major policy-making factor in democratic societies. The level of public awareness depends upon cultural traditions, the level of scientific knowledge, willingness and ability of the mass media to deliver this knowledge in an understandable form, the influence of civil society, and many other factors, including social status, education, and interests of individuals.

Side Payments: compensation to parties adversely affected by particular policies or private actions, offered either out of considerations of equity or because the agreement of potentially harmed parties is necessarily for the success of the policy or private action. The practical and theoretical possibility of side payments is important for assessing whether a policy is efficient in the sense of PARETO OPTIMALITY. In other words, to make a policy change that ultimately leaves no one worse off, the immediate winners generally have to provide side payments to the immediate losers.

Voluntary Agreements. See ENVIRONMENTAL VOLUNTARY AGREEMENTS.

5.2.4 ENVIRONMENTAL POLICY PRINCIPLES AND INITIATIVES

Accountability: in the environmental context, assigning responsibility for environmental damage to particular actors, and to the economic activities that cause the decline. See also POLLUTER PAYS PRINCIPLE.

Adaptation. See ENVIRONMENTAL POLICIES, TYPES OF.

Anticipatory Policy: a set of measures established before the emergence of an anticipated environmental problem. Anticipatory policies may be designed to avert or mitigate the consequences of such problems. *R. P.*

Best Environmental Practice (BEP): the application of the most appropriate combination of environmental protection measures. The CONVENTION ON THE PROTECTION AND USE OF TRANSBOUNDARY WATERCOURSES AND INTERNATIONAL LAKES (HELSINKI CONVENTION) stated that BEP should include at least the following graduated range of measures: a) provision of information and education to the public about particular activities and products; b) the development and application of CRADLE-TO-GRAVE "Codes of Good Environmental Practice"; c) availability of collection and disposal systems; d) saving of resources, recycling, recovery, and reuse; e) avoiding the use and generation of hazardous substances; f) application of economic instruments to activities and products; and g) a system of licensing. In contrast to the doctrine of BEST AVAILABLE TECHNOLOGY, which is aimed mainly at the reduction of emissions from point sources, BEP targets all kinds of sources.

See Convention on the Protection of the Marine Environment of the Baltic Sea Area, Helsinki, 1992, Annex II.

Best Practical Environmental Option (BPEO): a concept of ENVIRONMENTAL REGULA-TION suggested by the British Royal Commission on Environmental Pollution in 1976 and later developed into the concept of integrated pollution control, which has been under consideration by the EUROPEAN UNION (EU) since 1995. The basic idea of the BPEO is that the form and medium of disposal of pollutants are to be chosen to cause the least overall environmental damage. Integrated pollution control is more efficient than the traditional medium-to-medium approach to environmental management and it also matches the LIFE-CYCLE APPROACH of business management. In practice, however, it is more complicated than traditional schemes of regulation and requires sophisticated knowledge on the part of pollution inspectors as well as developing good rapport with the regulated businesses.

Bubble Policy: a type of environmental policy under which several existing sources of pollution are regulated as one source. The bubble policy ensures that environmental exposure to some pollutant is reduced or controlled overall, in contrast with the traditional approach of mandating a uniform maximum level of pollution for each emitter. Under this system, a firm can increase emissions at one source without incurring additional regulatory burdens, so long as it reduces emissions at other sources within the designated bubble area. Some regulators distrust the bubble approach because of their fear that businesses may use emission trading to evade emission controls. R. P.

COMMENTARY ON BUBBLE POLICY

An effective bubble policy must meet requirements similar to those of marketable permits. There must be an unambiguous baseline. Measurement of emissions within the bubble area must be accurate and reliable. Without these conditions, the possibility that emissions might actually increase in an area covered by a bubble policy can undermine environmentalists' support for the bubble approach. *Gerald Emison*

Cross-Media Approach: an integrated approach in addressing environmental problems that takes into account interrelationships among various media of the environment, such as air, water, or soil. The premise is that it is crucial to avoid improving quality standards in one environmental medium at the expense of other media. A cross-media approach recognizes the integrated nature of the environment, and seeks to reduce the overall pollution at the source, rather than eliminating waste from one medium by transferring it to another. Energy efficiency, clean technologies, and pollution prevention are obvious components of a cross-media approach.

See United Nations Economic Commission for Europe, *Guidelines on Environmental Management in CITs* (New York: United Nations, 1994), 12. R. P.

Environmental Policies, Types of: overall principles of response to environmental change. Environmental policies can be generally divided into four main groups, at least in terms of their objectives: a) prevention—actions aimed at preventing the change from occurring; b) MITIGATION—actions aimed at mitigating possible consequences of the change; c) adjustment—accepting the change temporarily and adjusting one's actions; and d) ADAPTATION—permanently adapting one's actions to the change. They can also be divided in terms of the kinds of instruments for inducing desired activities: regulation, economic instruments, voluntary agreements, etc. While obvious environmental policies often dominate the focus of attention of environmentalists, many other policies that are not so labeled often have highly significant impacts on the environment as well. For example, energy pricing policies ought to be considered as environmental policies because of their huge impact on fuel use.

Five "R" Principles: the environmental management maxim that emphasizes the importance of reduction, replacement, recovery, recycling, and reuse for sustainable industrial policy. *R. P.*

Full Cost Pricing: the principle calling for producers and resource users to pay for all of the costs involved in production, including the costs that government bears in providing natural and environmental resources and the damage caused by production. When environmental services and natural resources are enjoyed without full cost, users tend to overuse them. For example, if water charges do not include the replacement, investment, and maintenance costs of delivering water, as well as the costs of any contamination, more water will be consumed than is societally optimal. See also COST RECOVERY.

Keep Options Open Principle: one of the strategic principles suggested in the debates on global environmental management. The simple premise of the principle is that given the uncertainty of the future, it is wise not to foreclose options, and to consider the effects not only of changes most likely to happen, but other possibilities as well. Given the complexity of environmental phenomena and the high level of scientific uncertainty in this realm, in practice this principle can be very conservative in terms of actions that may affect the ecosystem.

Mitigation: measures taken to offset or ameliorate the adverse impact of human activities on environment. For example, according to U.S. legislation 40 (C.F.R. #1508.20), mitigation may include: a) avoiding impact altogether by not taking a certain action; b) minimizing impacts by limiting the degree of magnitude of the action and its implementation; c) rectifying the impact by repairing, rehabilitation, or restoring the affected environment; d) reducing or eliminating the impact over time by preservation and maintenance operations during the life of the action; and e) compensating for the impact by replacing or providing substitute resources or environment. Mitigation policies are often complementary to ADAPTATION and PREVENTION POLICIES. *R. P.*

Netting: a practice in environmental management in which all emission sources in the same area that are owned or controlled by a single company are treated as one source, thereby allowing flexibility in controlling individual sources in order to meet a single emission standard. The principle of netting allows businesses to modify existing facilities without undergoing the usual burdensome preconstruction review requirements and stringent emission controls, so long as there is no significant net increase in plant-wide emissions. In other words, netting relies entirely on internal (intrafirm) trades from the same facility. If a firm cannot meet the emission limits through netting, it must follow the OFFSET rules. *R. P.*

No-Regrets Principle: the principle of developing environmental policies that also improve the economic situation or achieve other policy objectives, such that citizens and policymakers have no basis for regretting a tradeoff between environmental protection and losses in other areas. In the case of global warming, for instance, a no-regrets approach might involve promoting energy efficiency to decrease carbon dioxide emissions, which at the same time reduces emissions of other trace gases and conserves coal, oil, and gas supplies for use by future generations. No one would dispute the attractiveness of such "win-win" strategies; the question is whether they even exist for particular instances.

Optimal Pollution Policy: the level of environmental protection that permits some pollution in order to find the optimal combination of benefits from production and costs from pollution. The concept is based on the premise that environmental policies must recognize that some pollution is inevitable, and therefore the task is to find the types and degrees of pollution that permit the most favorable tradeoffs between material improvements and environmental degradation.

Prevention. See ENVIRONMENTAL POLICIES, TYPES OF.

Subsidiarity: 1) in a general sense, the principle that decisions should be taken at the local level ("Think Globally, Act Locally"), except when environmentally relevant processes spill over political or geographical boundaries, which then calls for cooperation across jurisdictions. Therefore the subsidiarity principle requires strong local and other subnational institutions to address relatively confined environmental issues, but also calls for institutions of higher jurisdictional levels to address transboundary issues. On the international level, as well, subsidiarity means that decisions should be taken at the lowest level possible as long as spillover effects (i.e., negative EXTERNALITIES) can be addressed. Thus subsidiarity is a key principle for maintaining national sovereignty, while allowing for environmental decision making at the international level among adjacent nations and nations that share highly migratory species (such as sea turtles) or transboundary pollution problems. Examples of regimes based on the principle of subsidiarity include the climate change regime and international river basin regulations. Subsidiarity is a basic principle of environmental decision making in the EUROPEAN UNION (EU).

See Konrad von Moltke, *International Environmental Management, Trade Regimes and Sustainability* (Winnipeg: International Institute for Sustainable Development, 1996).

2) within the framework of the EUROPEAN UNION (EU), the principle that all European Union decisions should be taken as closely as possible to the citizens' level, while all-European institutions deal only with transnational matters. According to this principle, the EU does not take action (except for the areas that fall within its exclusive competence) unless its actions would be more effective than actions taken at national, regional, or local level. The principle of subsidiarity is closely bound with the principles of proportionality and necessity, which require that any action by the union should not go beyond what is necessary to achieve the objectives of the treaty. The basic principles of subsidiarity were laid down by the European Council in 1992. *R. P.*

Zero-Cost Improvements: actions that do not entail significant additional financial costs but improve the state of the environment, such as adequate management and the use of better information. In the spirit of the NO-REGRETS PRINCIPLE, environmental protection strategies should begin with an exploration of whether zero-cost improvements are possible, and the government's mandating of particular technologies or practices for environmental protection should not preclude reliance on "easy" zero-cost improvements. Critics of the zero-cost improvement approach argue that the search for easy, costless answers may delay or impede the adoption of tough policies to make necessary sacrifices for the environment. *R. P.*

5.2.5 POLICY FAILURES

Capture Theory: the argument that regulation may be rendered ineffective when the regulated industry gains influence and even control over the regulating agency. Developed by Grant McConnell and other political scientists, capture theory was introduced into economics by George Stigler. The means by which an industry can "capture" the regulators includes control over information and expertise, personal friendships, promise of future jobs for the regulators, and bribery. See also IRON TRIANGLES.

See Grant McConnell, *Private Power and American Democracy* (New York: Knopf, 1966); George J. Stigler, *The Citizen and the State: Essays on Regulation* (Chicago: University of Chicago Press, 1975).

COMMENTARY ON CAPTURE THEORY

With respect to natural resource management agencies in the United States, the Bureau of Land Management and the United States Forest Service have been heavily criticized as being captured by their respective constituencies: ranchers and the timber industry. *Toddi Steelman*

Corruption: abuse of public trust by government or state officials for the personal gain of the officials, their families, or associates. Thus the concept extends beyond individual enrichment, to include inappropriate behavior to support people of favored ethnic groups, political parties, and so on. The standard distinctions among types of corruption are: a) petty corruption (the bribery of low-level officials, or special treatment of associates by these officials); b) policy corruption (higher-level policymakers' choices that violate the public interest); and c) electoral corruption (electoral fraud and vote buying). Personal gain goes beyond financial gain, because corrupt practices can also be intended to provide political advantage through inappropriate and illegal behavior, such as electoral fraud.

COMMENTARY ON CORRUPTION

While corruption was once seen as a relatively small price to pay to overcome the rigidities of government and state procedures in developing countries ("greasing the wheels of government"), recent theory and findings show corruption often has enormous costs, including poor investment of government resources, distorted private investment, distorted prices, inefficient public services, lack of government accountability, and political cynicism. Squandering resources for personal or political gain is a major impediment to sustainable economic development.

See Robert E. Klitgaard, *Controlling Corruption* (Berkeley: University of California Press, 1987); Arnold J. Heidenheimer, Michael Johnston, and Victor T. LeVine, eds., *Political Corruption: A Handbook* (New Brunswick, NJ: Transaction Publishers, 1989). *William Ascher*

Economic Distortions: for neoclassical economics, the disruption of smoothly operating market systems brought on by inappropriate price setting, subsidies, rationing, exchange rates, etc., which create distortions that undermine economic efficiency. However, not all government interventions create distortions; some government interventions rectify market failures. Critics of government economic intervention link economic distortions on the macro and micro levels to the low economic growth rates suffered by many developing countries. The core of many economic distortions is the mispricing of inputs and outputs. Economic distortions in environmental and natural resource markets emerge when environmental services are underpriced and natural resources are either under- or overpriced. Pollution and resource depletion are often induced by market distortions that lower the costs that the producer has to bear.

Government Failure. See POLICY FAILURE.

Institutional Failure: a failure to achieve efficiency or a specified objective of equity due to distortions or rigidities in formal or informal institutions. Thus MARKET FAILURES may occur or persist because formal or informal institutions do not perform well in providing information, assuring competition, etc.; POLICY FAILURES may similarly be due to the inability of existing institutions to enact optimal policies. Identifying the institutional failure is the analytic task following the identification of market and policy failures in the effort to formulate institutional reforms. For example, weak community conflict resolution institutions may be found to be at the heart of the problem of reckless resource depletion.

Kleptocracy: ironic term used by some political scientists to describe political regimes characterized by extreme levels of corruption: governance by kleptomaniacs. When decisionmakers are primarily motivated by possibilities for financial gain for themselves, family members, or political cronies, large losses in economic efficiency and environmental degradation may result. This is particularly true in natural resource sectors, where access to minerals, land, or timber may be made available to favored persons or corporations (or those willing to pay bribes) at prices far less than those prevailing in the market. This in turn can lead to rates of extraction higher than would otherwise prevail, since the favored concessionaire usually wants to liquidate its holdings while the present political structure is in power. Kleptocracy can also direct foreign grants or loans away from their intended use, and can occasionally lead donors to suspend their activities. It can sometimes discourage foreign private investment, particularly by firms that are prohibited by their national law, or their internal business policy, from paying bribes. See also CORRUPTION.

See Robert Klitgaard, *Controlling Corruption* (Berkeley: University of California Press, 1987). *R. G. H.*

Note that in the United States, special interest groups, notably those that lobby for perverse subsidies, in Washington, D.C., spend $100 million per month in pressuring Congress to support subsidies that are harmful to the economy as well as the environment.

See A. Shuldiner and T. Raymond, *Who's in the Lobby?* (Washington, DC: Center for Responsive Politics, 1998). *Norman Myers*

Market Failure. See under SECTION 1.1.4.

Perverse Incentive: an incentive arising from government policy that prompts unintended negative behaviors. REGULATIONS and ECONOMIC INSTRUMENTS that create perverse incentives constitute a major category of POLICY FAILURES in that the responses to policies deviate from their intended effects. For example, antipollution prohibitions against driving automobiles with odd or even license numbers on given days may create an incentive for drivers to purchase additional cars, and often cheaper, more polluting ones. The insight that incentives are often difficult to predict and manage gives pause to those considering the use of elaborate regulation or economic instruments.

Policy Failure (Government Failure): government policies and actions that fail to produce optimal outcomes. Policy failures often entail the mispricing of goods (including natural resources) and services (including environmental services) that provoke inefficiency and the wasteful use of resources. The term *policy failure* was popularized in the 1970s as a counter to the concept of MARKET FAILURE, which implies that government intervention is justified when markets fail to produce optimal outcomes, for example because of imperfect competition, imperfect information, and unclear property rights. Policy failures, due to government ignorance, incompetence, or discrepancies between societal interests and the interests of government officials, create the possibility that government intervention may be counterproductive even when market failures exist. Yet policy failure also occurs when the policy interventions are *not* taken to correct market failures that could be addressed with less harm than the harm of the market failure itself. Policy failures also occur when government decisions lead directly to excessive resource depletion or greater pollution. This happens frequently when state enterprises are directly involved in natural resource exploitation or economic production that creates pollution.

COMMENTARY ON POLICY FAILURE

There are many causes of policy failure. The following represent some ways to deal with these causes: avoid simplistic problem definitions with a single objective; don't let a single organization dominate the policy process; don't assume the policy will benefit everyone; don't discount long-term consequences; and balance local and central (usually government) participation. Bring in experts from diverse fields right away and minimize their biases;

don't let expert opinions overrule preferences of those affected by decisions; don't exaggerate the expected benefits or discount the possible costs; and don't focus only on easily measured or easily understood data. Ask the hard questions. Coordinate governmental decision making, especially when there are stronger and weaker agencies; don't allow agency rivalries to block achievement of policy goals; and don't overcontrol participants or beneficiaries. Furthermore, avoid "benefit leakage," the extension of programs beyond their original targets to the point of wasting resources; make sure government management agencies do not succumb to the natural selfishness of the private sector; avoid intelligence failures and delays; ensure adequate coordination and appraisals; and insist on appropriate organizational arrangements. Furthermore, it is important to ensure that decisionmakers and managers are sensitive to criticism, learn explicitly and systematically from experience (strive to understand the complexity of causes and effects, consider long time spans and changing contexts, and conduct adequate follow-ups). Finally, consider the effects of termination and deal fairly with people affected by it, anticipate and manage antitermination coalitions, and realize that any program termination may be perceived as a policy failure.

See William Ascher and Robert G. Healy, *Natural Resource Policymaking in Developing Countries* (Durham: Duke University Press, 1990); Tim W. Clark, *Averting Extinction: Reconstructing Endangered Species Recovery* (New Haven: Yale University Press, 1997). *Tim W. Clark*

Rent-Seeking Behavior: efforts by private actors to qualify or obtain special privileges, such as exclusive licenses, tax exemptions, protectionism, etc., in order to obtain higher than normal profits ("rent" being the excess profit above the profit level prevailing in a competitive market situation). Rent-seeking behavior often involves providing benefits for the government officials with influence over rent granting, ranging from bribes to campaign contributions, as well as efforts to influence legislation and regulation through lobbying, and investments to make the rent seeker's firm more qualified to receive the rent. Rent seekers therefore often dissipate part of the rent by seeking it. When rent seeking involves activities that are not productive to the economy (e.g., lobbyists competing for the same special privilege or several firms investing in equipment necessary to qualify for a single license to produce a particular product), it causes a loss for the economy as a whole. Rent seeking opens up opportunities for corruption among public officials, and the persistence of rents often reflects the desire of public officials to provide the incentive for rent-seeking behavior that enriches these officials. Economic liberalization is often recommended as a strategy for reducing the opportunities for rent-seeking behavior. See also CORRUPTION; KLEPTOCRACY.

See Robert D. Tollison, "Rent Seeking: A Survey," *Kyklos* 35 (1982): 575–602.

Industry. See under SECTION 1.1.4.

5.3.1 TECHNOLOGICAL RESPONSES

Adapted Products: products that are designed to be less polluting in their production, use, and disposal. The production and consumption of adapted products often have to be encouraged by government incentives (on the grounds that they provide the positive EXTERNALITY of a cleaner environment) because they are usually more expensive than conventional products.

COMMENTARY ON ADAPTED PRODUCTS

Several companies are substituting less or nonpolluting products for more polluting ones. The process DuPont developed for its Petretec polyester film, for example, breaks down waste in a way that retains its original polyester properties, making it substitutable for virgin materials without depleting natural oil resources. Monsanto has developed genetically bioengineered plants such as potatoes and cotton that are protected against disease and insects rather than selling farmers the pesticides needed to prevent crop damage. Building protection into plants genetically obviates the need for millions of pounds of raw materials and enormous amounts of fossil fuels for energy to produce pesticides, hundreds of thousands of containers and packages that require disposal, thousands of gallons of fuel to distribute and apply the product, and millions of pounds of pesticide residue that pollute land and water. General Electric Plastics is testing ways of replacing metal components in automotive instrument panels with thermoplastic materials. The substitution allows suppliers to consolidate parts into fewer components for easier manufacturing. The thermoplastic panels will be more cost-effective and less environmentally damaging, since they will be easier to disassemble, recycle, and reuse. Similar product redesigns have been implemented by Korea's Samsung Electronics for refrigerators, washing machines, and televisions. *Dennis A. Rondinelli*

Ancillary Activities: supporting activities undertaken within any business to create the conditions for primary activities. Ancillary activities may take the form of pollution reduction in particular facilities in order to qualify for the expansion of primary activities under existing regulations that limit overall pollution to a particular level. A firm may also engage in significant environmental protection, individually or with other members of its industry, to maintain or improve its public image.

Appropriate Technology. See under SECTION 3.3.

Best Available Technology (BAT): technologies designated as most suitable for firms to adopt in order to comply with environmental regulation (e.g., particular types of scrubbers for reducing air pollutants; particular chemical processes for reducing noxious effluents). Identifying best available technologies is generally the responsibility of technical specialists in environmental agencies, with an obligation to assess the economic feasibility of new technologies as well as the costs for changing technologies when more effective ones become available. Nevertheless, regulations that require the adoption of a specified "best available technology" have been criticized as inefficient because they constrain the firm from adopting other technologies even when the specified best available technology is overly costly for a specific application or has become obsolete because of other technological advances. The business sector generally prefers environmental regulations that limit or charge for environmental damage, without dictating the technologies that the firms can choose to adopt.

Clean Technology. See under SECTION 3.3.

Five "R" Principles. See under SECTION 5.2.4.

Green Design: a concept of initially designing products and processes from a LIFE CYCLE PERSPECTIVE taking into consideration not only end uses but also product development, manufacturing, use, disposal, recycling, and/or reuse. It is closely associated with the concept of ADAPTED PRODUCTS. This approach is supported by governmental agencies of the United States, the United Kingdom, and Germany.

Product Stewardship: the concept of responsible and ethical management of a product throughout its life cycle—from the extraction of raw material inputs to final disposal. The concept was pioneered within the crop-protection chemical business in the 1980s and has been incorporated by some major chemical and energy businesses. In practical terms, product stewardship involves a systematic assessment of health, safety, and environmental hazards as well as the potential for exposure at each stage in the product's life and the implementation of appropriate risk reduction measures. *R. P.*

COMMENTARY ON PRODUCT STEWARDSHIP

Product stewardship is a concept taking hold in industrial countries seeking to curtail air pollution and solid and liquid wastes. Companies such as Dow Chemical, Procter and Gamble, and Scott Paper are responding by using product LIFE CYCLE ANALYSIS (LCA) to determine ways of reducing or eliminating wastes at all stages—from raw materials acquisition, production, distribution, and customer use to waste reclamation, recycling, reuse, and disposal. Japanese universities and research institutes are applying LCA to a wide range of products from aluminum cans, automobiles, and office buildings to vending machines, washing machines, and steel alloys. Firms serious about product stewardship seek alternative products and applications that are less polluting and alternative materials, energy sources, or processing methods that eliminate waste; compare the cost of managing for conformance versus for assurance; and adapt to customers' needs, preferences, and uses of products. *Dennis A. Rondinelli*

Remanufacturing: the RECYCLING of equipment or other products into the same or similar items. For example, the cases for computer-printer toner cartridges can be reused in the manufacture of new cartridges. As opposed to the recycling of the raw material components of used products, remanufacturing retains some of the value of the prior manufacturing, and reduces the energy costs as well as labor in the new manufacturing. Therefore this method, pioneered by Xerox Corporation, is believed to have higher environmental returns than simple recycling.

5.3.2 ECO-EFFICIENCY AND VOLUNTARY CODES

Business Charter for Sustainable Development: a voluntary code of environmentally sensitive behavior for corporations developed by a group of industry leaders in 1991 in preparation for the UNITED NATIONS CONFERENCE ON ENVIRONMENT AND DEVELOPMENT (UNCED) and currently endorsed by thousands of companies worldwide. The charter calls on industry to recognize environmental management as one of the highest corporate priorities and as a key to sustainable development. It contains sixteen principles, including integrated management, LIFE-CYCLE ANALYSIS, and the PRECAUTIONARY APPROACH.

Cradle-to-Grave Analysis. See LIFE-CYCLE ANALYSIS.

Eco-Efficiency: a term coined to describe business practices that highlight the positive connections between economic and ecological efficiency. Eco-efficiency is achieved when business efficiency is pursued through the reduction of ecological damage and resource inputs. It pertains to the entire product or service cycle. The concept challenges businesses to produce greater value for customers while minimizing resource use, pollution, and waste.

See Livio D. DeSimone and Frank Popoff, *Eco-Efficiency: The Business Link to Sustainable Development* (Cambridge, MA: MIT Press, 1997). R. P.

COMMENTARY ON ECO-EFFICIENCY

Advances in corporate environmental management require the adoption of a new philosophy that emphasizes what the World Business Council for Sustainable Development calls "eco-efficiency." Eco-efficiency encourages businesses to become more competitive, innovative, and environmentally responsible. It recognizes that economic growth and environmental quality are interdependent. Public policies can play a crucial role in encouraging businesses to integrate eco-efficiency practices into their overall business strategies, and in rewarding them for doing it. Firms attain eco-efficiency by reducing the energy and material intensity of goods and services, reducing toxic dispersion, enhancing material recyclability, maximizing sustainable use of renewable resources, extending product durability, and increasing the service intensity of goods and services. Developing policies that promote eco-efficiency will require not only reinventing environmental regulation but forging new partnerships among federal, state, and local governments and between the public and private sectors to discover, disseminate, and adapt innovative processes and technologies for improving environmental quality. See also SUSTAINABLE DEVELOPMENT. *Dennis A. Rondinelli*

Eco-Labeling. See Environmental Labeling.

Enviro-Capitalism: an emerging practice of using business tools to meet the growing demand for recreational and environmental amenities. Enviro-capitalists, for instance, encourage fee hunting to reward landowners for bearing the cost of providing habitat for wild animals and buy endangered species habitat instead of lobbying for regulations that restrict the use of private lands. Thus, they look for new opportunities to improve environmental quality by harnessing market forces at a lower cost. See also ecotourism.

See Terry Anderson and Donald Leal, *Enviro-Capitalists: Doing Good While Doing Well* (Lanham, MD: Rowman and Littlefield, 1997).

COMMENTARY ON ENVIRO-CAPITALISM

A typical example of enviro-capitalism is Ted Turner's Flying D Ranch. In return for a tax deduction, Turner granted conservation easements to the Nature Conservancy and converted what used to be a working cattle ranch into a bison ranch, which also provided a perfect habitat for elk, mule, deer, and other wildlife. The healthy size of each of these species' populations is controlled by selling hunting licenses. This system benefits the public in general (propagation of wildlife through better management), brings profits for the ranching enterprise, and provides employment for the guides and managers. In this case markets were used as a tool to turn wildlife into assets and then profits back into wildlife conservation. While enviro-capitalism is applicable in only a limited number of contexts, it may provide "win-win" results in some circumstances. *Natalia Mirovitskaya*

Environmental Accounting: the environmental auditing and environmental impact assessment framework used by corporations. Many corporations have developed increasingly elaborate methods of corporate environmental accounting. See also under section 5.1.1.

Environmental Audit. See under section 5.1.4.

Environmental Covenants: agreements with specific industries to set targets and design implementation instruments and procedures for monitoring environmental performance. Through environmental covenants, an industry sector agrees with government to meet certain environmental targets. Although the driving force for covenants may be the threat of regulation, it has been shown that in some circumstances environmental goals can be achieved—and even surpassed—more quickly through this mechanism than through legislation. A further benefit is that enforcement costs are typically lower than under a regulatory regime. Usually environmental covenants are voluntary agreements. However, in some countries (the Netherlands, for instance) they are legally binding contracts between government bodies and an industry association or group of companies, which specify commitments of businesses for reduction of environmental emissions and impact as well as government coresponsibilities. Some experts consider the use of covenants a major breakthrough in environmental policy. However, critics point out that the negotiation of a covenant acceptable to the industry may result in more modest targets than those set by the government alone. *R. P.*

Environmental Labeling. See under SECTION 5.2.3.

Environmental Voluntary Agreements: accords between private sector businesses or business associations and government environmental authorities that call upon businesses to undertake self-regulation and to meet particular environmental targets. These are often established through industry organizations such as manufacturers' associations. The main responsibility for determining how to meet these targets is left with the businesses themselves as much as possible. One form of voluntary agreement is the sector-wide ENVIRONMENTAL COVENANT; another is the pledge program that encourages action by individual corporations. Examples of voluntary agreements include energy-savings programs, pollution-prevention programs, and ENVIRONMENTAL LABELING programs. *R. P.*

Green Labeling. See ENVIRONMENTAL LABELING.

"Golden Carrot" Program: a market-oriented program of green technology innovation established in the United States to address the issues of climate change, ozone depletion, and energy efficiency together. The "golden carrot" idea was conceived in 1990 during discussions between Pacific Gas and Electric, the largest U.S. investor-owned electricity company, and the NATURAL RESOURCES DEFENSE COUNCIL on how utilities could get the maximum social benefits from their conservation programs, and was started in 1992 as a joint venture of twenty-four utility companies and the Environmental Protection Agency (EPA). In this program, utilities provide financial incentives ("golden carrots") for manufacturers to make major advances in energy efficiency and product performance. It is cheaper to finance technological innovations that save energy (e.g., in refrigerators, air conditioners, or lighting systems) than to build new power plants to run inefficient devices. The program leveraged greater private-sector investment in energy efficiency and pollution prevention, making the approach a valuable model for government leadership in promoting advanced technologies. See also DEMAND-SIDE MANAGEMENT.

Good Neighbor Agreements: a form of corporate-community relationship developed in the United States at the initiative of environmental groups. These agreements, negotiated between community representatives and businesses, stipulate actions that the companies must take. For instance, the Texas branch of Rhone Poulenc pays for independent environmental and safety audits, and funds community-conducted health studies and off-site monitoring. Entering into Good Neighbor agreements expedites the siting of facilities for companies and sometimes saves them from public litigation.

See Sanford Lewis, *The Good Neighbor Handbook* (Acton, MA: The Good Neighbor Project, 1992).

Life-Cycle Approach or Analysis (LCA; "Cradle-to-Grave" Analysis; Eco-Balancing): analysis and design of products and production to determine how to gain the greatest efficiency and minimize environmental damage, emphasizing all phases (raw material production, finished-product production, packaging, use, and disposal). The idea is to minimize energy use, emissions, and wastes in all phases. The life-cycle analysis is largely an engineering analysis rather than an economic analysis.

COMMENTARY ON LIFE-CYCLE ANALYSIS

Businesses are responding to environmental challenges by using product life-cycle analysis (LCA) to determine ways of reducing or eliminating wastes at all stages—from raw materials acquisition, production, distribution, and customer use to waste reclamation, recycling, reuse, and disposal. Using LCA, corporations can find alternative products and applications that are less polluting and alternative materials, energy sources, or processing methods that eliminate waste; compare the cost of managing for conformance versus for assurance; and adapt to customers' needs, preferences, and uses of products. Japanese universities and research institutes are applying LCA to a wide range of products from aluminum cans, automobiles, and office buildings to vending machines, washing machines, and steel alloys. Japan's Canon Corporation used life-cycle analysis to extend the life of its toner cartridges and make disposal less environmentally damaging. Procter and Gamble's Italian plant devised a method of printing directly on the plastic container of its dishwashing liquid product, eliminating the need for a shrink-sleeve label and saving ten tons of thin-film plastic a year. Kodak Pathé started collecting and recycling lead screens from its industrial film customers. In 1995, Kodak collected 4.1 tons of lead representing a 50 percent return on sales. By assessing its product and processing technologies, S. C. Johnson and Son was able to cut manufacturing waste nearly in half, reduce the use of virgin packaging materials by more than 25 percent, and reduce volatile organic compound (VOC) use by 16 percent between 1990 and 1995. The elimination of shrink wrap on auto aerosol products in Japan saved 4.2 tons of plastic. Development of lightweight shippers for liquid shoe polish in Mexico reduced corrugate by 125 tons and conversion to 70 percent recycled plastic in its five-liter pails in Brazil saved 110 tons of virgin plastic. Significant materials savings resulted from increasing recycled content of bulk delivery drums in Nigeria and adopting lightweight aerosol cans in Turkey.

See Michael A. Berry and Dennis A. Rondinelli, "Proactive Environmental Management: A New Industrial Revolution," *Academy of Management Executives* 12 (2) (1998): 38–50. *Dennis A. Rondinelli*

Natural Step: an international movement working to build an ecologically and economically sustainable society, founded in Sweden in 1989 by cancer researcher Dr.Karl-Henrik Robert. Natural Step disseminates a framework of scientifically based and easily understood principles of sustainable existence. In particular, "natural step" proponents argue that a) substances extracted from the earth should not accumulate in the ecosphere; b) society-produced substances should not accumulate systematically in the ecosphere; c) the physical conditions for production and diversity within the ecosphere must not be systematically eroded; and d) the use of resources must be efficient and fair with respect to meeting human needs. With the support of the king of Sweden, business executives, and political leaders, the Natural Step organization sent educational packages to every household and school in Sweden, outlining the steps needed to make Swedish society environmentally sustainable in the long term. In 1995, Natural Step began operation in the United States. Interface, Inc., a firm in the commercial interiors market with annual sales of $800 million, was the first American company adopting its framework. This movement is also popular in most European countries, Canada, and Australia.

See Natural Step's website at http://www.naturalstep.org.

Pollution-Prevention Pays (PPP) **Policy:** an approach developed by the Minnesota Mining and Manufacturing Company (3M Company) in the mid-1970s under the assumption that the reduction of pollution at the source rather than at the end of the production process was the key to sound environmental management. The PPP policy includes four interrelated processes: a) changing the resource basis of the process; b) modifying the production process to reduce harmful by-products; c) redesigning equipment to conform to more stringent operating conditions; and d) designing for recycling at the very start of a production process, including recovery of waste materials for reuse or sale. 3M business practices have been among the most successful examples of cost-efficient and sound environmental management in the private sector.

COMMENTARY ON POLLUTION-PREVENTION PAYS POLICY

Recent developments in the global economy are pushing firms in every industry to develop new strategies of competition and new processes for managing their environmental impacts. These trends include: a) a growing awareness of the relationships between economic and environmental sustainability; b) a better understanding of the business opportunities—both potential cost reductions and higher profits—in adopting quality environmental management practices; c) a growing realization in government and the private sector that regulatory controls, while necessary, are not sufficient to achieve pollution prevention; and d) growing international pressures on corporations to adopt voluntary standards for environmental management that go well beyond regulatory compliance as a precondition for participation in global trade and investment. Proactive corporate environmental policies reflect a growing realization that the objectives of sustainable environmental development and industrial progress are much the same—they both include maintaining growth and improving quality; satisfying the basic needs of life such as jobs, food, energy, water, and sanitation; conserving and enhancing the natural resource base from which many industries derive inputs; reorienting technology; and managing risk.

See Dennis A. Rondinelli and Michael A. Berry, "Industry's Role in Air Quality Improvement: Environmental Management Opportunities for the 21st Century," *Environmental Quality Management*, 7 (4)(1997): 31–44. *Dennis A. Rondinelli*

Primary Environment Care (PEC): an approach to empower communities to control the management of environmental resources, provide for the basic livelihood and health needs for all in the community, and enable all community members to pursue their self-directed development. PEC was modeled after the ideas and symbolism of primary health care. The approach was originally advocated by Canada's INTERNATIONAL INSTITUTE FOR SUSTAINABLE DEVELOPMENT (IISD) and is currently promoted by the United Nations Children's Fund (UNICEF) and some major NONGOVERNMENTAL ORGANIZATIONS. PEC has been credited with success in a few cases at the local level. However, skeptics argue that the PEC approach cannot operate without sound higher-level policies and practices, and can easily be undermined by economic restructuring and AUSTERITY MEASURES.

Responsible Care: a voluntary environmental program launched by the American chemical industry in 1988 in response to the challenge of PRODUCT STEWARDSHIP. The members of this program pledged to track their products through their life cycle (from manufacture to final disposal) and to adhere to a set of environmental standards higher than required by federal legislation. The Responsible Care program combines several newly emerging principles of business environmental management, including LIFE-CYCLE ANALYSIS and proactive communication with public. The Responsible Care program is currently used as a benchmark against which environmental performance of other businesses is compared. A similar initiative has been launched by the European Council of Chemical Manufacturers' Federations.

5.3.3 ENVIRONMENTALLY SENSITIVE BUSINESS MANAGEMENT

Corporate Environmental Reporting: a growing practice of businesses to publish information relating to environmental performance of the company. Currently, no uniform definition, much less standard, exists for such reporting; therefore existing practices vary greatly in their format, contents, and reliability. Some businesses disseminate carefully crafted reviews of the perceived environmental impacts of their businesses accompanied by numeric information on pollution from individual plants and its dynamics as well as details on pollution prevention and safety (the Anglo-American model of corporate reporting). In contrast, German companies use the "eco-balance" approach, which reflects the difference between the natural resources used by the company and the output of pollutants and useful products. Many companies prefer to do their environmental reporting simply through a statement of their "green" values.

See Frances Caincross, *Green, Inc.: A Guide to Business and Environment* (Washington, DC: Island Press, 1995).

Demand-Side Management: an approach to pollution prevention that originated in the utility industry, but has spread to other industries as well. It focuses on understanding customers' needs and preferences and on their use of products. It seeks to minimize or eliminate wasted product, to sell customers exactly what they demand, and to make the customer more efficient in the use of the product. Demand-side management forces an industry to look at itself in a new light, which often leads to the discovery of new business opportunities. In the case of the utilities, demand-side thinking emphasized that companies are not primarily in the business of selling electricity or gas, or even light or heat, but are really in the business of selling environmental conditions such as comfort, brightness, and conveyance. By looking at the market in terms of real demand, utilities can prosper by providing customers a variety of environmentally beneficial services and not just electricity. *D. A. R.*

Design for Environment (DFE): product design that minimizes pollution and waste, very similar to the LIFE-CYCLE APPROACH. DFE is becoming an integral part of pollution prevention in proactive environmental management. Corporations such as AT&T, Xerox, Hewlett Packard, and Baxter International are finding that it is far more efficient to design products for disassembly, modular upgradability, and recyclability at the outset than to deal with disposal problems at the end of a product's life. Procter and Gamble's objective, for example, is to design manufacturing waste out of business areas that account for at least 50 percent of its production volume by the beginning of 1998. DFE reduces reprocessing costs and returns products to market more quickly and economically. *D. A. R.*

Environmental Audit. See under SECTION 5.1.4.

Environmental Benchmarking: a method of comparing, ranking, or rating different business processes, units, or companies against environmental standards. The goal of environmental benchmarking is to make firms accountable for their environmental performance and to assess the effectiveness of alternatives for improving performance.

Environmental Options Assessment: an analytic framework that combines elements of LIFE-CYCLE ANALYSIS and ENVIRONMENTAL IMPACT ASSESSMENT. Using an environmental options assessment, business managers should be able to include environmental criteria in their analysis of product development and strategies, allowing them to generate strategies, assign priorities, and plan options for the four main aspects of material management: a) raw materials used; b) products consumed; c) substances emitted; and d) wastes recycled. This method provides managers with an environmentally sensitive investment system.

See Pieter Winsemius and Walter Hahn, "Environmental Option Assessment," *Columbia Journal of World Business* 27 (3–4) (1992): 248–66.

European Commission's Eco-Management and Audit Scheme (EMAS): a voluntary program of environmental certification that has been in effect in the EUROPEAN UNION (EU) since 1995. Its objective is to promote continuous improvement of corporate environmental performance. EMAS requires companies to establish and standardize environmental management and reporting systems and to publish detailed reports on their environmental management and performance.

Industrial Ecology: a new field in environmental studies that seeks to develop an ecological model of industrial systems. The industrial system under this approach is seen as a web of interconnected production units, linked by the flow of materials and energy determined by physical and ecological principles. The challenge is to manage the system in such a way that materials and energy are retained within the system, and that production systems and product cycles are integrated with natural ecosystems and material cycles. To ensure this, such techniques as LIFE-CYCLE ANALYSIS, FULL-COST ACCOUNTING, and DESIGN FOR THE ENVIRONMENT (DFE) are used. Principles of industrial ecology can be applied at different spatial levels: within and between companies, within industrial zones or communities, and at regional or national levels.

See Braden R. Allenby and Deanna J. Richards, eds., *The Greening of Industrial Ecosystems* (Washington, DC: National Academy Press, 1994).

COMMENTARY ON INDUSTRIAL ECOLOGY

As the need for proactive environmental management becomes clear, the search for innovative approaches to pollution prevention is moving beyond individual firms to intercorporate networks and strategic alliances. The concept of industrial ecology provides firms with new frameworks for working together to solve environmental problems. Industrial ecology models industries like living organisms where the byproduct of every metabolic process is food for another organism (see INDUSTRIAL METABOLISM). Industrial ecology and zero-pollution concepts are making it easier for corporations located in close physical proximity to design facilities that exchange energy and materials to develop eco-industrial parks where plants from different industries can exchange inputs and outputs with each other to their mutual benefit and to the benefit of the environment. *Dennis A. Rondinelli*

Industrial Metabolism: one of the basic notions of INDUSTRIAL ECOLOGY, first introduced by R. U. Ayres (1989). This concept views an industrial process as a parallel to the metabolic processes of a living organism. However, while the inputs and outputs of living organisms are in balance with the ECOSYSTEM, in the industrial process the inputs (energy and materials) and outputs (biomass and work) are not. Therefore the industrial metabolism model also encompasses a materials balance concept, which is the means of identifying the dissipative elements within an industrial process in order to reduce this dissipation.

See Robert U. Ayres, "Industrial Metabolism," in *Technology and Environment*, eds. Jesse H. Ausubel and Hedy E. Sladorich (Washington, DC: National Academy Press, 1989), 23–49.

ISO 4000. See under SECTION 4.3.2.

Porter Hypothesis: the argument that stringent environmental policies benefit the commercial interests of businesses and therefore may improve national competitiveness. This hypothesis, first presented by Michael Porter (Harvard Business School) assumes that businesses have large unexplored opportunities for improving products and saving money, that governmental regulations might force them to use these opportunities, and that subsequent product developments will pay better than the investments these companies would have made on their own. Extending this argument, Porter stated that European countries and Japan, by imposing tough environmental regulations, were wresting competitive advantage from the United States. This hypothesis is a logical extension of Porter's value chain analysis—a method of systematically analyzing and structuring all corporate activity and its relationships as a source of possible COMPETITIVE ADVANTAGES. Porter's hypothesis outlining the "win-win" strategy of clean environment and prospering economy has been a point of heated debate in environmental, economic, and political circles.

See Michael Porter, *Competitive Advantage: Creating and Sustaining Superior Performance* (New York: Free Press, 1986); Michael Porter, "America's Green Strategy," *Scientific American* (April 1991).

Portfolio Analysis: a management tool that examines all activities and risks holistically in order to determine the best balance between risk and opportunity. With a portfolio analysis applied to a firm's environmental strategy, the company can balance environmental risk and market opportunities.

See Kurt Fisher and Johan Schot, eds., *Environmental Strategies of Industry* (Washington, DC: Island Press, 1993).

Total Quality Environmental Management (TQEM): a business strategy that seeks to incorporate environmental protection and energy efficiency in all aspects of business operations and at all levels of a company, based on the assumption that both environmental quality and product quality are directly linked to long-term profits. TQEM was created by a group of large U.S. corporations as an extension of the Total Quality Movement introduced in Japan after World War II, to which some attribute Japan's extraordinary economic revival. It promotes continuous learning, assessment, data-driven decision tools, training, and other standard practices to integrate environmental management into a firm's business management. See also INDUSTRIAL ECOLOGY.

See Global Environmental Management Initiative, *Total Quality Environmental Management: A Primer* (Washington, DC: GEMI, 1992).

COMMENTARIES ON TOTAL QUALITY ENVIRONMENTAL MANAGEMENT

Corporations that adopt total environmental quality must develop a proactive environmental management system that not only sets out goals and objectives but that also measures performance aimed at continuous environmental improvement. Experience suggests that achieving total environmental quality requires companies to integrate several basic principles into their overall business strategy. These principles include: a) adopting an environmental policy that seeks to eliminate pollution based on life-cycle assessment of the firm's operations and communicating the policy throughout the company and to corporate STAKEHOLDERS; b) objectively assessing the effectiveness of environmental programs; c) comparing the company's environmental performance to that of the leading firms in the industry through BENCHMARKING and BEST-PRACTICES ASSESSMENTS; d) promulgating a company view that environmental performance is the responsibility of all employees; e) analyzing the impact of environmental issues on the future demand for products and the competitive economics of the industry; f) encouraging frequent discussion of environmental issues and activities at board meetings; g) developing and applying a formal system for monitoring proposed regulatory changes and for complying with changing regulations; h) routinely conducting environmental due diligence on potential acquisitions; i) developing budgets for environmental expenditures so that the firm does not incur surprise expenses that materially affect profitability; and j) identifying and quantifying environmental liabilities from past operations and developing plans for minimizing those liabilities.

See Micheal A. Berry and Dennis A. Rondinelli, "Proactive Environmental Management: A New Industrial Revolution," *Academy of Management Executives* 12 (2) (1998): 38–50. *Dennis A. Rondinelli*

Sometimes corporations profit from being environmentally responsible. But let's not lose sight of the fact that environmental protection is not intrinsically profit-maximizing. To a certain degree, "total environmental quality management" is happy talk implying that there is no conflict between profit and environmental protection. *William Ascher*

Voluntary Eco-Auditing. See ENVIRONMENTAL AUDIT.

5.3.4 BUSINESS ORGANIZATIONS AND INITIATIVES RELATED TO ENVIRONMENT

Business Council of Sustainable Development (BCSD): an international organization created in 1991 under the direction of Swiss industrialist Stephen Schmidheiny with the goal of bringing a business perspective to the UNITED NATIONS CONFERENCE ON ENVIRONMENT AND DEVELOPMENT (UNCED; RIO CONFERENCE; EARTH SUMMIT). BCSD members were forty-eight chief executive officers of major world corporations. Council members advocated changes in consumption patterns as well as price reform to reflect environmental costs of production use, recycling, and disposal. In 1995, the Geneva-based CBSD merged with the World Industry Council for the Environment, an initiative of the INTERNATIONAL CHAMBER OF COMMERCE, to form the WORLD BUSINESS COUNCIL FOR SUSTAINABLE DEVELOPMENT (WBSD).

CERES Principles: formerly the Valdez principles, a code of corporate environmental conduct in ten principles developed in the United States by the Coalition for Environmentally Responsible Economics (CERES) in 1990 after the notorious *Exxon Valdez* oil spill. CERES principles include much stronger obligations than any other attempts to establish standards for the international performance of corporations. Among the principles are provisions for the voluntary disclosure of pollution incidents and potential environmental hazards, performance of annual environmental audits, and the appointment of an environmental representative to each company's board of directors. To date, the CERES principles have been adopted by more than fifty corporations, including Bethlehem Steel, Ford Motor Company, and General Motors, but this is, of course, only a small fraction of U.S. companies.

Global Environmental Management Initiative (GEMI): a not-for-profit organization of major North American and Western European multinational companies that strives to foster environmental, health, and safety codes of business conduct. Its membership (including AT&T, Coca-Cola, Dow Chemical, DuPont, and Georgia-Pacific Corporation) represents a cross-section of industry with over one million employees and combined revenues exceeding $400 billion. GEMI promotes the use of TOTAL QUALITY ENVIRONMENTAL MANAGEMENT (TQEM) in a number of activities, including publishing a series of reports and primers on TQEM, BENCHMARKING, environmental reporting, ENVIRONMENTAL COST ACCOUNTING, ISO 14000, and environmental health and safety training. Though GEMI does not demand strong public disclosure policies or annual environmental audits, as called for by the CERES PRINCIPLES, it certainly goes beyond governmental regulatory standards.

International Chamber of Commerce (ICC): an international NONGOVERNMENTAL ORGANIZATION founded in 1919 to promote private investment, free trade, and the market economy and to represent business interests at governmental and international levels. As of 2000, the ICC had national committees and other members—individual companies, trade and industrial organizations—in some 130 countries, supported by over 7,000 corporations. The ICC Commission on Environment (ICCCE) was established in 1978 to devise environmental policies for businesses and to lobby governments on their behalf. In particular, the ICCCE provides business input into Climate Change Negotiations and into the work of the International Standards Organization on the elaboration of ISO 14000. The ICC adopted its Business Charter for Sustainable Development in 1990. The ICCCE was complemented in 1986 by the formation of the ICC's INTERNATIONAL ENVIRONMENT BUREAU (EIB).

International Environment Bureau (EIB): a specialized division of the INTERNATIONAL CHAMBER OF COMMERCE (ICC), set up in 1986 and funded by corporations from different nations. Its official goal is to promote the efficient management of the environment in such a way as to achieve sustainable economic growth. The IEB serves as a clearinghouse for technological and environmental expertise for businesses and other parties, especially in developing countries.

International Organization for Standardization (ISO). See under SECTION 4.3.2.

United Nations Industrial Development Organization (UNIDO). See under SECTION 4.2.4.

World Business Council for Sustainable Development (WBCSD): a coalition of international companies and regional and national business councils committed to economic growth and SUSTAINABLE DEVELOPMENT. By 2000 the WBCSD had 125 corporate members from 30 countries (largely from Western Europe, North America, and Japan), and 9 business council members. The council was founded in 1995 through a merger of the Business Council for Sustainable Development with the World Industry Council for the Environment. WBCSD is a major voice for the business community on issues of sustainable development. To promote environmental risk management in business, the WBCSD submits proposals to governments and business leaders relating to designing and implementing sustainable development policies. It also assists its members in technology and best-practices transfer. Radical environmental organizations such as GREENPEACE have criticized the WBCSD for minimizing the importance of environmental problems, exaggerating the effectiveness of business-sector environmental efforts, and supporting globalization initiatives.

World Economic Forum: an annual gathering of world political and business leaders that takes place in Davos, Switzerland. The forum is considered one of the most influential meetings of the world's decisionmakers. Traditionally its meetings were organized around macroeconomic issues. However, recently issues of SUSTAINABLE DEVELOPMENT have been included in the forum's agenda.

COMMENTARY ON BUSINESS SECTOR RESPONSES

The business responses to the challenge of SUSTAINABLE DEVELOPMENT have gone through roughly three phases. Initially, the focus was on "END-OF-PIPE" clean-up measures driven by the demands of regulation and public pressure. The second phase, which began in the 1980s, was focused on the concept of ECO-EFFICIENCY, in which "win-win" benefits were obtained through improved economic and environmental performance. In the third phase, into which some leading companies have been only recently entering, environmental performance is integrated into corporate strategic planning as a factor of competitive advantage. *Natalia Mirovitskaya*

5.4 Population and Consumption Responses

Cairo Conference on Population and Development. See UNITED NATIONS CONFERENCE ON POPULATION AND DEVELOPMENT.

International Organization of Consumer Unions (IOCU): an independent international organization that coordinates activities of over one hundred national consumer organizations. It was created in 1960 to foster cooperation in consumer protection and public education. It also acts as a clearinghouse and information center. The IOCU is headquartered in Malaysia and the Netherlands.

United Nations International Conference on Population and Development (ICPD; **Cairo Conference**): the United Nations International Conference on Population and Development was held 5–13 September 1994 in Cairo, Egypt. Representatives from more than 170 countries attended the conference. The ICPD Declaration was the first United Nations population conference document to officially recognize the interrelationship among population, environmental protection, and development, in contrast to the previously narrow focus on demographics. Attendees expressed satisfaction with the advances made in the Program of Action with regard to the role of nongovernmental organizations, the empowerment of women, reproductive health, reproductive rights, and treatment of unsafe abortion as a public health issue. The Cairo Conference was the third major intergovernmental meeting on population, following the 1974 Bucharest Conference and the 1984 Mexico City Conference. *T. S.*

United Nations Population Fund (formerly United Nations Fund for Population Activities [UNFPA]): the only multinational body that focuses exclusively on population issues. Created in 1969 as a subsidiary of the United Nations General Assembly, UNFPA coordinates the collection and dissemination of data on demographic trends around the world and promotes awareness of population problems and strategies that can be used to address them. Over the years it has provided assistance to most developing countries, in particular for maternal and child care and family planning. While the UNFPA provides no funding for abortions or abortion-related activities, its programs were hampered by a suspension of contributions from 1984 to 1993 by the United States, which opposed the fund's support for population programs in China that the United States considered coercive. UNFPA played a key role in planning the UN INTERNATIONAL CONFERENCE ON POPULATION AND DEVELOPMENT held in Cairo in 1994. It has taken the lead in facilitating the implementation of the plan of action adopted at the Cairo meeting and in preparing for a five-year progress report that culminated in a special session of the General Assembly in 1999.

See Barbara Crane, "International Population Institutions: Adaptation to a Changing World Order," in *Institutions for the Earth: Sources of Effective International Environmental Protection*, eds. Peter M. Haas, Robert O. Keohane, and Marc A. Levy (Cambridge, MA: MIT Press, 1993), 351–92. *M. S. S.*

Zero Population Growth: 1) the concept of stabilizing population at its current level, with birth and death rates in balance.

2) a U.S.-based nongovernmental organization created in 1968 by prominent environmentalists, demographers, and other scientists concerned with population pressure. It currently has over 300 local chapters whose primary mission is public education on global population issues. The group seeks to stabilize the population of the United States, which it considers to be crucial for control over global social, environmental, and resource problems.

5.5 International Environmental Management

5.5.1 BASIC CONCEPTS AND ANALYTICAL APPROACHES

Blocking (Veto) State: in international negotiations a nation-state that by virtue of its influence on a particular issue is able to block or weaken an international agreement. Blocking power enhances the bargaining leverage of any state and may give it the chance to come out of negotiations with the best deal. For instance, the Indian delegation to the 1990 London Conference of the Parties to the MONTREAL PROTOCOL at first seemed to oppose any binding agreements but eventually settled for a compromise that provided India with large-scale financial and technical assistance. If India, a major user of ozone-damaging aerosols, had not agreed to the protocol, it would have been difficult to secure the agreement of other countries.

Environmental Diplomacy: diplomatic practices by governments involved in global, transboundary, and transnational environmental and resource issues. Environmental diplomacy is a relatively new branch of international relations and is the main form of decision process through which international environmental policy is made. In distinction from other branches of international relations, environmental diplomacy has a strong transboundary dimension, comparatively high degree of issue complexity, and a high level of scientific and technical uncertainty. It is also strongly influenced by interest groups, media, scientific, and economic considerations.

See Gunnar Sjostedt, ed., *International Environmental Negotiations: Process, Issues and Contexts* (Stockholm: FRN, 1991).

Hegemonic Power: in environmental diplomacy the dominant power of a nation that has significant control over international arrangements and receives sufficient benefits to warrant that nation's willingness to underwrite the continuation of the arrangements. The theory of hegemonic stability argues that a hegemon is necessary for stable international regimes. Some observers argue that strong international environmental regimes cannot be established, either because there is no sufficiently hegemonic power to underwrite a set of strong environmental agreements, or because the hegemon (often the United States on the global level) does not have an interest in establishing a strong environmental order.

See Robert Pahre, *Leading Questions: How Hegemony Affects the International Political Economy* (Ann Arbor: University of Michigan Press, 1999).

Human Needs Theory: a conceptual approach to environmental conflict analysis and resolution, rooted in hierarchical needs-based theory of self-actualization and research on the physiological and sociobiological causes of conflict. In this approach, environmental conflicts are seen as a form of disagreement between parties ascribing different values to environmental goods and services. Resource disputes, in contrast, are regarded as conflicts of interest in which parties may disagree on the distribution of a resource or environmental goods, while basically assenting on the value of this resource or goods. The distinction between value-based and distribution-based disputes has important implications for the conflict management strategy.

See John Burton, ed., *Conflict: Human Needs Theory* (London: Macmillan, 1990).

Lead State: in environmental diplomacy a nation-state that sponsors and asserts leadership on behalf of the most advanced proposal for international regulation. The lead state has a strong commitment to effective action on the issue, moves the process of negotiations forward by proposing its own negotiating formula as the basis for an agreement, and attempts to get the support of other actors. Lead-state influence can take many forms: identification and definition of the particular environmental problem, raising public awareness, unilateral action, use of diplomatic clout, threatening trade sanctions and/or pledging financial resources to sway other states. The role of lead state is not static nor is it completely solitary. Over time, the role of a lead state may shift from country to country or groups of countries. For instance, in the case of the ozone regime, Sweden and Finland first took a lead role only to be replaced by the United States when they proposed actual target reductions. The lead state designation differs from HEGEMONIC POWER in that the hegemon is defined as the most powerful nation, which is prepared to underwrite the new arrangements; the lead state refers to a highly active, initiating nation, which may or may not be hegemonic.

See Gareth Porter and Janet Brown, *Global Environmental Politics* (Boulder, CO: Westview Press, 1996).

Supporting State: in environmental diplomacy, a state that is willing to support and work for the most far-reaching proposal in international environmental negotiations.

See Gareth Porter and Janet Brown, *Global Environmental Politics* (Boulder, CO: Westview Press, 1996).

Swing State: a nation-state that has the flexibility to take different positions on a given international issue, allowing it to gain significant concessions in negotiations to establish an INTERNATIONAL REGIME. These concessions frequently involve financial and technical assistance, to support the country's economic and institutional transformation in order to comply with the rules of the new regime. India and China were engaged in swing-state diplomacy during the 1990 negotiations of the ozone-producing countries that signed the MONTREAL PROTOCOL. On that occasion both nations abstained from participation until they were given guarantees of technology transfer.

See Gareth Porter and Janet Brown, *Global Environmental Politics* (Boulder, CO: Westview Press, 1996).

Veto State. See BLOCKING (VETO) STATE.

5.5.2 GLOBAL CONFERENCES ON ENVIRONMENT AND DEVELOPMENT

Cairo Conference. See UNITED NATIONS INTERNATIONAL CONFERENCE ON POPULATION AND DEVELOPMENT (ICPD; CAIRO CONFERENCE).

Earth Summit. See UNITED NATIONS CONFERENCE ON ENVIRONMENT AND DEVELOPMENT.

Rio Conference. See UNITED NATIONS CONFERENCE ON ENVIRONMENT AND DEVELOPMENT (UNCED).

Stockholm Conference. See UNITED NATIONS CONFERENCE ON HUMAN ENVIRONMENT (UNCED).

United Nations Conference on Environment and Development (UNCED; Rio Conference; Earth Summit): the highest-profile international environmental conference ever held, with a huge official conference and an even bigger unofficial conference of NONGOVERNMENTAL ORGANIZATIONS and other interested parties. UNCED was held in June 1992 in Rio de Janeiro, Brazil. Nearly one hundred leaders from around the world met to discuss the issues related to development and the environment. Four preconference sessions of the Preparatory Committee of the Earth Summit were held to set the agenda for the meeting. AGENDA 21 evolved out of the Earth Summit as the central agreement and guiding document to tackle the world's ecological and economic problems. In addition to Agenda 21, several other documents emerged from the Earth Summit, including the CONVENTION ON BIODIVERSITY, the CONVENTION ON CLIMATE CHANGE, the RIO DECLARATION ON ENVIRONMENT AND DEVELOPMENT (also known as the EARTH CHARTER), and the Statement of Forest Principles. *T. S.*

United Nations Conference on the Human Environment (Stockholm Conference): the first worldwide environmental conference, held in Stockholm in June 1972, attended by 114 states (not including, however, the Soviet Union and the other socialist-bloc countries) and over 400 intergovernmental and NONGOVERNMENTAL ORGANIZATIONS. The Stockholm Conference is frequently described as a watershed in the development of international environmental policy and law. It also became the first international forum to expose confrontational approaches of developing and industrialized countries toward environment and development issues. The Stockholm Conference approved a DECLARATION ON THE HUMAN ENVIRONMENT containing twenty-six principles on the management of global environment, an Action Plan with 109 recommendations on international environmental cooperation, and the Resolution on Institutional and Financial Arrangements. The first four global environmental agreements (the WORLD HERITAGE CONVENTION, MARPOL, CITES, and the CONVENTION ON THE PREVENTION OF MARINE POLLUTION BY DUMPING OF WASTES AND OTHER MATTER) were negotiated at the Stockholm Conference or on its recommendation. Also on the recommendation of the Stockholm Conference, the UN General Assembly created the UNITED NATIONS ENVIRONMENT PROGRAMME (UNEP).

United Nations Conferences on Human Settlements (Habitat I and II): international conferences on urbanization held in Vancouver, Canada, in June 1976 and Istanbul, Turkey, in June 1996, respectively. They addressed a wide range of problems related to urban areas where a rapidly growing proportion of the world's population lives, especially in developing countries. Habitat I led to the establishment in 1978 of the UN Center for Human Settlements, based in Nairobi, Kenya, and the adoption of the Global Strategy for Shelter to the Year 2000 by the General Assembly in 1988. Habitat II, which was the last major UN conference of the 1990s, was notable for inviting the participation of representatives of local governments and grassroots organizations. Considerable attention was given to strategies for alleviating the growing shortages of housing in major cities and to the question of whether international law recognizes a basic human right to housing. Emphasis was also placed on creating environmentally sustainable cities by avoiding destructive and wasteful patterns of natural resource use.

See Gail V. Karlsson, "Habitat II," in *A Global Agenda: Issues Before the 51st General Assembly of the United Nations,* eds. John Tessitore and Susan Woolfson (Lanham, MD: Rowman and Littlefield, 1996), 253–59. *M. S. S.*

5.5.3 INTERNATIONAL AGENCIES AND COMMISSIONS

Arctic Council: a high-level intergovernmental forum established on September 19, 1996, by the eight Arctic states: Canada, Denmark, Finland, Iceland, Norway, the Russian Federation, Sweden, and the United States. In addition, the status of permanent participants is granted to three organizations that represent the majority of Arctic indigenous people: the Inuit Circumpolar Conference, the Saami Council, and the Association of Indigenous Minorities of the North, Siberia, and the Far East of the Russian Federation. The council meets at the ministerial level biennially and operates on a consensus basis. The chair and secretariat of the council rotate concurrently every two years among the eight Arctic states, beginning with Canada in 1996, the United States in 1998, and Finland in 2000. The main activities of the organization are the protection of the Arctic environment and SUSTAINABLE DEVELOPMENT as a means of improving the economic, social, and cultural well-being of the North. The council coordinates implementation of the programs established under the Arctic Environmental Protection Strategy, in particular the Program for the Conservation of Arctic Flora and Fauna, the Arctic Monitoring and Assessment Program, the Program for the Protection of the

Arctic Marine Environment, the Program on Emergency Response, and the Program on Sustainable Development and Utilization.

Canada/USA International Joint Commission (ICJ). See under SECTION 6.8.

Commission on Sustainable Development (CSD): a body emerging from the 1992 UNITED NATIONS CONFERENCE ON ENVIRONMENT AND DEVELOPMENT (UNCED) to oversee and implement the UNCED agreements. These include AGENDA 21; the RIO DECLARATION ON ENVIRONMENT AND DEVELOPMENT; and the nonbinding Authoritative Statement of Principles for a Global Consensus on the Management, Conservation, and Sustainable Development of All Types of Forests (also known as the STATEMENT OF FOREST PRINCIPLES). The commission is composed of fifty-three members elected from UN member states for terms of three years and meets annually for a period of two to three weeks. Thirteen members are elected from Africa; eleven from Asia; ten from Latin America and the Caribbean; six from Eastern Europe; and thirteen from Western Europe. *T. S.*

Global Environment Facility (GEF). See under SECTION 4.2.4.

United Nations Commission on Human Settlements (UNCHS): the intergovernmental policy-making body established by the UN General Assembly to coordinate human settlements activities in the UN system. It meets biennially with the UNITED NATIONS CENTER FOR HUMAN SETTLEMENTS (HABITAT), with UNCHS serving as the secretariat. The commission's activities include research and publication in: a) settlements policies and strategies; b) settlement planning; c) shelter and community services; d) development of the indigenous construction sector; e) low-cost infrastructure for human settlements; and f) institutions and management. The commission's work is also connected with the activities of United Nations Habitat and Human Settlements Foundation, set up in 1975 to assist developing nations in strengthening their environmental programs relating to all aspects of human settlements. The foundation provides seed capital for "habitat" projects and facilitates technical cooperation.

United Nations Commission on Sustainable Development. See COMMISSION ON SUSTAINABLE DEVELOPMENT.

United Nations Educational, Scientific and Cultural Organization (UNESCO): the UN specialized agency focusing on education, science, culture, and communication. With 186 member states as of 1999, UNESCO undertakes research on education, science, and culture; provides technical and advisory services to its members through consulting and training; disseminates information; oversees standards setting; and provides financial support to relevant nongovernmental organizations through fellowships and travel grants. Recent conferences of UNESCO have focused on the role and importance of education in developing countries, reflecting UNESCO's commitment to education as a human right in accordance with UN human rights definitions.

United Nations Environment Programme (UNEP): a UN program created by the UN General Assembly in December 1972 in accordance with the recommendations of the UNITED NATIONS CONFERENCE ON THE HUMAN ENVIRONMENT (STOCKHOLM CONFERENCE). From a political perspective, UNEP's powers, functions, structure, budgetary arrangements, and location (Nairobi, Kenya) have been a compromise between developed and developing countries. It was established not as a specialized agency, but as a program within the UN system. UNEP does not have executive powers and is controlled by a fifty-eight-member governing council. Administrative expenses come from the UN general budget, but all program activities are funded from a voluntary fund; therefore insufficient funds has been a limiting factor throughout UNEP's history. Despite the limitations in its mandate, powers, funding, and geographical isolation, however, UNEP's achievements are considerable. It has created a comprehensive system of global environmental assessment (see GEMS, INFOTERRA, IRPTC), has been instrumental in negotiating over fifty multilateral environmental agreements and, in particular, in creating international agreements on the OZONE LAYER, BIODIVERSITY, and HAZARDOUS SUBSTANCES. Of all the UN organizations and programs, UNEP is probably the most involved in cooperation with nongovernmental actors.

United Nations Non-Governmental Liaison Office (UN-NGLS): an interagency unit of the United Nations system established in 1975 to support interaction between the UN and NONGOVERNMENTAL ORGANIZATIONS, mainly on international issues of SUSTAINABLE DEVELOPMENT. This has included the coordination of nongovernmental organization participation in major UN conferences as well as providing them with outreach and publications on a wide range of environment- and development-related issues. During the last fifteen years, UN-NGLS efforts have greatly strengthened the nongovernmental organization role in environmental policy making, by helping these organizations to inject their views, technical abilities, and experiences into major international action plans and strategies.

United Nations Revolving Fund for Natural Resources Exploration: a special-purpose fund, set up in 1974 and directed by the UNITED NATIONS DEVELOPMENT PROGRAMME with the main task of facilitating exploration for natural resources in developing countries. The fund provided the expertise and technical equipment for exploration programs and also helped to mobilize the resources for feasibility studies. An initial belief that the fund would be replenished by recycling funds from commercially profitable projects was not realized; instead it depended on contributions from governments, which kept its operations to a relatively small scale. The fund was phased out by the end of 2000. Some observers point to this as an example of the withdrawal of the official international community from direct economic activity.

United Nations Specialized Agencies: task-oriented organizations affiliated with the United Nations but autonomous in their operations. These agencies deal with economics (WORLD BANK GROUP, INTERNATIONAL MONETARY FUND [IMF], UN INDUSTRIAL DEVELOPMENT ORGANIZATION [UNIDO]); labor and social issues (INTERNATIONAL LABOUR ORGANISATION [ILO]); cultural and educational issues (UN EDUCATION, SCIENTIFIC, AND CULTURAL ORGANIZATION [UNESCO]); agriculture (UN FOOD AND AGRICULTURAL ORGANIZATION [FAO], INTERNATIONAL FUND FOR AGRICULTURAL DEVELOPMENT [IFAD]); health (WORLD HEALTH ORGANIZATION [WHO]); and other issues (International Civil Aviation Organization [ICAO], the INTERNATIONAL TELECOMMUNICATIONS UNION [ITU], the Universal Postal Union [UPU], WORLD METEOROLOGICAL ORGANIZATION [WMO], Intergovernmental Maritime Consultative Organization, and the World Intellectual Property Organization). Each agency has been established by an international treaty and has its own statutes, budget, governing body, and staff. In some cases, such as the World Bank Group and the International Monetary Fund, the connection with the United Nations is practically in name only. These organizations are not bound by the principles established by the UN General Assembly or other UN bodies; neither are obligated to take human rights considerations into account in their decisions—indeed, their charters preclude taking any sort of political considerations into account. Specialized agencies derive about 40 percent of their operational funds from the United Nations programs and funds. Member states also provide them with resources for particular projects. Currently, almost every specialized agency has a department or program relating to SUSTAINABLE DEVELOPMENT.

World Health Organization (WHO): the largest of the United Nations specialized agencies, WHO was established in 1948 to pursue "the attainment by all peoples of the highest possible level of health," which the organization defines as a state of complete physical, mental, and social well-being and not merely the absence of disease or infirmity. To reach this goal, the WHO emphasizes eight main elements: public education on health problems; proper food supply and nutrition; safe water and sanitation; maternal and child health, including family planning; immunization against major infectious diseases; prevention and control of local diseases; appropriate treatment of common diseases and injuries; and provision of essential drugs. The WHO's functions consist of: a) acting as the directing and coordinating authority on international health work; b) promoting technical cooperation; c) assisting governments in strengthening health services; d) furnishing appropriate technical assistance and, in emergencies, necessary aid; e) stimulating and advancing work on the prevention and control of epidemic, endemic, and other diseases; f) promoting the improvement of nutrition, housing, sanitation, recreation, economic, and working conditions and other aspects of environmental hygiene; g) promoting coordinated biomedical and health services research; h) promoting improved standards for biological, pharmaceutical, and similar products and procedures; and i) fostering activities in the field of human health, especially activities affecting the harmony of human relations.

The general record of the WHO is widely regarded as successful, not just in the control and eradication of such life-threatening diseases as smallpox, bubonic plague, malaria, and typhoid but also in assisting developing countries in the creation of health services and training facilities. In addition to combating disease, the WHO has launched diverse environmental projects. Recently, the WHO adopted a new policy, called "Health for All for the 21st Century," to provide a bridge between its "Health for All" policy and SUSTAINABLE DEVELOPMENT. The four main components of this new policy are to a) combat poverty; b) promote health in all settings; c) align sectoral policies that affect health; and d) include health issues in planning for sustainable development. These components are compatible with WHO's Global Strategy, ratified in 1992, which emphasized the new perspective that health was an essential component of sustainable development.

5.5.4 INTERNATIONAL ENVIRONMENTAL AND SUSTAINABLE DEVELOPMENT RESEARCH

Brundtland Commission. See WORLD COMMISSION ON ENVIRONMENT AND DEVELOPMENT.

Climate Variability and Predictability (CLIVAR): an international research program launched in 1995 by the WORLD CLIMATE PROGRAM and cosponsored by THE INTERNATIONAL COUNCIL OF SCIENTIFIC UNIONS, INTERGOVERNMENTAL OCEANOGRAPHIC COMMISSION, and UNITED NATIONS EDUCATIONAL, SOCIAL, AND CULTURAL ORGANIZATION (UNESCO). The main objectives of the program are: a) to describe and understand the physical processes responsible for climate variability and predictability on seasonal, interannual, decadal, and centennial time scales; b) to extend the record of climate variability over the time scales of interest through the assembly of quality-controlled paleoclimatic and instrumental data sets; c) to extend the range and accuracy of seasonal to interannual climate prediction through the development of global coupled predictive models; d) to understand and predict the response of the climate system to increases of radioactive gases and aerosols; and e) to compare these predictions to the observed climate record in order to detect the anthropogenic modification of the natural climate signal. The program is headquartered at the Southampton Oceanography Centre in Great Britain.

See www.dkrz.de/clivar/hp.html

DIVERSITAS: a partnership of intergovernmental and NONGOVERNMENTAL ORGANIZATIONS established in 1991 to promote and catalyze knowledge about biodiversity, including its origin, composition, ecosystem function, maintenance, and conservation. It is cosponsored by the International Union of Biological Sciences (IUBS), the SCIENTIFIC COMMITTEE OF PROBLEMS OF THE ENVIRONMENT (SCOPE), UNITED NATIONS EDUCATIONAL, SOCIAL, AND CULTURAL ORGANIZATION (UNESCO), the INTERNATIONAL COUNCIL OF SCIENTIFIC UNIONS (ICSU), the INTERNATIONAL GEOSPHERE-BIOSPHERE PROGRAM (IGBP), and the International Union of Microbiological Societies (IUMS). The main goals of DIVERSITAS are to provide accurate scientific information and predictive models of the status of biodiversity and sustainability of the earth's biotic resources, and to build a worldwide capacity for the science of biodiversity. The Scientific Steering Committee in which each partner is represented manages the program. The research effort of DIVERSITAS focuses on the origins, maintenance, and change of biodiversity; systematics, inventorying, and classification; monitoring of biodiversity; and the conservation, restoration, and sustainable use of biodiversity. The six Special Target Areas of Research focus on problems of special concern within biodiversity science and issues that are often neglected: soil and sediment biodiversity, marine biodiversity, microbial biodiversity, inland water biodiversity, global mountain biodiversity, invasive species, and human dimensions of biodiversity. Currently this is the only umbrella program that coordinates a broad research effort in the biodiversity sciences at the global level.

See www.icsu.org/DIVERSITAS/

European Centre for Nature Conservation (ECNC): an international research center established in 1993 with the purpose of furthering European nature conservation by providing innovative and integrated multidisciplinary expertise through a proactive and pan-European approach. It is located in Tilburg, the Netherlands, and is financed by several Dutch governmental agencies and Tilburg University.

See www.ecnc.nl

European Environment Agency (EEA): an agency formed in 1993 as a research and development facility for the EUROPEAN UNION (EU). Located in Copenhagen, the EEA is intended to provide the European Union and its member states with objective, reliable, and comparable environmental information at the European level, enabling them to take the necessary measures to protect the environment, assess the results of these measures, and inform the public.

See Stanley Johnson and Guy Corcell, *The Environmental Policy of the European Communities*, 2d ed. (London: Kluwer, 1995), 363–64.

Global Change in Terrestrial Ecosystems (GCTE): an international research effort coordinated through the INTERNATIONAL GEOSPHERE/BIOSPHERE PROGRAM (IGBP). Its scientific objectives are to: a) predict the effects of changes in climate, atmospheric composition, and land use on terrestrial ecosystems, including agriculture, forestry, soils, and BIODIVERSITY; and b) determine how these effects lead to feedbacks to the atmosphere and the physical climate system. GCTE research is focused on four main areas (ecosystem physiology; changes in ecosystem structure; global change and impacts on agriculture, forestry, and soils; and global change and biodiversity) and thus far has involved over one thousand research scientists and projects worth $47.5 million. The results of the first six years of the program's implementation are summarized in B. H. Walker et al.

See B. H. Walker et al., eds., *The Terrestrial Biosphere and Global Change: Implications for Natural and Managed Ecosystems* (London: Cambridge University Press, 1999); http://GCTE.org

Global Environmental Monitoring System (GEMS). See under SECTION 5.1.5.

Global Monitoring System. See under SECTION 5.1.5.

Global Ocean Ecosystem Dynamics (GLOBEC): an international research program started in 1996 as a core project of the INTERNATIONAL GEOSPHERE/BIOSPHERE PROGRAM and cosponsored by the Scientific Committee on Oceanic Research (SCOR) and INTERNATIONAL OCEANOGRAPHIC COMMISSION. The main objective of GLOBEC is to advance understanding of the structure and functioning of the global ocean ecosystem, its major subsystems, and its response to physical forcing so that a capability can be developed to forecast the responses of the marine ecosystem to global change. GLOBEC research activities consist of regional programs, national programs, and crosscutting activities that seek to: a) understand how multiscale physical environmental processes force largescale marine ecosystem changes; b) determine the relationships between structure and dynamics in oceanic systems that typify significant components of the global ocean ecosystem, with emphasis on trophodynamic pathways, their variability, and the role of nutrition quality in the food web; c) determine and predict the impacts of global change on stock dynamics using coupled physical, biological, and chemical models; and d) determine how changing marine ecosystems will affect the global system by quantifying feedback effects. The program is headquartered at Plymouth Marine Laboratory in Great Britain.

See http://www.pml.ac.uk/globec

Human Dimensions of Global Change Program (HDGCP): an international research program that brings together social scientists, natural scientists, and those involved in the management of human activities. HDGCP seeks to: a) improve scientific understanding of the complex dynamics governing human interactions with the total earth system; b) strengthen efforts to explore and anticipate social change affecting global environment; c) identify broad social strategies to prevent or mitigate undesirable impacts of global change; and d) provide a basis for policy issues that arise from global environmental change. The HDGCP operates through a series of working groups, monitored by an International Scientific Committee and under overall direction of a Steering Committee, made up of the International Federation of Institutes for Advanced Study, the International Social Science Council, the UNITED NATIONS EDUCATIONAL, SCIENTIFIC AND CULTURAL ORGANIZATION (UNESCO), and the UNITED NATIONS UNIVERSITY (UNU). The main areas of interdisciplinary research within the HDGCP framework are global climate change and global risk assessment.

International Development Research Center (IDRC): a Canadian government-supported research institute established in 1970 by an act of the Canadian Parliament. The IDRC's main goal is to initiate, support, and conduct research into the issues of development, often in conjunction with counterparts from developing countries. Projects supported by IDRC are usually identified and managed by researchers from recipient nations. The IDRC is considered one of the world's most influential think tanks in the sphere of international development.

International Geosphere-Biosphere Programme (IGBP): a research program launched in 1986 by the INTERNATIONAL COUNCIL OF SCIENTIFIC UNIONS (ICSU) with the principal objective of describing and understanding the interactive physical, chemical, and biological processes that regulate the total earth system, the changes occurring in this system, and the impact of human activities. The research agenda of the program has been shaped by its steering committee, which is composed of high-level scientists from different disciplines and countries. The steering committee identifies priority issues and tries to devise the most efficient ways of addressing them through a combination of international and national research. Currently, the research effort within the IGBP framework is focused on a set of specific global change problems that are believed to have the greatest significance for the next ten to one hundred years, such as global ocean ecosystem dynamics and land-ocean interactions in the coastal zone. Funds for program coordination are provided by the ICSU and other international organizations, national governments, academia, and businesses, while the research itself is generally funded at the national level.

International Human Dimensions Program on Global Environmental Change (IHDP): an international research effort launched in 1990 as a joint effort of the International Council of Scientific Unions (ICSU) and the International Social Science Council (ISSC). The main objectives of the IHDP are to identify and generate research activities in priority areas of human interaction with the environment, promote international collaboration, and link policymakers and researchers. Four core projects have been conducted under IHDP auspices: a) the Land-Use and Land-Cover Change (LUCC) project explores land-use and land-cover dynamics through comparative case studies analysis, empirical observations, and diagnostic models in order to define links between land-use and land-cover change and other critical global environmental change issues such as climate change, food production, health, urbanization, coastal zone management, transboundary migration, and availability and quality of water; b) the Global Environmental Change and Human Security (GECHS) project focuses on defining the roles that institutions play as drivers of global environmental changes and exploring institutional responses to changes in biophysical systems; c) the Institutional Dimensions of Global Change (IDGC) project explores the multiple linkages between global environmental change and human security; d) the Industrial Transformation (IT) project explores the transformation of the industrial system toward sustainability, and how to decouple industrial activities from their environmental impacts.

See http://www.unibonn.de/ihdp/

Man and the Biosphere Program (MAB): an interdisciplinary program of research and training launched by the UNITED NATIONS EDUCATIONAL, SCIENTIFIC AND CULTURAL ORGANIZATION (UNESCO) in 1971. The MAB is designed to initiate and coordinate international research essential to ensure the rational use and conservation of the world's living resources. The program emphasizes an integrated ecological approach to the study of human-nature relationships and is organized around fourteen major research themes, including perceptions of environmental quality and environmental education. The MAB functions through over a thousand field research projects in more than one hundred countries. These projects are planned and implemented under the direction of national committees overseen by an international coordinating committee. A major achievement of the program has been the establishment of over three hundred BIOSPHERE RESERVES all over the world. Given the organizational constraints and financial limitations of UNESCO, there are different estimates of MAB's effectiveness. According to Lynton Caldwell, the MAB "presents an appropriate model for an international agency in environmental policy and administration" and "has broken ground in international cooperative environmental investigation upon which its own future efforts may build."

See Lynton Caldwell, *International Environmental Policy*, 2d ed. (Durham, NC: Duke University Press, 1996), 337–39.

Management of Social Transformations (MOST): an international research program launched in 1992 by the UNITED NATIONS EDUCATIONAL, SCIENTIFIC AND CULTURAL ORGANIZATION (UNESCO) and hosted and largely financed by that organization. The objective of the program is to promote international comparative social science research with its main emphasis on large-scale, long-term autonomous research and to establish sustainable links between scientific and policy communities. MOST research aims at surveying the effects of global changes on local and regional levels in order to support the effort made by local population and authorities to cope with them. The program operates on three high-priority research areas: a) multicultural and multiethnic societies; b) cities as arenas of accelerated social transformation (sustainable management of cities); and c) coping locally and regionally with economic, technological, and environmental transformations.

See http://www.unesco.org/most

Past Global Changes (PAGES): an international research project launched by the INTER-NATIONAL GEOSPHERE-BIOSPHERE PROGRAM (IGBP) to provide a quantitative analysis of the earth's past climate and environment. It is supported by the Swiss and U.S. National Science Foundations and the U.S. National Ocean and Atmosphere Administration (NOAA). The primary task of PAGES is to organize the international scientific effort to produce a coherent and quantitative record of the earth's natural history through multiproxy studies. The program focuses on: a) dynamics of the global climate and the earth's natural environment and factors responsible for these changes; b) the extent of anthropogenic impact over climate and the global environment; c) the limits of natural greenhouse gas variation and the natural feedbacks to the global climate system; and d) the important forcing factors that produce climate change on societal time scales. The project is administered from Bern, Switzerland.

Scientific Committee on Problems of the Environment (SCOPE): a special analytic body of the INTERNATIONAL COUNCIL OF SCIENTIFIC UNIONS (ICSU) created in 1960. SCOPE does not undertake research itself, but assesses global environmental problems on the basis of scientific information drawn from different disciplines and different countries. Within SCOPE, international teams of scientists analyze available information, identify knowledge gaps, and propose research priorities. Many SCOPE reports turned out to be pioneering efforts. For instance, the 1979 SCOPE report on "Environmental Impact Assessment" was instrumental in promoting EIA as an instrument of environmental policy in many countries. Similarly, the 1986 SCOPE report, "The Greenhouse Effect," helped to place global climate change on the international political agenda.

See E. R. Munn et al., eds., *Policy Making in an Era of Global Environmental Change* (London: Kluwer Academic, 1996).

South Commission: a commission set up in 1987 by the "nonaligned movement" (largely DEVELOPING COUNTRIES without strong ties to either the Western powers or the Soviet bloc). The South Commission was founded to pursue three main goals: investigate the common problems of developing countries, examine the possibilities of their working together to solve these problems, and develop a new dialogue with developed countries. The report of the South Commission, entitled "The Challenge to the South," and its abridged version "Environment and Development: Towards a Common Strategy of the South in the UNCED Negotiations and Beyond" (1990) might be considered an answer from the developing countries to the BRUNDTLAND REPORT. These materials demonstrated that members of the South Commission were largely concerned with the industrial development and economic growth of their countries and intended to use Northern environmental concerns as leverage to ensure further assistance to developing countries' economic development (see also NORTH VS. SOUTH).

United Nations University (UNU): an autonomous university within the UN framework established in 1975 and based in Tokyo. Its mandate is to facilitate international collaborative research, mainly in the areas of human and social development, as well as the use and management of natural resources. According to its charter, UNU is "an international community of scholars engaged in research, postgraduate training and the dissemination of knowledge" to further the purposes and principles of the United Nations. UNU promotes cooperation among scientists and scholars by organizing multinational and multidisciplinary networks of researchers that cooperate in research and training projects on a wide range of global issues. The university also has several research and training centers such as the World Institute for Development Economics Research (UNU/WIDER) located in Helsinki, Finland, and the UNU Institute for Natural Resources in Africa, based in Côte d'Ivoire and Zambia. UNU research covers natural resource management and development, the causes and consequences of tropical deforestation, management of coastal and water resources, arid land management and DESERTIFICATION, and AGROFORESTRY.

World Climate Programme (WCP): an international scientific program established following the First World Climate Conference in Geneva, Switzerland, in February 1979. Its major sponsoring organizations are the WORLD METEOROLOGICAL ORGANIZATION (WMO), the UNITED NATIONS ENVIRONMENT PROGRAMME (UNEP), the Intergovernmental Oceanographic Commission of the UNITED NATIONS EDUCATIONAL, SCIENTIFIC AND CULTURAL ORGANIZATION (UNESCO), and the International Council for Scientific Unions (ICSU). The objectives of the WCP are to use existing climate information to improve economic and social planning; improve the understanding of climate processes through research; and detect and warn governments of impending climate variations or changes, either natural or humanmade, which may significantly affect critical human activities. The World Climate Programme comprises the following components: a) World Climate Data and Monitoring Program (WCDMP); b) World Climate Applications and Services Program (WCASP); c) World Climate Impact Assessment and Response Strategies Program (WCIRP); d) WORLD CLIMATE RESEARCH PROGRAMME (WCRP); and e) Global Climate Observing System. The Coordinating Committee for the World Climate Programme (CCWCP) provides overall coordination between the components of the WCP as well as coordination with other related international climate activities. The WCP is headquartered with the WORLD METEOROLOGICAL ORGANIZATION in Geneva, Switzerland.

World Climate Research Programme (WCRP): a major international research effort initiated in 1980 as part of the WORLD CLIMATE PROGRAMME (WCP) with the objective of "develop[ing] the fundamental scientific understanding of the physical climate system and climate processes needed to determine to what extent climate can be predicted and the extent of man's influence on climate." Currently the WCRP is one of the few major programs focusing on global change. WCRP research has been focused on two principal issues: a) short-term climate processes that determine the equilibrium climate and the long-term response to a change in environmental conditions (e.g., the atmospheric concentration of greenhouse gases); and b) long-term responses of oceanic circulation and ice to changes in atmospheric forcing, predictions of natural climatic variations, and predictions of the transient response of climate to changing environmental conditions. The WCRP has initiated a number of formal studies, experiments, and research projects, such as CLIMATE VARIABILITY AND PREDICTABILITY (CLIVAR).

See World Climate Research Programme, *Draft of the WCRP Long-term Plan 1996–2005*, 1994.

World Commission on Environment and Development (WCED; Brundtland Commission): a highly prominent international commission convened in 1987 to consider environmental and economic growth issues. With Norwegian Prime Minister Bro Harlem

Brundtland as chair, and prominent commissioners from developing countries (such as Guyana's Sir Shridath Ramphal, secretary-general of the British Commonwealth), the WCED produced the document *Our Common Future*, which brought attention to quality-of-life issues in development and focused on how the world could make the transition to sustainability. The WCED effectively established the present generation's responsibility for safeguarding future generations' development options and opportunities by protecting the planet's environment and natural resources. The definition of SUSTAINABLE DEVELOPMENT advanced by the WCED is "development that meets the needs of the present without compromising the ability of future generations to meet their own needs." The alleviation of poverty in the developing world was tied to sustainability. Attention was paid to the international economy by recognizing the need to reorder patterns of international trade and flows of capital. *T. S.*

5.5.5 INTERNATIONAL LEGAL INITIATIVES

Additionality, Principle of: the principle proposed by developing nations that the additional expense for addressing the environmental problems in developed countries should not detract from the development assistance provided to the developing world. This principle is one of the most controversial issues in North-South relations (see NORTH VS. SOUTH). It was first put forward at regional meetings of the G-77, then officially stated in the STOCKHOLM DECLARATION OF PRINCIPLES and later reaffirmed by resolution of the UN General Assembly. The principle of additionality implies that "the preoccupation of developed countries with their own environmental problems should not affect the flow of assistance (to developing nations) . . . and that this flow should be adequate to meet the additional environmental requirements of such countries." In practical terms this means that existing development funds should not be diverted to environmental purposes and that increases in the foreign aid budget of most developed countries should cover the additional costs imposed by environmental protection measures on development projects. Though developing countries were able to ensure official adoption of this concept at the international level, its implementation has always depended entirely upon the discretion of donors.

Common but Differentiated Responsibilities: a legal principle elaborated to link the relative contributions of different countries to global environmental threats and the distribution of obligations to mitigate these threats. The essence of the principle is a compromise between Northern concerns that all countries should participate in global environmental policies and Southern arguments that global environmental threats are largely byproducts of Northern development and consumerism and therefore it is the obligation of developed countries to deal with them. The principle of "common but differentiated responsibilities" is incorporated into the RIO DECLARATION, the CLIMATE CHANGE CONVENTION, and most other environmental agreements negotiated in the 1990s. However, attempts to implement the principle demonstrate that representatives of the developed and developing countries interpret it differently. Governments of developing nations tend to assume that the principle confirms an obligation on the part of the North to take the lead in reversing environmental degradation, while in the view of many developed countries, this principle does not absolve any country of commitments to mitigate global environmental degradation.

Common Concern of Mankind: an emerging principle of international law that stipulates the obligation of all nations to address environmental issues. It was first applied in 1988 when the UN General Assembly declared that the global climate was the "common concern of mankind." The VIENNA OZONE CONVENTION and the GLOBAL CLIMATE CONVENTION have accepted this principle in lieu of the COMMON HERITAGE OF MANKIND. However, its legal significance and consequences are still not clear.

Common Heritage of Mankind: a proposed principle of international law that calls for sharing the benefits from international commons. First proposed in 1967 by Dr. Arvid Pardo, the Maltese ambassador to the United Nations, in relation to deep seabed resources, the principle suggests that the resources of the area "cannot be appropriated to the exclusive sovereignty of states but must be conserved and exploited for the benefit of all without discrimination." The concept was applied in the 1982 UN CONVENTION ON THE LAW OF THE SEA (UNCLOS) (which was a partial explanation of why it did not enter into force in its original form) and the 1979 MOON AGREEMENT. The practical implication of this principle would be the widespread sharing of benefits from exploiting international commons among the developing countries, even if investments and operations were undertaken by the developed countries. To date, this principle is not universally accepted and has been recently substituted by the principle of COMMON CONCERN OF MANKIND.

Compensation, Principle of: the proposed principle in international law that calls for developed nations to assist poor nations to reconcile their development efforts with environmental quality objectives. The principle of compensation is one of the most debated issues in contemporary international environmental law and politics. The principle was initially put forward at the 1972 UNITED NATIONS CONFERENCE ON THE HUMAN ENVIRONMENT (STOCKHOLM CONFERENCE) and was later formalized by the UN General Assembly in the Declaration of a NEW INTERNATIONAL ECONOMIC ORDER. Most developed nations, however, still oppose this principle.

Convention, International: an agreement between two or more states that establishes both rights and obligations for the parties and usually results from an international conference. Adherence to a convention is governed by international law and is binding only upon the parties to the particular agreement unless it represents or comes to represent customary law. The term *convention* is usually used interchangeably with *international treaty*, though traditionally the term *treaty* was used to define international agreements of the utmost importance for its members. Often the rights and obligations specified in conventions are very general (therefore they are sometimes called FRAMEWORK CONVENTIONS); conventions are frequently followed by more specific (and more costly) obligations specified in PROTOCOLS.

Crimes against Environment: a concept introduced to international legal discourse by the International Law Commission (ILC) in its Draft Articles on State Responsibility. The ILC defined international crime as "an internationally wrongful act which results from the breach of the State of an international obligation so essential for the protection of fundamental interests of the international community that its breach is recognized as a crime by that community as a whole" and included in this category breaches of the rules prohibiting massive pollution of the atmosphere or sea. This concept was also promoted by the Canadian Parliament in 1985 and became a subject of heated discussion after the 1990–91 Persian Gulf War created substantial international environmental damage.

See Law Reform Commission of Canada, *Protection of Life: Crimes against the Environment* (Ottawa: Law Reform Commission, 1985); Patricia W. Birnie and Alan Boyle, *International Law and the Environment* (Oxford: Clarendon Press, 1992).

Draft International Covenant on Environment and Development: a document prepared in 1995 by the WORLD CONSERVATION UNION (IUCN) in cooperation with the International Council of Environmental Law, proposing a strong and broad international treaty on nations' environmental responsibilities. The Draft Covenant, which includes a proposed treaty text and detailed commentary, is intended to stimulate consideration of a global instrument on environmental conservation and SUSTAINABLE DEVELOPMENT. The document was meant to meet the needs emphasized by the UNITED NATIONS CONFERENCE ON ENVIRONMENT AND DEVELOPMENT (UNCED), such as elaborating the balance between environmental and developmental concerns; clarifying the relationships between the various existing treaties; and ensuring national participation in both developing and implementing these legal measures, with particular focus on developing countries. The Draft Covenant has been an influential source of the UN ENVIRONMENT PROGRAMME and the UN General Assembly's processes of identifying principles of environmental law for sustainable development.

Environmental Justice: concepts of, and criteria for, fairness in the distribution of environmental quality. Environmental justice is a relatively recent addition to the theories and movements on social justice, particularly stimulated by studies demonstrating that the poor often lack environmental amenities and are frequently subjected to efforts to strip away their resource rights as well.

Framework Convention: in international environmental diplomacy, a negotiated treaty that lays the basis for international efforts to address an environmental problem. These agreements typically acknowledge the existence and potential seriousness of the problem, call for international cooperation in monitoring and scientific research as well as the exchange of pertinent information among the parties, and establish procedures for periodic conferences of the parties to consider supplemental agreements, in particular in the form of protocols. Most framework conventions encourage the parties to adopt appropriate measures to ameliorate the problem, but do not bind them to specific actions or outcomes by a certain date. Framework treaties are normally adopted when there is widespread, if not universal, concern about a problem, but there is not enough consensus on the immediacy or severity of the problem or on the concrete steps that should be taken to address it. Examples include the CONVENTION FOR THE PROTECTION OF THE MEDITERRANEAN SEA AGAINST POLLUTION (1976), the LONG-RANGE TRANSBOUNDARY AIR POLLUTION CONVENTION (1979), the VIENNA CONVENTION ON PROTECTION OF THE OZONE LAYER (1985), and the FRAMEWORK CONVENTION ON CLIMATE CHANGE (1992). Specific obligations sometimes follow in the form of PROTOCOLS.

See Lawrence Susskind, *Environmental Diplomacy: Negotiating More Effective Global Agreements* (New York: Oxford University Press, 1994), 30–37. *M. S. S.*

Framework conventions came into fashion in the 1970s as an early step in the development of environmental regimes that can be taken even before there is a consensus on the seriousness of a problem and on what regulatory measures should be taken to mitigate it. Critics argue that framework conventions have become a convenient way of delaying needed responses to environmental problems and falsely leave the impression that significant progress has been made in addressing them. *Marvin S. Soroos*

Framework conventions do not provide the details for concrete commitments, but they often increase the pressure on governments to follow up with specific protocols. Some government officials may think that environmental framework conventions let them off the hook by making nonspecific promises of future action, yet environmental groups can often mobilize around the opportunity to push governments to sign protocols. The upcoming battle over the KYOTO PROTOCOL will be a crucial test of this logic. *William Ascher*

Incremental Costs, Principle of: the principle that developed countries should pay for the additional costs of environmental protection in the developing countries, which exceed the amount already borne by a developing country under a business-as-usual strategy. The principle calls for partial burden-sharing in relation to responsibility for global environmental problems. The principle of incremental costs is one of the most contentious issues in North-South relations on SUSTAINABLE DEVELOPMENT and global environmental protection.

COMMENTARY ON THE PRINCIPLE OF INCREMENTAL COSTS

The pitfall of applying the incremental costs principle is the assumption that the policies already adopted by a developing country are efficient from that country's perspective, and therefore it is possible to define the exact amount of additional costs; practice often contradicts this assumption. *Natalia Mirovitskaya*

Intergenerational Equity, Principle of: fairness in the distribution over time of income, wealth, access to social services and natural resources, or other desired outcomes. SUSTAINABLE DEVELOPMENT, particularly as defined in the BRUNDTLAND COMMISSION REPORT, requires the current generation to desist from depleting natural resources and environmental services for its own consumption, unless investments are made with the proceeds to provide future benefits of equal or greater benefit.

International Court for Environmental Arbitration and Conciliation (ICEAC; International Environmental Court): an independent, international judicial mechanism set up in 1994 for the settlement of environmental disputes. In accordance with its charter, the ICEAC addresses disputes through: a) legal assessment of the facts and consultation with the parties concerned at their request; b) conciliation of disputing parties, that is, through recommending a compromise which may be worked out in a voluntary agreement; c) arranging a full judicial hearing upon mutual agreement of the parties, in which case decision is mandatory.

The ICEAC attends to a broad range of cases, including disputes on compensation for transboundary environmental pollution, the use and protection of shared natural resources and complexes, compensation on military-induced environmental damage, and the protection of environmental rights of individuals. The ICEAC's powers are limited by its incapacity to impose a decision unless all parties agree to come under the court's judgment.

International Court of Justice (ICJ; World Court): the principal judicial organ of the United Nations, founded in 1945, and until now the most far-reaching attempt to establish the rule of law in the settlement of international disputes. Located in the Hague, the Netherlands, the ICJ consists of fifteen judges elected by the General Assembly and the Security Council on the basis of their qualifications, political sponsorship, and representativeness of the world's principal legal systems. It has a double jurisdiction: settling international disputes between nation-states and giving advisory opinions to governments as well as to nonstate actors, such as international organizations and individuals. Three environmental treaties identify the ICJ as a source for arbitration and meditation: the MONTREAL PROTOCOL ON SUBSTANCES THAT DEPLETE THE OZONE LAYER; the UN CONVENTION ON THE LAW OF THE SEA; and the Resolution on the Permanent Sovereignty over Natural Resources.

COMMENTARY ON THE INTERNATIONAL COURT OF JUSTICE

Though all UN members are automatically members of the World Court and, according to its charter (article 36), must recognize the compulsory nature of its jurisdiction, in practice the ICJ's effectiveness is low. For instance, when in May 1973 the ICJ voted to ask France to stop any further testing in French Polynesia that would deposit radioactive fallout on Australian territory, France did not recognize the ICJ's jurisdiction in the case and ignored its ruling. This case highlights the basic impediment facing the ICJ: lack of an adequate enforcement mechanism, even though in theory if a party fails to comply with a judgment rendered by the ICJ, the other party may have recourse to the UN Security Council. *Natalia Mirovitskaya*

Intragenerational Equity, Principle of: fairness in the distribution of income, wealth, access to social services and natural resources, or other desired outcomes across current populations. The existing distribution can be measured by determining the share of the outcome received by the lowest to highest segments of the population (e.g., the poorest quintile (20 percent) of the population may receive 5 percent of the total national income). The most desirable intragenerational equity is a matter of values; it is a separate question from the distribution that would create the most productive economy. While one can compare a given intragenerational distribution to total equality (the GINI INDEX OF INEQUALITY measures the discrepancy between an existing distribution and the line of compete equality), ideas of economic justice may not hold equality as ideal; many believe that effort, willingness to save, difficulty of work, market demand for skills, or other factors justify rewarding some people more than others.

COMMENTARY ON THE PRINCIPLE OF INTRAGENERATIONAL EQUITY

People will disagree about the most desirable income distribution in any given country. Therefore, debates over income distribution are often short-circuited when the defenders of the status quo challenge the advocates of redistribution to specify the optimal distribution. Yet this disagreement about the optimal should not distract us from realizing that most developing countries, and a number of developed and transitional countries as well, have unacceptably unequal distributions of income by anyone's standards, apart from those with a personal stake in maintaining their own income levels. In many developing nations, a strategy of devoting greater tax revenues to invest in the health and education of low-income families is the best way to improve the distribution of income in the long run. *William Ascher*

Joint Implementation (JI; Activities Implemented Jointly): transfer of technological, intellectual, and/or financial resources between parties to international environmental agreements to assist each other in meeting environmental targets of these agreements. For instance, in the case of the FRAMEWORK CONVENTION ON CLIMATE CHANGE, joint implementation involves funding from wealthier countries to countries with lower marginal ABATEMENT costs, so that the reduction in GREENHOUSE GASES can be accomplished where it is less expensive. If the investment results in greenhouse gas reductions, the credits can be divided between host and investing parties to help them in meeting their respective obligations under the convention. For investor nations, joint implementation can reduce both costs and risks; for the host country, the benefits are technological and financial assistance as well as indirect environmental and economic benefits. Despite its obvious economic and environmental advantages, the joint implementation approach has been criticized for allowing investing countries to take advantage of the market distortions in developing and transitional countries, which might result in their decreased economic welfare. Within the discourse on ENVIRONMENTAL COLONIALISM, joint implementation has been represented as one of the tools for the North to shift its burdens to the South and to perpetuate inequities (see also NORTH VS. SOUTH).

See Joyeeta Gupta, *The Climate Change Convention and Developing Countries: From Conflict to Consensus?* (Dordrecht: Kluwer, 1997); Onno Kuik, Paul Peters and Nico Schrijver, eds., *Joint Implementation to Curb Climate Change: Legal and Economic Aspects* (Dordrecht: Kluwer, 1994). *R. P.*

Liability for Environmental Harm, Principle of: a basic though still developing principle of international environmental law that holds that a state is responsible for damage caused by its subjects beyond its borders. Many international agreements include this principle. However, there is as yet no consensus regarding the details for when and how compensation for damage is to be determined and paid. Most judicial claims based on this principle were settled directly among the private individuals involved.

Permanent Sovereignty over Natural Resources, Principle of: a fundamental principle of international law that refers to the right of peoples and nations to exercise permanent sovereignty over their natural wealth and resources. The principle was initially established by the 1962 United Nations General Assembly resolution "Permanent Sovereignty over Natural Resources" in response to demands of the developing countries for economic self-determination. The principle of permanent sovereignty is the basis of other state rights, such as the right to regulate the entry and operations of foreign investors and the right of a nation to pursue its own economic and environmental policies. The principle of permanent sovereignty over natural resources was later reaffirmed in many international agreements and declarations. However, there are ambiguities between this concept and international implications of resource management. Therefore, it is increasingly recognized that the principle of permanent sovereignty, as well as other state rights rising from it, are coupled with duties, such as a duty not to harm the interests of other states, a duty to cooperate with other countries in equitable use and management of transboundary resources, etc.

Precautionary Principle (PP): a newly emerging principle of international environmental law that calls for nations to take preventive measures when there is reason to assume that substances or energy introduced into the ecosystem may cause damage, even when there is no conclusive evidence of a causal relationship between such inputs and their alleged effects. The potential damages include hazards to human health, harm to living resources and ecosystems, damage to amenities, or interference with other legitimate uses of the environment. The principle has been included in numerous international documents, such as the European Bergen Declaration of 1990, the RIO DECLARATION (PRINCIPLE 15) and the 1992 RIO FRAMEWORK CONVENTION ON CLIMATE CHANGE. The techniques of implementing the precautionary principle include quantitative RISK ASSESSMENT, minimal regret criteria, safe minimum standards, and ENVIRONMENTAL BONDS. Within Europe, the precautionary principle has emerged as a central, though contested, concept in addressing SUSTAINABLE DEVELOPMENT issues, and is included in the environmental strategies of most European countries.

COMMENTARIES ON THE PRECAUTIONARY PRINCIPLE

The contested nature of the term arises from the means to avoid future environmental degradation. Here a nested set of acronyms become central. These include ALARA—as low as reasonably achievable; ALARP—as low as reasonably practicable; BATNEEC—best available technique not entailing excessive cost; BPEO—best practical environmental option; and BPM—best practicable means. The balance between economic costs and the reduction of environmental risks is frequently resolved by cost-benefit analysis, a reductionist methodology that ultimately reduces everything to a market price subverting the potential for precautionary action.

See Timothy O'Riordan and J. Cameron, eds., *Interpreting the Precautionary Principle* (London: Earthscan, 1994). *Ian Welsh*

Though of recent origin, the precautionary principle has taken a focal place in the discussions of international environmental and resource management regimes. Despite the fact that it has been adopted by a number of conventions and other international agreements, there is still no clarity about its meaning and extent as well as whether this principle should be considered merely a guideline or a substantive mandate for action. The extension of the precautionary principle to the realm of resource management resulted in the elaboration of the concept of the "precautionary approach." The latter has been designed as a tool that provides managers of renewable resources with the power to continue harvesting or engage in other types of resource extraction along with the flexibility to introduce stringent cutbacks or prohibitions only when a threshold has been reached that would threaten a stock's sustainability. The experience with this approach thus far has been limited to several cases of international fisheries management. For instance, the CONVENTION ON THE CONSERVATION OF ANTARCTIC MARINE LIVING RESOURCES (CCAMLRA) has applied the precautionary approach to multispecies interaction, particularly in regulating bycatches and prohibiting some fisheries in order to protect non-target species. The precautionary approach under CCMLRA also covered the protection of seabirds, the prohibition of the disposal of synthetic material, and the curtailment of large-scale pelagic driftnet fishing. The initiation of new fisheries is subject to review and authorization of the scientific committee. *Natalia Mirovitskaya*

Prevention of Environmental Harm, Duty of: a fundamental principle of international law concerning the responsibility of a polluting state for transboundary pollution. The earliest affirmation of this principle was given by the 1941 TRAIL SMELTER CASE. It was later stated in Principle 21 of the STOCKHOLM DECLARATION ON THE HUMAN ENVIRONMENT as the state's "responsibility to ensure that activities within their jurisdiction or control do not cause damage to the environment of other States or of areas beyond the limits of national jurisdiction." The principle has also been supported by several United Nations General Assembly declarations and was incorporated into major international agreements, such as the UN CONVENTION ON THE LAW OF THE SEA (UNCLOS) and the CLIMATE CHANGE CONVENTION.

Protocol: a negotiated international agreement that supplements a previously adopted TREATY or CONVENTION, commonly known as a FRAMEWORK CONVENTION. Protocols are ratified separately, so it is not uncommon for states to be parties to framework conventions, but not to the protocols that are concluded later. In contrast to framework conventions, which set the stage for international cooperation in addressing a problem, protocols normally bind the parties to specific actions, such as a percentage reduction of certain air pollutants within a set time period, which will begin to mitigate the problem at hand. Protocols have proven to be flexible instruments for strengthening international environmental regimes as more is known about a problem and a consensus emerges on the next steps that should be taken to address it. Notable examples of protocols are the Helsinki Protocol on Reduction of Sulfur Emissions (1985) and the MONTREAL PROTOCOL ON SUBSTANCES THAT DEPLETE THE OZONE LAYER (1987).

See Lawrence E. Susskind, *Environmental Diplomacy: Negotiating More Effective Global Agreements* (New York: Oxford University Press, 1994), 30–37. M. S. S.

Right to Development: one of the principles of international cooperation that affirms the individual's right to the benefits of development. The 1986 United Nations Declaration on the Right of Development (UNDRD) was the first document to recognize development as a legal right with correlative legal duties and responsibilities at the global level. This document focused on the "human person" as the central subject of development and sought to base the international obligation of states and the international community to promote development on the universal recognition accorded to basic human rights of all persons. It identified basic elements that comprise the concept of development: a) equality of opportunity for all in the access to basic resources, education, health services, food, housing, employment, and fair distribution of income; b) an active role for women in the development process; c) adoption of economic and social reforms to remove social injustices; and d) encouragement of popular participation in all spheres relating to development. The UNDRD reaffirmed the responsibility of the international community to promote more rapid development for developing countries. Basic elements of the UNDRD were strengthened by the 1992 RIO DECLARATION ON ENVIRONMENT AND DEVELOPMENT, which proclaimed the necessity for fulfilling the right to development in order to "equitably meet developmental and environmental needs of present and future generations." However, despite ongoing discourse, the contents and structure of the right to development have not been resolved.

Soft Law: common norms or rules of conduct that do not have the binding status of treaties on participating states. These rules (guidelines, codes of conduct, plans of action, declarations), though explicitly drafted as nonbinding, may still have significant impact on the behavior of states on environmental issues. Soft laws are usually proposed by international organizations in anticipation of voluntary compliance by states. The generality and ambiguity of the soft law allows states to proceed with their implementation at their own pace and therefore brings to the agreement actors that might otherwise abstain from participation. The UNITED NATIONS ENVIRONMENT PROGRAMME's Action Plan for Biosphere Reserves and UN FOOD AND AGRICULTURE ORGANIZATION (FAO) Codex Alimentarius exemplify soft law. Soft law is considered a special feature of international environmental and economic law and in these areas it develops more rapidly than formal international conventions. Some soft law agreements can eventually "harden," either by their absorption into the treaty law, or by renegotiation.

Subsidiarity: See under SECTION 5.2.4.

Treaty: a written contract or agreement between two or more parties that is considered binding by international law. A treaty might be concluded between states, heads of states, governments, and international organizations. It is the most formal and highest instrument among international agreements, which include, among others, CONVENTIONS, PROTOCOLS, charters, declarations, and undertakings. There are several stages before a treaty becomes international law: a) adoption, b) signature, c) ratification/approval/acceptance, d) formal confirmation, and e) entry into force. After ratification, the ratifying parties are obligated to make any necessary changes in domestic legislation. Some states use the practice of acceptance or approval as synonymous with ratification. International organizations express their consent to be bound by a treaty by "formal confirmation." States that did not participate in treaty negotiation can later accede to it (stage of accession) in accordance with the provisions of the treaty. The controversies that arise in establishing treaties include questions about which government authorities can speak for the government as a whole, as well as what the constitutions of each country require in terms of compliance. A treaty enters into force after a specified number of ratifications is achieved. Therefore some treaties are adopted by several countries but go into effect only many years later, if at all.

5.5.6 MAJOR NONGOVERNMENTAL ORGANIZATIONS AND MOVEMENTS

Big 10: a collective term for the major U.S. environmental lobbying organizations: the SIERRA CLUB, the National Audubon Society, the National Parks and Conservation Association, the Izaak Walton League, the Wilderness Society, the National Wildlife Federation, the Defenders of Wildlife, the Environmental Defense Fund, FRIENDS OF THE EARTH, and the NATURAL RESOURCE DEFENSE COUNCIL. The Big 10 are considered "mainstream" environmental organizations in that they are less confrontational than organizations like GREENPEACE. The Big 10 devote much of their effort to lobbying U.S. government agencies and Congress, mainly with respect to nature protection and environmental conservation. They are also referred to as the "Eco-Establishment."

COMMENTARY ON THE BIG 10

The Big 10 are often criticized for being out of touch with their grassroots membership and for being coopted into the traditional political environment of Washington, D.C. In the 1980s, the Big 10 saw marked increases in their membership and fundraising activities as the Reagan administration abandoned environmental protection in favor of less stringent regulatory policies. Following the boom of the 1980s, the Big 10 have seen their membership drop, along with their budgets. *Toddi Steelman*

Chipko Movement: a largely women-led movement in northern India opposing commercial logging. In Hindi, *Chipko* literally means "to hug," i.e., "tree-huggers." The Chipko Movement mobilized women and their children of the villages in northern India's Himalayas, where deforestation is a severe problem and especially affects women because of fuel-wood scarcity. Recognizing trees as valuable, living beings, members of the Chipko Movement literally embraced the trees, thereby preventing commercial timber harvesting. The Chipko Movement was instrumental to the elaboration of the WOMEN IN DEVELOPMENT concept. *T. S.*

Club of Rome: a NONGOVERNMENTAL ORGANIZATION and think tank founded by the Italian industrialist and environmentalist Aurelio Peccei in the 1960s. The Club of Rome sponsored the 1971 *Limits to Growth* study, which predicted environmental and resource collapse if economic growth persisted. Despite widespread criticism of the model used for this study, the Club of Rome sponsored a series of other complex-model-based forecasting studies, also predicting dire environmental consequences.

Committee of International Development Institutions on the Environment (CIDIE): a committee set up in 1981 following the 1980 UN DECLARATION OF ENVIRONMENTAL POLICIES AND PROCEDURES RELATING TO ECONOMIC DEVELOPMENT. The task they set for themselves was to ensure that UN member countries take into account environmental considerations while formulating their policies of economic development. The secretariat for the CIDIE is provided by the UN ENVIRONMENT PROGRAMME.

Development Alternatives with Women for a New Era (DAWN): a women's network from developing countries involved in research and political action to develop an alternative development paradigm. Since its founding in 1984, DAWN has emphasized the political nature of development as well as the imbalances of power within and between nations, rather than the presence or absence of resources. DAWN criticizes the dominant model of development based on economic growth, and advocates "people-centered" and equitable development based on the values of cooperation, resistance to hierarchies, accountability, and commitment to peace. Above all, DAWN insists that women's work should be validated and accounted for in national and international accounting systems. Currently, DAWN's agenda is focused on the topics of environment, population, and alternative economic frameworks. DAWN sees the real causes behind environmental crisis as "maldevelopment" and economic injustices within the global economic system; therefore it calls for linking environmental conservation to the issues of health, technology, income distribution, power, and social organization. Members of DAWN have been extremely active in bringing their perspective to recent international negotiations on global issues.

See Gita Sen and Caren Grown, *Development Crises and Alternative Visions: Third World Women's Perspectives* (London: Earthscan, 1988); Rosi Braidotti, Ewa Charkiewicz, Sabine Hausler, and Saskia Wieringa, *Women, the Environment and Sustainable Development: Towards a Theoretical Analysis* (London: Zed Books, 1994).

Earth Council: an international NONGOVERNMENTAL ORGANIZATION created in 1992 to promote the implementation of the Earth Summit agreements (see UNITED NATIONS CONFERENCE ON ENVIRONMENT AND DEVELOPMENT (UNCED; Rio Conference; Earth Summit). It is led by a body consisting of eighteen prominent political, business, scientific, and civil leaders, such as Maurice Strong (secretary general of the STOCKHOLM and RIO CONFERENCES), Jim MacNeil (former secretary general of the BRUNDTLAND COMMISSION), Klaus Schwab (president of the WORLD ECONOMIC FORUM), and Princess Basma Bint Talal of Jordan. The Earth Council is funded by the private sector and its work has been guided by three fundamental objectives: awareness for the needed transition to more sustainable and equitable patterns of development, public participation in decision making processes at all levels of government, and cooperation between various actors of civil society and governments worldwide. In pursuit of these objectives, the Earth Council is networking with other nongovernmental organizations, business, financial, and professional organizations from over seventy countries. One of its most important initiatives is a program called "Making Sustainability Work," the goal of which is to promote a holistic and integrated approach to operationalizing SUSTAINABLE DEVELOPMENT.

Earth First! (E1!): a radical environmental movement of the 1980s. First started in the U.S. Southwest under the slogan "rednecks for wilderness" (inspired by Edward Abbey's novel *The Monkey Wrench Gang*), E1! rapidly spread all over the country and to many other nations. Its militant tactics in defense of wilderness and native bio-diversity comprised guerrilla theater, media stunts, and civil disobedience, including the so-called ecotage: sabotaging inanimate objects like whaling ships, bulldozers, and road-building equipment. E1!, however, is known not only for its so-called aggressive nonviolence, considerable following, and ardent government opposition, but for de-veloping extensive wilderness inventories and innovative suggestions for preserving BIODIVERSITY. In the 1990s, the U.S.-based organization splintered into several rival groups.

See David Foreman, *Confessions of an Eco-Warrior* (New York: Crown Books, 1991); Riley E. Dunlap and Angela G. Mertig, eds., *American Environmentalism* (New York: Taylor and Francis, 1992). *T. S.*

Earthwatch Institute (Earthwatch): a NONGOVERNMENTAL ORGANIZATION founded in 1972 to improve understanding of the planet and the processes that affect the quality of life, to monitor global change, and to conserve endangered species and habitats. Membership is open to individuals, organizations, businesses, and education centers. Earthwatch arranges volunteer (Earth Corps) help for field projects pertaining to envi-ronmental issues, such as surveying unique lakes in Kenya and tracking endangered rhinos in Zimbabwe. Since 1972 40,000 volunteers from the Earth Corps contributed over $35 million to environmental field projects that have also provided training for nationals from developing countries.

Environmental Defense Fund (EDF): a U.S.-based mainstream environmental organiza-tion that supports a broad range of regional, national, and global environmental efforts. It was founded in 1967 by volunteer conservationists on Long Island in New York as "a partnership between scientists and lawyers" concerned mainly with the use of the pesticide DDT. The initial composition of the group made it possible to use litigation, "armed with scientific and technical arguments" as its primary mechanism in protect-ing the environment. In a few years EDF became a major litigator in such different areas as lead toxicity, the protection of sperm whales, and pesticide hazards. It has been re-garded as one of the major centers of environmental expertise and policy action in the United States. EDF brings polluters to court for breaking environmental laws and challenges the lack of enforcement of environmental legislation by the government. The organization advocates the use of economic incentives to achieve environmental protection and pollution clean up. Within the environmental movement, EDF stands out for its position that environmental change is a matter of reinforcing rather than restructuring a market-driven industrial order.

See Robert Gottlieb, *Forcing the Spring: The Transformation of the American Environ-mental Movement* (Washington, DC: Island Press, 1993).

Environment Liaison Centre International (ELCI): based in the UNITED NATIONS ENVI-RONMENTAL PROGRAMME (UNEP) headquarters (Nairobi, Kenya), the ELCI was founded in 1974 to provide liaison between NONGOVERNMENTAL ENVIRONMENTAL ORGANIZA-TIONS worldwide and the UN system. Recently, the ELCI functions were redefined to emphasize socially, economically, and environmentally SUSTAINABLE DEVELOPMENT by facilitating the voice and collective action of grassroots organizations, and by promoting voluntary community action as a strategy for achieving sustainable development. ELCI serves as a networking mechanism for nongovernmental organizations working on the CONVENTION TO COMBAT DESERTIFICATION, the CONVENTION ON BIOLOGICAL DIVERSITY, the MONTREAL PROTOCOL, the UN CONFERENCE ON HUMAN SETTLEMENTS, among others. As the core of a global network of environmental voluntary organizations, ELCI is instrumental in stimulating critical civic support for UNEP activity. It has a membership of about 730 environment and development NGOs in 114 countries and maintains contacts with many others.

European Environment Bureau (EEB): a confederation of 120 environmental organizations (mainly European) with a combined membership of 20 million. It was formed in 1974 to lobby European institutions to "promote an equitable and sustainable life-style, to undertake and coordinate research on European environmental issues, to serve as coordinator for European environmental groups, to promote the protection and conservation of the environment and to educate the public on European environmental problems."

Friends of the Earth International (FOEI): one of the largest international federations of autonomous national environmental advocacy organizations, founded in 1971. Its objectives are to: a) protect the earth against further deterioration and restore damage inflicted upon the environment by human activities and negligence; b) preserve the earth's ecological, cultural, and ethnic diversity; c) increase public participation and democratic decision making; d) achieve social, economic, and political justice and equal access to resources and opportunities; and e) promote environmentally sustainable development. As of 1999, FOEI encompassed fifty-four national member groups with over one million individual members.

Global Legislators for a Balanced Environment (GLOBE International): an international organization founded in 1989 as a forum in which democratically elected and environmentally committed parliamentarians from around the world can collaborate on policy responses to global environmental challenges. GLOBE International has over five hundred members in more than one hundred countries and four member organizations in the European Union, Japan, Russia, and the United States. By stressing particular environmental problems, urging effective action by governments and private sector leaders, and suggesting alternative approaches, GLOBE has had a direct and significant impact on the international environmental agenda. A flexible, streamlined structure allows GLOBE members to act quickly and informally when they perceive a need for concerted action. When GLOBE members reach consensus in particular issue areas, commitments and guidelines for legislative action are drafted for ratification by the full membership. These guidelines serve as models for GLOBE members and their counterparts in parliaments around the world as they develop environmental policies.

Grassroots Organizations: nongovernmental groups whose members are organized to pursue their own interests, as distinct from NONGOVERNMENTAL ORGANIZATIONS whose mission is to assist in providing for the welfare of others or for public goods. Referring frequently to small groups of low-income people, grassroots organizations often receive the assistance of nongovernmental organizations to mobilize their members, either to defend their rights or to organize for more effective economic cooperation. The flowering of grassroots organizations is often seen as an indication of the strengthening of civil society as distinct from state-sponsored organizations.

Green Cross International (GCI): a global, nonaligned organization that brings together decisionmakers and grassroots movements from all sectors of society (nongovernmental organizations, business and industry, government, academia, science and the arts, religious, and other groups) for a dialogue on SUSTAINABLE DEVELOPMENT. The organization was initiated at the UNITED NATIONS CONFERENCE ON ENVIRONMENT AND DEVELOPMENT (UNCED; Rio Conference; Earth Summit) in 1992 and officially launched in Kyoto, Japan, on April 20, 1993 under the leadership of GCI president Mikhail S. Gorbachev. GCI is headquartered in Geneva and supported by regular contributions of national organizations and the government of Switzerland. GCI works mainly through its national organizations, whose effectiveness and visibility depend largely upon domestic leadership. Most branches have programs on: a) disaster prevention and response (aimed to reduce the risks of human-induced environmental incidents in the chemical industry or through military activities); b) international environmental codes (designed to promote environmentally sound legislation); c) changing consumption patterns; and d) environmental education. The projects are tailored to specific local needs and capacities. For instance, the GCI environmental education initiatives include such different programs as creation of ecological battalions in the Bolivian army and "Education through Entertainment" in the United States, which encourages leading Hollywood directors, producers, writers, and performing artists to incorporate environmental themes into entertainment.

Greenpeace International: a major radical international NONGOVERNMENTAL ORGANIZATION. Founded in 1971 with the goal of creating a "green and peaceful world" by drawing attention to environmental abuses through direct confrontation with governments and corporations, Greenpeace has offices in thirty-two countries and maintains its independence by relying almost entirely on contributions from about five million supporters in 158 countries and by the sales of Greenpeace merchandise. Greenpeace has supported a wide variety of initiatives, ranging from the banning of nuclear testing to a global warming awareness campaign. Recent actions by Greenpeace activists include protesting clearcutting of ancient forests in the Russian Far East by the Korean Hyundai Corporation. By using blockades and generating a government inquiry, this effort helped to stop several logging operations. Greenpeace has gained much of its visibility through highly publicized efforts by its ships to disrupt nuclear testing, certain types of fishing, and other activities regarded by Greenpeace chapters as environmentally threatening. Greenpeace International campaigns have focused on the GLOBAL CLIMATE AGREEMENT, the PVC ban, nuclear waste dumping, and OVERFISHING.

International Academy of the Environment: an independent foundation organized by several European nations and headquartered in Geneva. Among its objectives is "to provide high-level decision-makers with the basic knowledge and management principles to enable them to take decisions that are consistent with SUSTAINABLE DEVELOP-

MENT." Its budget, around $8 million annually, has been largely provided by the Swiss Confederation.

International Council of Scientific Unions (ICSU): a confederation of national and international scientific organizations, founded in 1919 as the International Research Council, adopting its current name in 1932. Its aim is to encourage scientific research, and so promote the cause of world peace and security, by coordinating the activities of international scientific unions and of its national members. Membership includes academies and research councils in 71 countries, 20 international units and 245 scientific associates. One ICSU program, SCOPE, has provided a worldwide system of monitoring changes in the environment caused by human activity. Because of inadequate funding, however, the program was not fully implemented.

International Environment Bureau (IEB). See under SECTION 5.3.4.

International Institute for Applied Systems Analysis (IIASA): a nongovernmental, interdisciplinary research institute founded in 1972, notable for the participation of scientists from the Soviet Union and Eastern Europe as well as from Western countries. IIASA is also notable as a leading proponent of systems analysis and complex modeling approaches to environmental and natural-resource studies. With its main research site in Laxenburg, Austria, IIASA's institutional membership now comprises scientific and professional bodies in sixteen countries. IIASA's objectives are to initiate and support individual and collaborative research on problems associated with social, economic, technological, and environmental change. During the last decade, the IIASA environment program has included projects on ecologically SUSTAINABLE DEVELOPMENT of the BIOSPHERE, acid rain, decision-support systems for managing large international rivers, future environments of Europe, and analysis of alternative development paths. The institute also pursues clearinghouse and networking activities and special projects or risk technologies, processes of international negotiations, food, and agriculture. The principal consumers of IIASA's research are the policymakers at the national and international levels of government and industry. Currently, IIASA's research program is concentrated on three central core themes: energy and technology, population and society, and natural resources and environment.

International Institute for Environment and Development (IIED): a London-based research institute financed by private funds, UN agencies, and governments, founded in 1971 as the International Institute of Environmental Affairs. IIED studies environmental and developmental issues of concern to governments, international institutions, and NGOs; and acts as a clearinghouse for ideas on environment and development. It operates the Joint Environmental Union for the Conservation of Nature and Natural Resources and Earthscan, an information service.

International Union for the Protection of Nature (IUPN). See WORLD CONSERVATION UNION.

Natural Resources Defense Council (NRDC): an environmental NONGOVERNMENTAL ORGANIZATION formed in the United States in the 1970s. The NRDC lobbies the U.S. government in support of environmental protection laws, and litigates with industry and the government over how those laws are to be interpreted and implemented. The

NRDC is one of the more active users of citizen-suit litigation to achieve its policy goals. *D. S.*

Nature Conservancy: a U.S.-based NONGOVERNMENTAL ORGANIZATION dedicated primarily to establishing nature reserves and other forms of protected areas. Founded in 1951 by the Ecologists Union (an activist split-off of the Ecological Society of America), the Nature Conservancy raises money for land purchases or other support for protected areas, concentrating mainly on areas of greatest risk of loss of BIODIVERSITY. Through a membership of approximately 900,00 individuals, organized in state chapters, and 1,850 corporate associates, the Nature Conservancy has helped to place more than one and a half million acres into conservation easements in the United States, in roughly 1,600 preserves. Internationally, the Nature Conservancy has contributed to placing 60 million acres into protected status outside of the United States, particularly in developing countries. The Nature Conservancy is also involved in inventorying biodiversity and lobbying on conservation issues.

Nongovernmental Organizations (NGOs): voluntary, nonprofit organizations that have at least an ostensible commitment to contributing to the public good, whether by fostering economic development, encouraging education, protecting the environment, etc. The term "private voluntary organization" is essentially interchangeable with this definition of the NGO. NGOs often attract relatively well-educated activists and professional administrators, who sometimes focus on supporting GRASSROOTS ORGANIZATIONS. NGOs have become the bulwark of "civil society" as the counterweight to the state. Increasingly, governments, international organizations, and international NGOs have channeled resources for environmental and developmental programs through local NGOs in developing countries. However, while the explicit mandate of most NGOs is to serve the common interest, over time some NGOs become bureaucratic and their leadership comes to be more interested in their own remuneration and standing.

Private Voluntary Organization (PVO). See NONGOVERNMENTAL ORGANIZATIONS (NGOS).

Rainforest Action Network. See under SECTION 6.3.2.

Sierra Club: a U.S.- and Canada-based nongovernmental conservation advocacy organization founded in 1892 by John Muir, a leading U.S. environmentalist who took the more extreme, preservationist position in his famous debate with Gifford Pinchot (see PRESERVATIONIST VS. CONSERVATIONIST DEBATE). Through state and local chapters, the Sierra Club promotes environmental awareness and activism, largely through education programs and wilderness "outings." Its 600,000 members worldwide are organized to exert pressure on government to influence policies on conservation, nuclear safety, and so on. The Sierra Club International has taken a combative stance on many issues, including population, trade, global warming, and the intersection of human rights and environmental issues. An example is the successful effort in the 1980s to block WORLD BANK support for Brazilian activities in the Amazon. Among the "establishment" environmental groups in the United States, the Sierra Club remains one of the most confrontational vis-à-vis government and industry.

United Nations Non-Governmental Liaison Office (UN-NGLS). See under SECTION 5.5.3.

World Conservation Union (IUCN): one of the world's oldest international conservation organizations, established in Fontainbleau, France, in 1948 as the "International Union for the Protection of Nature." Today, headquartered in Gland, Switzerland, it is an association of 74 governments, 105 government agencies, and over 700 NONGOVERNMENTAL ORGANIZATIONS working at the field and policy levels, together with scientists and experts. Such a unique membership allows the IUCN to work at all levels of the environmental debate, from the local and community level up to the global level. The IUCN's activities include protection of endangered animal and plant species, the creation of national parks, assessing the status of ecosystems, and such recent projects as the preparation of the "Red List of Threatened Species" and protecting the coral reefs of Tanzania. A particular focus of current activities is organizing the Global Biodiversity Forum.

World Resources Institute (WRI): an independent, nonprofit center for policy research and technical assistance in the areas of global environment and development. It was founded in 1982 with the mission "to move human society to live in ways that protect the Earth's environment and its capacity to provide for the needs and aspirations of current and future generations." WRI's research focuses on resource economics, forests, biodiversity, climate change, energy, sustainable agriculture, and the interaction of global environmental concerns with economic development and social equity. WRI conducts research on international environmental policy, and gathers and disseminates information among government and private sector decisionmakers, NONGOVERNMENTAL ORGANIZATIONS, and educators. Through its publications and special outreach programs, WRI influences international environmental policy and institutional change. For instance, through its climate protection initiative designed to assess the climate-related activities of various countries, it plays a major technical role in the ongoing climate negotiations. WRI's annual publication *World Resources* traces the status and trends of natural resources worldwide. *T. S.*

Worldwatch Institute: a NONGOVERNMENTAL ORGANIZATION established in 1974 by Lester Brown to analyze and disseminate information on global environmental issues. Since its inception, Worldwatch has been committed to "fostering the evolution of an environmentally sustainable society—one in which human needs are met in ways that do not threaten the health of the natural environment or the prospects of future generations." The Worldwatch publications (e.g., *Worldwatch* bi-monthly magazine, *State of the World* annual report on the status of the environment, *Vital Signs* and *Environmental Alert*—both annual volumes of environmental statistics) raise public awareness and are influential in environmental debates. Lester Brown, still president of Worldwatch Institute, is widely regarded as "one of the world's most influential thinkers" and "the guru of the environmental movement." *T. S.*

World Wildlife Fund for Nature (WWF): formerly known as the World Wildlife Fund, the world's largest and most accomplished independent conservation organization. The WWF was founded in 1961 to protect the world's endangered wildlife and habitat and since then has completed over 11,000 extensive research and conservation programs worldwide. Its current membership exceeds 4,700,000 individual members worldwide and a global network of 24 national organizations. The overreaching goal of the organization is to "stop, and eventually reverse, the accelerating degradation of our planet's environment," while also addressing ways in which humans can best serve their natural environments to promote long-term sustainability. By calling attention to the plight of "flagship species," such as the panda, the wwf has been able to generate public support for its programs and to enjoy considerable influence with governments around the world. To accomplish its global environmental initiatives, the wwf seeks partnerships with international governmental organizations, nongovernmental organizations, and community groups. The wwf has also helped establish new civic and environmental groups, such as the TRADE RECORDS ANALYSIS OF FAUNA AND FLORA IN COMMERCE (TRAFFIC). wwf's publication *Caring for the Earth: A Strategy for Sustainable Living* has been accepted as a national strategy document in over sixty countries.

Major Problems of Environmental Degradation and Development

6.1 Agricultural Decline and Pollution

6.1.1 AGRICULTURAL TRENDS, PRACTICES, AND PROBLEMS

Agricultural Collapse: the severe decline of agriculture in many developing countries, usually brought about by government policies that discourage agricultural investment and encourage able-bodied people to leave rural areas for the cities. These policies include price ceilings on foodstuffs, overvalued exchange rates that discourage agricultural exports, easy entry of free or heavily subsidized food imports, and credit policies that favor industry at the expense of agriculture. Nonagricultural commodity booms (e.g., oil booms that occur when world oil prices are very high) can also trigger an agricultural collapse by strengthening the local currency and thereby make agricultural exports too expensive to be competitive. In some countries, agricultural collapse is also due to environmental deterioration, such as DESERTIFICATION, SOIL EROSION, salinization, and soil or water contamination.

See C. Peter Timmer, *Agriculture and the State: Growth, Employment, and Poverty in Developing Countries* (Ithaca: Cornell University Press, 1991).

Agricultural Extensification: the approach of increasing agricultural production by expanding the area under cultivation or grazing. Agricultural extensification pushes agriculture into frontier areas, risking the ecosystems of those areas. The virtue of extensification is that it increases production without subjecting existing cropland or pasture to greater pressures, yet it often faces the prospect of disappointing productivity, reflecting the fact that the areas opened up for agriculture often have adverse conditions that discouraged agriculture previously. Extensification is often the "path of least resistance" for resettlement programs.

Agricultural Intensification: the approach of increasing agricultural production by using already exploited agricultural land more intensively. Agricultural intensification can be accomplished through higher investments, technological innovations, and more efficient farm systems, but often also entails greater inputs of fertilizer and pesticides. The GREEN REVOLUTION, utilizing new grain varieties and greater inputs of water, fertilizer, and often pesticides, permitted much more intensive agriculture in many countries, and is often credited with averting starvation. However, while agricultural intensification spares nonagricultural areas from direct incursions, the more intensive use of pesticides and fertilizers, and the greater volume of waste products (such as pig waste), may have general polluting effects.

Declining Relative Prices: the tendency of particular types of goods or services to be priced progressively lower in terms of the prices of other types of goods and services. In particular, the specter of declining relative prices of raw materials produced by developing countries, especially but not exclusively agricultural outputs, led prominent development theorists (most notably Raul Prebisch, the Argentine economist who served as secretary general of the UN Economic Commission for Latin America) to predict that developing countries were at a progressive disadvantage in world trade, and therefore to argue for industrialization. This diagnosis was one of the major rationales for draining capital from the agricultural sector to finance the growth of the industrial sector. The evidence that raw materials face declining relative prices is mixed, and the industrialization efforts rationalized on the basis of the expectation of declining relative prices were often disastrous.

Dirty Dozen: a list of the world's most toxic pesticides that have been linked to widespread human poisonings, severe health effects, and environmental damage. In reality the "dirty dozen" list includes more than twelve pesticides: aldicarb, camphechlor (toxaphene), chlordane, heptachlor, chlordimeform, DBCP, DDT, aldrin, dieldrin, endrin, EDB, HCH/BHC, lindane, paraquat, parathion, methyl parathion, pentachlorophenol (PCP), and 2,4,5-T. Despite demonstrated harm to human health and the environment, these toxic chemicals are still widely used in households and in agriculture, and are still exported from the United States and other countries even if their domestic use is banned. The dirty dozen were first targeted in the late 1980s by the Pesticide Action Network (PAN), an international coalition of over four hundred environmental, labor, consumer, farmer, and women's groups. PAN groups worldwide educate governments and population on the danger of these pesticides. Seventy-two countries have now banned or taken other legislative action against DDT, while at least eighty countries have banned or severely restricted other dirty dozen pesticides such as aldrin, dieldrin, and endrin. In many countries, growing rejection of toxic pesticides has been accompanied by increased reliance on INTEGRATED PEST MANAGEMENT and other nontoxic alternatives.

Engel's Law (and Engel's Curve): the theory that the proportion of national income spent on food declines as a nation's income rises. Engel's curve is the relationship between income and the consumption of a given good. The slope of the curve at any point represents the income elasticity of demand for food, or the change in demand induced by a change in income. Applied to food demand, the curve flattens because the increase in demand for food approaches zero after food requirements are well met. Thus, additional income will be spent on other goods, lowering the proportion of income spent on food. *R. P.*

Food Self-Sufficiency: the doctrine of ensuring that a nation's internal capacity to produce food is maintained so that the country can provide all or most of its own foodstuffs. Food self-sufficiency generally arises from a combination of nationalism and fears that international conflicts might result in deliberate cut-offs of food to that nation, or that world food shortages would leave scant supplies to import. The problem for many nations espousing a food self-sufficiency doctrine is that they lack COMPARATIVE ADVANTAGE in producing such volumes of food; they pay far more for domestically produced food than they would to import certain types of food, while land and capital are diverted from more productive economic activities. Domestic production can distort the economy in serious ways, as occurs in Japan because of the long-standing policy of subsidizing domestic rice production. Food self-sufficiency thus differs from the less problematic objective of food security—to ensure that people have enough to eat, whether or not the food is produced domestically.

Green Revolution: agricultural innovations undertaken during the 1960s and 1970s in response to the problem of feeding a rapidly increasing population. Based on research largely funded by the Rockefeller Foundation and associated with geneticist Norman Borlaug, the UN FOOD AND AGRICULTURAL ORGANIZATION (FAO) launched the Indicative World Plan for Agricultural Development (labeled the "Green Revolution") in 1963, involving the introduction of high-yield varieties (HYVs) of wheat, corn, and rice, along with regimens of greater fertilizer, pesticide, and water inputs, to the developing world agriculture. By the early 1980s over half of the wheat and rice areas of developing countries were planted with HYVs. The Green Revolution has been credited with spectacular success in saving millions from famine and changing several developing countries from grain importers into occasional exporters. Yet because of the costs of inputs needed to sustain the HYVs, the Green Revolution has also been accused of worsening the rural income distribution, increasing the number of landless and unemployed peasants, increasing regional disparities, reducing the genetic diversity of cultivated plants, fostering dependence on multinational corporations, and harming the environment.

Land Rotation Agriculture. See SWIDDEN AGRICULTURE.

Shifting Cultivation. See SWIDDEN AGRICULTURE.

Slash-and-Burn Agriculture. See SWIDDEN AGRICULTURE.

Soil Erosion: the deterioration of soil, often caused by wind and water effects abetted by inappropriate agricultural techniques. Soil erosion reduces agricultural productivity unless inputs such as fertilizers are provided, and is often a key aspect of DESERTIFICATION. In some regions, soil erosion has rendered huge areas unfit for any sort of agriculture. Soil erosion may be caused by overly intensive agriculture, pasturage that leads to soil compaction, poor agricultural techniques that expose soil to wind and water erosion, and the removal of trees and other groundcover that hold topsoils in place and reduce heavy winds and water run-off. Agricultural techniques such as terracing, crop rotation, and no-till planting (i.e., planting that does not involve tilling the soil) can reduce soil erosion.

Swidden Agriculture (Shifting Cultivation; Land Rotation Agriculture; Slash-and-Burn Agriculture): the agricultural practice of burning trees and brush in a wooded area in order to plant crops. Because the nutrients from the burning are only sufficient to permit cultivation for a limited period (e.g., one or two years), the cultivators must then shift to another site and repeat the process.

COMMENTARY ON SWIDDEN AGRICULTURE

As long as the area available for swidden agriculture is sufficiently large to permit a long enough cycle before returning to a given site, this form of agriculture can be sustainable. However, increased populations and decreasing areas can cause shifting cultivators to reduce the cycle and return to earlier sites before they have had time to regenerate sufficiently. Many governments in the past have looked upon swidden agriculture as destructive, and have tried to prohibit it. The amount of deforestation apparently due to swidden agriculture often seems high, but this perspective reflects a lack of appreciation for the regenerative potential of the lands subjected to swidden agriculture, as well as ignorance about the interaction between commercial logging and shifting cultivators (the latter often follow logging roads to open up new areas). Yet there is increasing evidence that swidden agriculture is less harmful environmentally than was believed, and that alternative uses of the land—for example, leaving it to commercial logging—may be even worse. *William Ascher*

6.1.2 AGRICULTURAL RESPONSES AND STRATEGIES

Agrarian Reform: policy changes in the agricultural sector intended to improve agricultural productivity, alleviate poverty, redistribute assets, or meet some combination of these objectives. Agrarian reform subsumes *land reform,* which typically emphasizes redistribution of land and clarification of land titles. Yet agrarian reform goes beyond land reform by including reforms of agricultural credit, improvement of agricultural infrastructure, reform of pricing mechanisms, and provision of agricultural extension. Recognition of the importance of these additional components has grown as the experiences in many developing countries have shown that simple reform often does not increase productivity or allow new smallholders to keep their land. Agrarian reform can also encompass policies to enhance the protection of rural ecosystems. Pricing policy reforms are particularly important for sending farmers appropriate signals to avoid overuse of inputs such as fertilizers and pesticides and to encourage investments in the maintenance of land quality.

Agroecology: an area of interdisciplinary research that applies general ecological principles and socioeconomic perspectives to the design and management of agricultural systems and emphasizes the connections of these systems with the broader ecological, economic, and political systems. Agroecology integrates INDIGENOUS KNOWLEDGE with modern technology to ensure not only production but also ecological and socioeconomic sustainability of the agricultural system.

Agroforestry: a combination of land use practices that integrates the management of agricultural crops, both plant and animal, with the practice of forestry. Different species and combinations of trees, crops, and /or livestock are used, depending on whether agroforestry is practiced in a temperate or tropical climate and other ecosystem characteristics. As a land use management philosophy, agroforestry seeks to actively manipulate the biological and physical interactions between trees, crops, and animals while maximizing environmental protection and satisfying human needs. *T. S.*

Alternative Farming: a set of farming practices that restrict the use of pesticides, chemical fertilizers, and other artificial inputs typical of traditional farming. These practices might include systems with only slightly reduced dependence on these inputs, such as INTEGRATED PEST MANAGEMENT, as well as systems that avoid them completely, such as ORGANIC FARMING.

Biodynamic Agriculture: a concept of farming based on the notion that cosmic and terrestrial forces tend to enhance the growth and quality of plants and animals. It draws heavily from the work of Austrian philosopher and scientist Rudolf Steiner (1861–1925), who argued that the physical bodies of human beings were to be effective vehicles for the growth and intentions of the human spirit. Biodynamic agriculture as a practice is based on the "biodynamic preparations" that enhance soil fertility through applications of a variety of common herbs, silica, and cow manure, as well as on the "biodynamic sowing calendar" that defines appropriate timing to sow specific plants based on the relative positions of different cosmic objects.

See Mary Gold, *Sustainable Agriculture: Definitions and Term* (Beltsville, MD: National Agricultural Library, 1994), 4–5.

Codex Alimentarius: an international system of standards for various food products and their components, developed in 1962 by the joint efforts of the UN FOOD AND AGRICULTURE ORGANIZATION (FAO) and the WORLD HEALTH ORGANIZATION (WHO) to protect the health of consumers and to ensure fair practices in international trade. As of January 2000, the Codex Alimentarius comprised 237 food standards for commodities, 41 codes of hygienic or technological practice, and 25 guidelines for contaminants, etc. Codex standards for raw and processed food commodities that are internationally traded include uniform labeling requirements, the presence of additives and contaminants, sanitary requirements, and composition of any foodstuff. Currently there are more than 160 member countries that accept Codex Alimentarius standards partially or fully. The CODEX ALIMENTARIUS COMMISSION is an institutional mechanism for this international system of standards.

COMMENTARY ON THE CODEX ALIMENTARIUS

The positive role of such standards in the harmonization and development of international trade is indisputable. However, in practice it is difficult for many countries to accept Codex standards in the statutory sense due to differing legal formats and administrative systems, varying political systems, and sometimes the influence of national attitudes and concepts of sovereign rights. Despite these difficulties, however, Codex standards have become the benchmarks against which national food measures and regulations are evaluated and an increasing number of countries align their national food standards (especially those relating to safety such as additives, contaminants, and residues) with them. On the other hand, some critics believe that the Codex Alimentarius negotiations are heavily influenced by the large multinational food and agrochemical companies whose prevalence within governmental delegations results in scaling standards downward. Greenpeace USA research has revealed that Codex safety levels for eight widely used pesticides were lower than the U.S. standards by a factor of twenty-five.

See "Power: The Central Issue," *The Ecologist* 22 (4) (July/August) 1992: 159. *Natalia Mirovitskaya*

Codex Alimentarius Commission: an intergovernmental body established in 1962 by the UN FOOD AND AGRICULTURE ORGANIZATION (FAO) and the WORLD HEALTH ORGANIZATION (WHO) to set international guidelines for the control of food quality and safety. The commission is responsible for compiling the standards, codes of practice, guidelines, and recommendations that constitute the CODEX ALIMENTARIUS and that can be used by countries to facilitate trade. It meets every two years, with plenary sessions attended by senior governmental officials, representatives of industry, consumers' organizations, and academic institutes. Between these plenary sessions, the commission performs its work through committees composed of delegates from member countries, including subject committees such as food labeling and commodity committees such as milk products. The secretariat of the commission is headquartered in Rome.

Food Aid: the practice of providing foodstuffs from national and international sources, either as grants or on concessional terms, to nations that are unable to meet their domestic food requirements. Massive food aid began in the 1950s with the programs relying on U.S. grain surpluses (such as Public Law 480) and currently averages about 9 percent of all U.S. FOREIGN ASSISTANCE. UN food assistance is provided under the auspices of the UN FOOD AND AGRICULTURE ORGANIZATION (FAO). Food aid has been justified as crucial for improving the nutrition of people in food-scarce countries, and for averting famine during natural and social disasters. However, long-term food aid (or bulk transfer of food to governments to be sold on the local market) has also been heavily criticized for undermining the rationale for agricultural production within the recipient countries, relegating them to food, economic, and political dependence, exacerbating rural poverty, distorting the eating habits and traditions of local population, and contributing to environmental degradation. Given the possibility of major climatic change, the fact that most food aid comes from the American plains is disconcerting as well. Another line of criticism against food aid comes from the members of THE ENVIRONMENTAL FUND and their ideological successors, who have argued that food aid violates the principle of carrying capacity by "artificially allowing more people to live on the land that can live from it."

See The Environmental Fund, *Behind the "Food Crisis"* (Washington, DC: The Environmental Fund, 1977); World Food Program, *World Food Aid: Experiences of Recipients and Donors* (London: James Currey and Heinemann, 1993).

Food Security: access to adequate food supplies. The UN FOOD AND AGRICULTURE ORGANIZATION (FAO) defines this as access by all people at all times to the food needed for healthy and active life. Achieving food security means ensuring that sufficient food is available, that supplies are relatively stable, and that those in need of food can obtain it. However, as long as a nation has the capacity to purchase and distribute food imports, food security does not necessarily mean FOOD SELF-SUFFICIENCY in the sense of the capacity of a nation to produce its own food. Currently, some 800 million people in developing countries are chronically undernourished and at least 86 countries of the world are believed to lack food security. This issue was a major focus of the 1996 WORLD FOOD SUMMIT.

Integrated Pest Management (IPM): 1) methodology of pest control that seeks to combine all available techniques of pest suppression, crop rotations, resistant crop varieties, and encouragement of natural enemies and diseases, together with the selective and judicious use of pesticides, into a sound ecological framework. Rather than trying to eradicate the entire pest population with heavy and expensive doses of pesticides, integrated pest management tries to find the optimal level of pesticide and other pest-control interventions that allow for high agricultural productivity with reasonable economic costs and minimized environmental damage, even if the pest population is not completely eradicated. In some circumstances, conventional pesticides are replaced by the introduction of animals or even plants that prey on or repel harmful pests or disrupt their reproduction.

2) a program developed by the UN FOOD AND AGRICULTURE ORGANIZATION (FAO) in the early 1980s specifically for South and Southeast Asia. The program was supposed to coordinate economically and environmentally acceptable methods of pest control with judicious and minimal use of toxic pesticides.

3) a facility created in 1994 by the WORLD BANK, the FAO, the UN ENVIRONMENT PROGRAMME and the UN DEVELOPMENT PROGRAMME to promote international cooperation in reducing chemical pesticides used in developing countries and combine their use with natural methods of pest control.

COMMENTARY ON INTEGRATED PEST MANAGEMENT

The idea of IPM originated in the 1980s following unsuccessful efforts to regulate international pesticide trade due to pressures on developing countries from the pesticide exporters. The pressures concentrated on preventing the inclusion of the "prior informed consent" from the regulations, so the recipient countries were not able to get information about pesticides that were banned, withdrawn, or restricted for use in other countries before sanctioning their import. *Natalia Mirovitskaya*

Nature Farming: an approach to agriculture based on the writings and practices of the Japanese philosopher Mokichi Okada (1882–1955). In its ideal variant, it is an integrated system of production, marketing, and consumption that builds upon the inherent life-giving powers of the soil, fire (sunlight), and water. Nature farming takes into account the interdependence of human beings, nature, and the earth and it should therefore protect the natural environment, sustain human health and security, and contribute to an "ideal civilization." The techniques of nature farming include soil evaluation, appropriate tillage practices, crop rotation, cover crops, composting, nonchemical pest and weed control, and other environment-friendly agricultural practices and are to be adapted to each farm's unique conditions. Nature farming is practiced by many Japanese farmers and has been promoted by the Japanese NONGOVERNMENTAL ORGANIZATION, MOA International, in several other countries. See also ORGANIC AGRICULTURE.

Organic Agriculture: a production system that avoids or largely excludes the use of synthetically compounded fertilizers, pesticides, growth regulators, and livestock feed additives. To the maximum extent feasible, organic farming systems rely upon crop rotations, crop residues, animal manure, legumes, green manure, off-farm organic wastes, mechanical cultivation, mineral-bearing rock, and biological pest control to maintain soil productivity and control pests. See also BIODYNAMIC AGRICULTURE; NATURE FARMING.

See Mary Gold, *Sustainable Agriculture: Definitions and Terms* (Beltsville, MD: National Agricultural Library, 1994), 6.

Permaculture (Permanent Agriculture): a term coined in the late 1970s by Australian scholar Bill Mollison to define "a design system for creating sustainable human environments . . . applied at the scale of a home garden, all the way through to a large farm." One of the many alternative practices described as sustainable agriculture, permaculture is distinct by its emphasis on design, i.e., the location of each element in a landscape, and the evolution of landscape over time.

See Mary Gold, *Sustainable Agriculture: Definitions and Terms* (Beltsville, MD: National Agricultural Library, 1994), 6.

Sustainable Agriculture: the aspiration to design agricultural systems that could provide the food and fiber needs of the population without sacrificing health, environmental quality, and a sound natural resource endowment. The sustainability of modern, increasingly industrialized agriculture is challenged by the growing use of fertilizers, pesticides, and scarce water required by the introduction of higher-yield varieties and more intensive cultivation. Sustainable agriculture may require reliance on a broader set of crops, greater use of local knowledge concerning cultivation that is compatible with ecosystems, full pricing of agricultural inputs in order to reduce inefficiency-producing subsidies, and full charges for environmental damage. *T. S.*

Urban Agriculture: the practice of crop cultivation, livestock raising, and aquaculture within the boundaries or the immediate periphery of a city. The dramatic rise in urban populations, and growing populations of the urban poor, have contributed to the rapid spread of urban-based food production. The UN DEVELOPMENT PROGRAMME identifies five main systems of urban agriculture: horticulture (cultivation of fruits, vegetable, and ornamental plants); animal husbandry; AGROFORESTRY; AQUACULTURE; and other farming activities (including beekeeping, worm-raising, and the cultivation of beverage, culinary, and medicinal crops). The choice of a particular pattern of urban agriculture depends on a combination of social, physical, and economic factors, including culture, traditions, market, land and water availability, climate, soil condition, etc. For instance, in the face of the economic crisis following the dissolution of the Soviet Union, dachas (small houses and land plots on the outskirts of cities traditionally used for recreation) evolved into a major source of food security for millions of people.

6.1.3 ORGANIZATIONS RELATED TO AGRICULTURE

Consultative Group on International Agricultural Research (CGIAR): the body that directs sixteen specialized international agricultural research centers. CGIAR was formed in 1971 as the "informal consortium of governments, international and regional organizations, and private foundations . . . to nurture agricultural research to improve the quantity and quality of food production in the developing countries." It is headquartered in the WORLD BANK and is sponsored jointly by the World Bank, UN FOOD AND AGRICULTURE ORGANIZATION (FAO), and the UN DEVELOPMENT PROGRAMME. As an umbrella organization, CGIAR does not have a formal governing charter and its activities are based on consensus decisions made at the semiannual meetings of members and on recommendations from its technical advisory committee. CGIAR research centers, which are to a large degree autonomous and dispersed throughout the world, are dedicated to a broad range of agricultural and forestry research topics (e.g., rice research, AGROFORESTRY, tropical agriculture, agricultural policy, forestry policy, corn and wheat research, drylands agriculture). The CGIAR research centers have been major contributors to the development of GREEN REVOLUTION plant varieties. Recently CGIAR's research has been reoriented toward SUSTAINABLE AGRICULTURE. Some critics have questioned the need for internationally funded agricultural research centers when private corporate research centers are investing heavily in research.

Food and Agriculture Organization (FAO). See UN FOOD AND AGRICULTURE ORGANIZATION (FAO).

International Fund for Agricultural Development (IFAD): a specialized UN agency founded in 1977 and headquartered in Rome. IFAD, which emerged out of the 1974 World Food Conference, aims to mobilize funds from the Western nations and OPEC countries for agricultural and rural projects in developing countries with special emphasis on the small farmers, landless poor, and impoverished rural women. Since its establishment, IFAD has financed nearly 500 projects in 111 countries, involving over US$5 billion of multilateral and bilateral loans at zero interest rates and very long payback periods. As of early 2000, IFAD had 161 member states. Being dependent on voluntary contributions rather than annual fixed payments, IFAD is therefore less financially secure than other UN financial agencies.

COMMENTARY ON THE INTERNATIONAL FUND FOR AGRICULTURAL DEVELOPMENT

Despite its financial precariousness, IFAD is considered one of the most efficient UN agencies. Its small size, emphasis on providing assistance to those who would have no other source of loans, and its ideological commitment to helping the poor to help themselves, make IFAD an archetype of the "barefoot" approach to development. *Natalia Mirovitskaya*

Special Program for Food Security (SPFS): a UN FOOD AND AGRICULTURE ORGANIZATION (FAO) initiative launched in 1994 to focus on the food security problems of the eighty-six "Low-Income Food-Deficit Countries" (LIFDCS)—the countries least able to meet their food needs. The main goal of this initiative is to help LIFDCs improve their national food security through rapid increases in productivity and food production and by reducing year-to-year variability in production. The program is based on the assumption that most countries have viable and sustainable means of increasing food availability but that these means are not realized due to technical, institutional, or policy constraints. By helping farmers and other STAKEHOLDERS to identify these constraints, the program would enhance food security. A distinctive feature of the SPFS is its South-South Cooperation Scheme, which provides an opportunity for strengthening partnerships among developing countries at different stages of development. Within this scheme, a large number of experts from more advanced developing countries work for two to three years in the recipient countries. The SPFS has been operating in fifty-four countries—roughly two-thirds of the LIFDCs—with a biennial budget for its own operations (not including the foreign assistance that it brokers) of approximately $10 million.

UN **Food and Agriculture Organization** (FAO): the most prominent international body devoted to agriculture, forestry, fisheries, and related issues. As the largest self-governing organization within the United Nations system, in terms of both operating budget and staff size, the FAO has a total membership of 174 countries plus the EUROPEAN UNION (EU) and a biennial budget of $US650 million (as of 1997). The FAO was founded in 1945 with a mandate to raise levels of nutrition and standards of living, to improve agricultural productivity, and to better the conditions of rural populations. FAO offers direct development assistance, collects, analyzes, and disseminates information, coordinates international food assistance during emergencies, provides policy and planning advice to governments, and acts as an international forum for debate on food and agriculture issues. It has been involved in many projects related to land and water development, forestry and fisheries, plant and animal production, food standards, commodities and trade, as well as in several UN environmental programs. FAO also serves as a lead agency in THE TROPICAL FORESTRY ACTION PLAN and in the UN'S CONVENTION TO COMBAT DESERTIFICATION.

COMMENTARY ON THE FOOD AND AGRICULTURE ORGANIZATION

The immensity and multiplicity of its tasks make the FAO an obvious target for criticism. Environmentalists attack the agency for promoting commercial development of tropical forests rather than conservation or sustainable use, as well as for its support for large-scale irrigation projects, its orientation on export crops, and its alleged advocacy of heavy use of chemical inputs. First World countries regard the FAO as dominated by developing countries (its governance is based on the same one-member, one-vote principle as the UN General Assembly), while developing countries reproach it for its low efficiency in food assistance. In the early 1990s, FAO was restructured to integrate sustainability into its programs and activities and to focus more on the world's shrinking natural resource base, sustainable agriculture, forestry, and fisheries. The results of this reorientation are yet to be seen. *Natalia Mirovitskaya*

World Food Council (WFC): an international forum that advocates, catalyzes, and co-ordinates international policies aimed at alleviating world hunger and improving the global food system. The WFC is comprised of ministers from thirty-six nations representing all regions of the world who are assisted by a small staff of food policy specialists. It works together with the UN ENVIRONMENT PROGRAMME to assess policy issues and advance action toward achieving ecologically sustainable FOOD SECURITY.

World Food Programme (WFP): an intergovernmental organization founded in 1961 to provide food aid to developing countries both in emergencies and as means to promote economic and social development. The WFP is the world's largest international food aid organization, with cumulative expenditures (including food purchases) of over $25 billion since its inception. It currently operates in eighty-two countries. WFP provides nearly one third of all global food aid and targets primarily the least developed, low-income, food-deficit countries. Assistance is delivered through three main programs: Food-For-Life (emergency aid to victims of natural or humanmade disasters); Food-For-Growth (food aid to needy children, pregnant and breastfeeding women, and the elderly); and Food-For-Work (promoting self-reliance by paying chronically hungry people with food for their work in social programs). WFP is also one of the world's largest donors of resources for environmental activities, spending some $1 million a day on projects for afforestation, soil conservation, and activities to promote SUSTAINABLE AGRICULTURE.

COMMENTARY ON THE WORLD FOOD PROGRAMME

Despite the general ho-hum performance of UN organizations, the WFP has been doing very good work in Central America, where it has been able to provide safety-net assistance through well-designed programs such as food-for-work and school lunch programs. Compared to other donors, the WFP has been effective in providing assistance with a minimum of disruption of local market incentives and of policy formulation. *Gustavo J. Arcia*

World Food Summit: the first meeting of top government leaders to focus on the food security issue. The summit, convened by the UN FOOD AND AGRICULTURE ORGANIZATION (FAO) in Rome in November 1996 was in reaction to widespread malnutrition and growing concern about the capacity of agriculture to meet future food needs. Among various issues discussed at the summit were biodiversity concerns and the connection between food security and environmental degradation. The meeting adopted the Declaration on World Food Security and the World Food Summit Plan of Action. It was not designed as a binding conference, nor was it aimed at creating new mechanisms or institutions. Each participant state was to consider how and what it might do to contribute to the implementation of the policies, strategies, and the plan of action adopted by the summit.

Acid Deposition (Acid Rain): the phenomenon by which air pollutants, in particular sulfur dioxide (SO_2) and nitrogen oxides (NO_x), form air- or rain-borne acids that are damaging to forests, freshwater aquatic life, and stone surfaces of buildings and monuments. Acid deposition occurs in a dry form when the pollutants gravitate out of the atmosphere and form acidic solutions when they combine with moisture on the earth's surface. Alternatively, the pollutants may undergo further oxidation in the atmosphere in the presence of photo-oxidants such as low-level ozone to form tiny droplets of sulfuric and nitric acids. These acids dissolve in water vapor and eventually precipitate to the earth in rain, snow, or mist. Most dry deposition occurs soon after the pollutants are released and near the pollution source, while wet deposition normally takes place several days later. Winds may carry the pollutants hundreds or even thousands of kilometers from their source. Wet deposition became an increasingly serious problem in heavily industrialized regions, such as Europe and North America, after tall smokestacks were mandated as a strategy for dispersing pollutants that were threatening human health in heavily polluted areas.

See Chris C. Park, *Acid Rain: Rhetoric and Reality* (New York: Routledge, 1987), 22–50. *M. S. S.*

COMMENTARY ON ACID DEPOSITION (ACID RAIN)

The controversy over acid deposition in the United Stated led to the passage of the Clean Air Act Amendments of 1990. The CAAA 1990 mandated reductions in coal-fired power plant emissions of sulfur dioxide and nitrogen oxides and set caps on the amount of sulfur dioxide that could be emitted. To accomplish these goals, a market-based system of sulfur dioxide emission allowances was established. *Toddi Steelman*

Air Pollution: an unwanted change in the atmosphere that can harm humans or other organisms and can damage their surroundings. It results both from human activities (industrial, agricultural, and domestic) and natural sources, such as volcanic eruption. Air pollution usually comes in four basic forms—particulate, gaseous, photochemical, and radioactive—and in various combinations of these. It can also be classified on the basis of its origin (stationary or point sources, mobile or nonpoint sources, and domestic sources, such as home heating), geographical distribution (local, transboundary, and global), and physical or chemical complexity (primary or secondary). In the United States, the term is usually applied to six conventional pollutants identified and measured by the Environmental Protection Agency: carbon monoxide, lead, nitrogen oxides, ozone, particulate matter, and sulfur oxides. Toxic effects of air pollution damage both natural and human-dominated ecosystems, human health, and human-made capital. It also affects atmospheric chemistry and physics, therefore contributing to global climate change. The ecological and economic costs of such damage are extensive, diffuse, and long-term. This makes the control of air pollution a multidisciplinary and expensive effort, the benefits of which may be slow to be realized.

Air Pollution Index (API; Air Quality Index): a quantitative measure for gauging ambient air quality. The API is obtained by combining data for various air pollutants into an aggregate measure.

Air Quality Standards: standards that establish the targets for limiting specific air pollutants. Various jurisdictions, from cities and counties to nations, set these targets as a first stage in regulating or otherwise discouraging activities that produce air pollution. Air quality standards thus prescribe maximum and/or permissible concentrations of substances, such as carbon monoxide, fine particulates, sulfur dioxide, etc. The level of pollutants prescribed by regulations may not be exceeded during a specified time in a defined area. There are two debates concerning air quality standards: a) how to balance the benefits of strict standards with the economic costs of restricting activities in order to meet these standards; and b) whether the same standards ought to be applied across a wide variety of locations, with different environmental and economic characteristics. See also AMBIENT STANDARDS.

Ambient Standards. See under SECTION 1.2.7.

Long-Range Transboundary Air Pollution Convention (LRTAP): the first treaty to address the problem of ACID DEPOSITION caused by air pollutants that cross international frontiers. The convention was concluded at a ministerial-level meeting convened in Geneva in November 1979 under the auspices of the United Nations Economic Commission for Europe, whose membership includes most Eastern and Western European countries, as well as the United States and Canada. A typical FRAMEWORK CONVENTION, the agreement did not bind the parties to a schedule for specific reductions of acid-forming pollutants, but called upon them to endeavor to limit, reduce, or prevent air pollution using the BEST AVAILABLE TECHNOLOGY that is economically feasible. The agreement has been supplemented by a series of protocols that impose limits or reductions by specified dates on sulfur dioxide (1985), nitrogen oxides (1988), and volatile organic compounds (1991). A revised sulfur protocol was adopted in 1994, based on calculations of CRITICAL LOADS, the amount of acid deposition that an area can absorb without serious environmental damage.

See John McCormick, ed., *Acid Earth: The Politics of Acidification* (Washington, DC: Island Press, 1997). *M. S. S.*

COMMENTARY ON LONG-RANGE TRANSBOUNDARY
AIR POLLUTION CONVENTION

While the Geneva Convention of 1979 has the typical weaknesses of FRAMEWORK CONVENTIONS, the series of protocols contain path-breaking regulatory mechanisms that have set a precedent for further protocols that limit other pollutants, such as the MONTREAL PROTOCOL on ozone-depleting substances. The shift from across-the-board reductions in pollutants to differentiated reduction targets based on CRITICAL LOADS was an especially significant regulatory refinement. *Marvin S. Soroos*

Trail Smelter Case: an international dispute that arose in 1927 when the United States formally complained that fumes from a large and recently expanded lead and zinc smelter in Trail, British Columbia, were causing significant damage to agricultural operations across the border in the state of Washington. The case was initially taken up in the INTERNATIONAL JOINT COMMISSION, which in 1931 awarded the United States $350,000 to compensate its farmers for damages. In 1935 the two disputants agreed to submit the remaining issues to a three-person arbitral panel, which in an interim ruling awarded the United States an additional $78,000. The tribunal's final opinion, issued in 1941, is notable for the statement that "no state has the right to use or permit the use of its territory in such as a manner as to cause injury by fumes in or to the territory of another or the properties or persons therein, when the case is of serious consequence and the injury is established by clear and convincing evidence."

See John D. Wirth, "The Trail Smelter Dispute: Canadians and Americans Confront Transboundary Pollution, 1927–41," *Environmental History* 1 (2) (April 1996): 34–51. *M. S. S.*

COMMENTARY ON TRAIL SMELTER CASE

The Trail Smelter case is a frequently cited landmark case in international law not only for being the first to address TRANSBOUNDARY AIR POLLUTION, but also for affirming the principle that states have a responsibility to ensure that activities taking place within their territories do not cause significant damage to people or property of other countries. Its applicability is limited, however, because the links between pollution sources and specific environmental damages are usually not as clear-cut as in the Trail Smelter case. *Marvin S. Soroos*

Transboundary Air Pollution: the spread of airborne pollution across national boundaries or other jurisdictional borders. For instance, ACID DEPOSITION may be transported in the form of "fog, dew, rain, frost, sleet, hail, snow, and dry deposits" over a thousand kilometers and eventually affect areas in excess of a million square kilometers in other countries. Over the past thirty years, long-range air pollution has become a major environmental concern due to its impact on various life forms, especially human health (e.g., the increased cancer incidence due to the 1986 Chernobyl nuclear disaster) and deforestation resulting from acid rain. Transboundary air pollution has also become a major concern because of the difficulty of resolving disputes over damages when the pollution extends across international boundaries.

6.3 Biodiversity Loss and Conservation Efforts

6.3.1 BIODIVERSITY CONCEPTS, TRENDS, AND CONDITIONS

Biocolonialism. See under SECTION 4.1.

Biodiversity Loss: the loss of the earth's biological resources, encompassing both the mass extinction of species and the loss of genetic diversity within species. It is widely believed that BIODIVERSITY loss is taking place at a historically unprecedented and alarming rate. Unlike previous epochs of biodiversity loss, which were probably caused by dramatic climate change (e.g., the ice ages), the current decline in biodiversity is mainly caused by the shrinking of habitats. The current destruction of the world's tropical forests, which hold an estimated 50 to 90 percent of all species, could lead to the loss of one-fourth to one-half of all species within a few decades. This would constitute a sixth extinction spasm in the history of the planet. Other major causes for biodiversity loss are pollution and the introduction of non-native species. Biodiversity loss is addressed by several international treaties like the LONDON CONVENTIONS, the BONN CONVENTION, the CONVENTION ON INTERNATIONAL TRADE IN ENDANGERED SPECIES (CITES), and the 1992 CONVENTION ON BIODIVERSITY.

See Reed E. Noss and Allen Y. Cooperrider, *Saving the Nature's Legacy: Protecting and Restoring Biodiversity* (Washington, DC: Island Press, 1996); G. K. Meffe, C. R. Carroll et al., *Principles of Conservation Biology* (Sunderland, MA: Sinauer Associates, 1997).

Biodiversity Indices. See under SECTION 1.2.7.

Biosafety: in the context of the CONVENTION ON BIOLOGICAL DIVERSITY, safety procedures to regulate the potential transboundary effects of harmful biological agents, including genetically altered organisms. Public perceptions on GENETIC ENGINEERING, especially the potential for human cloning, along with developing nations' concern over the exploitation of their BIODIVERSITY resources, gave an impetus to negotiations on internationally binding regulations on the transboundary movements of modified organisms. The Biosafety Protocol to the CONVENTION ON BIOLOGICAL DIVERSITY is supposed to address these concerns, although until now its draft remains rather vague. In anticipation of this protocol, some countries have developed their own guidelines and agreements on biosafety while the UN FOOD AND AGRICULTURE ORGANIZATION (FAO), the EUROPEAN UNION (EU), the UN ENVIRONMENT PROGRAMME (UNEP), the UN INDUSTRIAL DEVELOPMENT ORGANIZATION (UNIDO) and the ORGANIZATION FOR ECONOMIC COOPERATION AND DEVELOPMENT (OECD) introduced various SOFT LAWS on biotechnology.

Conservation versus Preservation: the dispute over approaches to maintaining BIODIVERSITY and landscapes, centered on the dilemma of whether to try to preserve (see PRESERVATION) as many ecosystem elements as possible or to conserve (see CONSERVATION) the general configuration of the ecosystem while allowing ecosystem changes that are very expensive or difficult to prevent. Preservation involves maintaining protected areas unchanged from the conditions they were in when their importance was first recognized. Conservation recognizes that natural communities of plants and animals exist in dynamic equilibrium, and it aims to prevent any development that would alter or destroy natural habitat but without interfering with natural ecological changes. The conservation approach has now become much more widely accepted than preservation.

COMMENTARIES ON PRESERVATION VS. CONSERVATION

In the United States, the debate over preservation and conservation was played out initially by John Muir and Gifford Pinchot in the late 1800s. Muir, an avid preservationist, valued resources for their intrinsic worth and advocated for their wholesale protection. Muir was responsible for the establishment of protecting the Yosemite Valley in the 1880s. In contrast, Pinchot, the founder of the U.S. Forest Service, valued resources for their instrumental worth and advocated using those resources for the greatest good of the greatest number in the long run. *Toddi Steelman*

The idea of preservation was easier to maintain in turn-of the-century America than in Europe where the entire landscape had long been transformed by human action. For example in Britain, preservation is a concept associated with that of the countryside—an idyllic but far from natural state of affairs. *Steve Rayner*

In the traditions of the Western world, "wise use" conservation and preservation are two fundamentally different kinds of conservation. Modern conservation combines elements of both. It differs from "wise use" conservation in its rejection of utilitarianism and instrumental values (nature-as-commodity) and from preservationism in its rejection of a pure hands-off approach to nature. Modern conservation seeks to sustain species and ecosystems, and has come to focus on biodiversity as an overarching goal. Recognizing that much of the world's biodiversity is found in landscapes occupied and used by people, international biodiversity conservation increasingly accepts the notion of sustainable human use of the environment. The role of indigenous peoples is of special interest in biodiversity conservation and very controversial. One side argues that the agendas of indigenous peoples are not always consistent with biodiversity conservation, especially at a time when many of these groups are engaged in a market economy and use nontraditional technologies. The other side argues that indigenous peoples' worldviews and practices do not match preservationist goals but do match the conservation-through-use approach of sustainability, and recommends a partnership of indigenous peoples and conservationists as the best long-term option for biodiversity conservation.

See UN Environmental Programme, *Global Biodiversity Assessment* (New York: UNEP, 1995); Fikret Berkes, *Sacred Ecology* (Philadelphia: Taylor and Francis, 1999). *Fikret Berkes*

Endangered Species: plant and animal species threatened with extinction and whose survival is unlikely if causal factors continue operating. At this point, there is no universally accepted criterion for considering particular species endangered and for taking policy actions. For example, in the United States, the 1995 Endangered Species Conservation Act (ESA) defines endangered species as any biological species that, without any action, would be placed on an irreversible course to extinction within two human generations, or forty years. This act classifies species as critical (those that probably cannot survive without direct human intervention), threatened (abundant in parts of their range but declining in total numbers), and rare (exist in relatively low numbers over their ranges but are not necessarily in immediate danger of extinction). The WORLD

CONSERVATION UNION (IUCN) expands this classification to: a) extinct (not seen in the wild for at least fifty years); b) critically endangered (facing an extremely high risk of extinction in the wild in the immediate future); c) endangered (facing a very high risk of extinction in the wild in the near future, as defined by such criteria as an observed, estimated, inferred, or suspected reduction of at least 50 percent over the last ten years or three generations); d) vulnerable (facing a high risk of extinction in the wild in the medium-term future, defined by such criteria as an observed, estimated, inferred or suspected reduction of at least 20 percent over the last ten years or three generations); and e) indeterminate and insufficiently known. Since the 1960s, the process of extinction has accelerated rapidly through the impacts of population growth and technology on natural ecosystems. The first global convention on wildlife conservation with both strong legal commitments and an enforcement mechanism was the CONVENTION ON INTERNATIONAL TRADE IN ENDANGERED SPECIES (CITES).

COMMENTARY ON ENDANGERED SPECIES

The loss of endangered species is not fully appreciated by the public and many decisionmakers, although there has been growing awareness worldwide over the last two decades. Usually the problem is perceived as a biological one— wherein species are reduced first by habitat loss and conversion, and when the species has shrunk to only a few small populations, a host of random demographic, genetic, and environmental variations and catastrophes pushes the species to extinction—and biologists are expected to address it. However, there is also a host of professional, organizational, and policy dimensions to endangered species conservation that are typically overlooked or underappreciated. A species can be recovered from endangerment if these elements can be configured effectively to address the entire problem and if the species's biology permits recovery. The challenge is to understand the interaction of both biological and nonbiological dimensions in actual endangered species cases and how to configure species recovery and conservation efforts in the future.

See Tim W. Clark, *Averting Extinction: Reconstructing Endangered Species Recovery* (New Haven: Yale University Press, 1997); Tim W. Clark, Richard P. Reading, and Alice L. Clarke, eds., *Endangered Species Recovery: Finding the Lessons, Improving the Process* (Washington, DC: Island Press, 1994). *Tim W. Clark*

Extinct Species: species that no longer exist, either because no survivors were left to reproduce, or because descendants evolved into separate species. Often extinction is the result of the failure of the species to adapt to environmental change, such as climate change or loss of habitat. Though extinction is a natural process and has occurred throughout geological history, the current rate of extinction is believed to be unprecedented and mainly human-induced.

Genetic Engineering: a biotechnical process by which the insertion of new genetic material into cells results in the modification of the characteristics of an organism. By endowing plants and animals with the characteristics of the species from which the genetic materials came, genetic engineering can accelerate the development of desired living forms far more rapidly than conventional breeding methods. However, genetic engineering bears the risk that dangerous organisms may escape, or be intentionally released, into the environment. There is much debate about the seriousness of the risk of accidental escape of dangerous genetically engineered organisms, with some defenders of genetic engineering arguing that numerous genetic changes would have to be introduced to create dangerous organisms.

Keystone Species: a term used loosely to designate species that are especially important to the structure and resilience of their communities. Often keystone species are not obvious and emerge as such only after their disappearance from a habitat, whose dynamics then become radically different. Identifying keystone species before their disappearance is crucial for conservation efforts.

Riparian Habitat. See under SECTION 1.2.4.

Spasms of Extinction: periods of mass species extinction, during which more than half and up to 90 percent of all existing species disappeared in a short period of time. The fossil records attest to at least five spasms of extinction. Short-term results of spasms of extinction for biodiversity are dramatic; however, from the long-term perspective these periods are usually followed by phases of rapid speciation, during which surviving species rapidly evolve to occupy vacated ecological niches. Some ecologists think that human action has already provoked a sixth spasm of extinction.

Vavilov Centers: areas of the most valuable concentrations of genetic resources. Russian scientist Nikolai Vavilov suggested that the center of origin of a cultivated plant is to be found in the regions where its wild cousins show maximum genetic diversity. Through his extensive worldwide expeditions, Vavilov was able to locate many of these centers that now are prime targets for on-site conservation.

6.3.2 RESPONSES AND STRATEGIES

Bioprospecting: exploration of wild plants and other organisms located mainly in the tropics (and therefore mainly in the developing world) for commercially valuable genetic and biochemical resources. Hopes that bioprospecting will yield more pharmacological or other biological discoveries (such as the rosy periwinkle, which has been the base for highly successful anti-leukemia drugs) have prompted a large number of bioprospecting ventures. Prior to the CONVENTION ON BIOLOGICAL DIVERSITY (CBD), most countries considered genetic resources to be the COMMON HERITAGE OF MANKIND, and practically no precedent existed for national policies and legislation to govern and regulate their exploration, or to ensure that the country in which the bioresource was found would share in the benefits of its commercialization. Carried out as traditional resource-exploitation ventures, bioprospecting has often been referred to as "biopiracy" and was just another example of ENVIRONMENTAL COLONIALISM. The convention, by asserting the sovereignty of nations over their biodiversity, explicitly recognizes the right of countries to establish legislation regulating access to genetic resources and to require payment for that access. This might turn bioprospecting into a positive force for SUSTAINABLE DEVELOPMENT of the host countries. A major limitation on bioprospecting is that the United States has not ratified the CBD and therefore American companies are not officially subject to its requirements.

Biosafety Protocol (Cartagena Agreement): a legally binding agreement that regulates the transboundary transport of living modified organisms (LMOS), such as tomatoes, grains, corn, soybeans, and other products that have been genetically modified for greater productivity or nutritional value, or for resistance to pests or diseases. Agricultural LMOS form the basis of a multibillion-dollar global industry. The Biosafety Protocol was approved after five years of negotiations in January 2000 by representatives from 174 countries, and will enter into force for its members after ratification by fifty signatories. In response to a growing public concern about the potential risks of biotechnology, the protocol specified that governments were to signal whether or not they are willing to accept imports of agricultural commodities that include LMOS by communicating their decision to the world community via an Internet-based Biosafety Clearing House. In addition, shipments of these commodities that may contain LMOS are to be clearly labeled. Stricter "Advanced Informed Agreement" procedures will apply to seeds, live fish, and other LMOS that are to be intentionally introduced into the environment. In these cases, the exporter must provide detailed information to each importing country in advance of the first shipment, and the importer must then authorize the shipment. The objective is to ensure that recipient countries have both the opportunity and the capacity to assess risks involving the products of modern biotechnology.

Biosphere Reserves: terrestrial and coastal systems of ecological significance that are internationally protected under the auspices of the UN EDUCATIONAL, SCIENTIFIC AND CULTURAL ORGANIZATION (UNESCO). More than 350 biosphere reserves exist in 90 countries, collectively constituting a world network. The system is administered within the framework of UNESCO's MAN AND THE BIOSPHERE (MAB) PROGRAM. They are nominated by national governments and must meet a minimal set of criteria and adhere to management conditions. Each biosphere reserve is intended to fulfill three basic functions: a) contribute to the conservation of landscapes, ecosystems, species, and genetic variation; b) foster economic and human development that is sociocultural ly and ecologically sustainable; and c) provide support for research, monitoring, education, and information exchange related to local, national, and global issues of conservation and development. The reserves represent a major tool for implementing the concerns of AGENDA 21, the CONVENTION ON BIOLOGICAL DIVERSITY, and other international agreements. Individual biosphere reserves remain under the sovereign jurisdiction of the countries in which they are situated. Since many of the existing reserves include human settlements inside their boundaries, they are also managed for SUSTAINABLE DEVELOPMENT. Examples of biosphere reserves are the Sierra de las Minas Biosphere Reserve in Guatemala and Mount Culal in Kenya. *T. S.*

Buffer Zone: the area surrounding the most-restricted-use ("core") area of a park, reserve, or protected area, where less-restrictive-use controls pertain in order to provide economic activities for local populations. Buffer zones represent the recognition that the local population's needs must be addressed, and that the violation of core-area limitations is likely to be greater if the local population is denied the opportunity to exploit resources throughout the protected area. The buffer zone strategy is sometimes criticized for being ineffective (i.e., local populations exploit the buffer zones and still encroach on the core areas) or for limiting the local population's uses of the buffer areas themselves. *T. S.*

Collective Intellectual Property Rights (CIPRS). See under SECTION 3.4.2.

Ecotourism: a type of rural tourism based on NONCONSUMPTIVE USE of natural resources. Ecotourism has been defined in dozens of ways since the term was introduced in the mid-1980s by Mexican architect Hector Ceballos. The Ecotourism Society, a U.S.-based international group, defines it as "responsible travel to natural areas which conserves the environment and improves the welfare of local people" (Lindberg and Hawkins 1993). Implicit in nearly all definitions are a type of place visited (a natural area, which may or may not be officially protected), a motivation for travel (to experience nature and sometimes cultural heritage features as well), and a set of positive impacts (which may include both protection of the resource at the destination and generation of economic benefits for local people.)

See Kreg Lindberg and Donald Hawkins, *Ecotourism: A Guide for Planners and Managers* (North Bennington, VT: Ecotourism Society, 1993); Hector Ceballos-Lascurain, *Tourism, Ecotourism and Protected Areas* (Gland, Switzerland: International Union for Conservation of Nature, 1996). *R. G. H.*

COMMENTARIES ON ECOTOURISM

Accurate figures on the current level of ecotourism are not available—indeed the multiple criteria of the definition make one suspect that true ecotourism is more a goal to be sought than an actual phenomenon, since few if any sites meet all the criteria. Tourist visitation to natural areas, including many remote tropical and mountain destinations, has increased greatly in recent years. Moreover, the governmental and nongovernmental entities that support many protected areas have come to regard tourism as a source of political support and economic revenue, both for the protected areas and for adjoining communities that might otherwise threaten them. *Robert G. Healy*

One of the most prominent oxymoronic expressions masquerading under the banner of sustainability. The "benefits" of ecotourism within "ecologically managed" destinations have to be balanced against the greenhouse impacts of jet aircraft travel. In the early 1990s it was estimated that jet flights released 550 tonnes of carbon dioxide, 220 million tonnes of water, and 3.5 million tonnes of nitrogen oxides into the atmosphere. By the start of the next century the number of arrivals will have doubled compared to 1990. The idea that ecotourism benefits local peoples is also called into question by reports that as little as twenty percent of the cost of a holiday remains within the destination. Recent reports argue that indigenous peoples involved in "authentic" meetings with ecotourists are, in some instances, enslaved (George Monbiot, "The Tourist Trap," *The Guardian Supplement* December 12, 1998, 7). Other reports suggest that "nature watch" trips are reportedly threatening populations of dolphins and whales (D. Harrison, "Whales under Stress as Man Crowds Sea," *The Observer*, October 18, 1998, 7). Meanwhile, space tourism gathers momentum, promising even more atmospheric pollution while the rich indulge their "tourist gaze." Local low-mileage holidays are the only form of ecotourism apart from the emergent virtual tourism industry being marketed in Japan. *Ian Welsh*

Ex Situ Preservation: a CONSERVATION method that entails the removal of germplasm resources (seed, pollen, sperm, individual organisms) from their natural environment and maintaining them outside of their original habitat, most often in GENE BANKS. Ex situ preservation outside of the country of origin may arouse objections that the country's natural heritage—and possibly commercial opportunities from it—have been appropriated by foreigners.

Gene Banks: facilities established for EX SITU CONSERVATION of seed specimens, tissues, or reproductive cells of plants or animals in order to preserve genetic diversity. Specimens from these collections are distributed as needed either to farmers to replenish a crop or to scientists for study. The creation of gene banks was pioneered by Russian scientist Nikolai Vavilov in 1920s. Vavilov demonstrated the importance of genetic diversity in plant breeding and developed a systematic action plan for collecting genetic material from the "primary centers of origin." Currently, most gene banks are located at the territories of a few developed countries and are under the auspices of the CONSULTATIVE GROUP ON INTERNATIONAL AGRICULTURAL RESEARCH (CGIAR). However, the WORLD BANK and transnational corporations have been exerting greater influence over CGIAR and the issue of location of the gene banks has become an additional point of North-South environmental debate (see NORTH VS. SOUTH). Developing countries object to the possibility of "Northern" control over genetic material that has been largely collected from the developing world.

See Nikolai Ivanovich Vavilov, *The Origin, Variation, Immunity and Breeding of Cultivated Plants: Selected Writings* (Waltham, MA: Chronica Botanica Co., 1951).

COMMENTARY ON GENE BANKS

The idea that biodiversity can be protected by freezing seed stocks as patent regimes reduce the range of plant varieties actively grown is contentious. No one can be sure that frozen seedstocks will a) remain viable indefinitely; and b) not be lost due to some accident—power failure being a prime candidate. In Europe a diverse network of organizations, such as the Soil Association and the Henry Doubleday Research Association (HDRA), actively campaign on this issue. The HDRA "Seed Guardians" undertake a "grow-out," i.e., to cultivate varieties no longer commercially available.

See Derrick Purdue, "Winds of Change: Conserving Biodiversity and Social Movements," Doctoral dissertation, University of the West of England, Bristol, England, 1998. *Ian Welsh*

Global Biodiversity Strategy: a 1992 report prepared jointly by the WORLD RESOURCES INSTITUTE, the WORLD CONSERVATION UNION (IUCN), and the UN ENVIRONMENT PROGRAMME (UNEP). It addressed such issues as the nature and value of biodiversity, and the dynamics and causes of its loss. The report also outlined a systematic agenda of policy reforms and conservation action at local, national, and international levels necessary to protect biodiversity over the long run while using its resources for food, medicine, and other necessities.

Habitat Management Plan. See under SECTION 5.1.3.

In Situ Preservation: a conservation method used to preserve the genetic integrity of gene resources by conserving them within the ecosystems of the original habitat or natural environment. See also EX SITU CONSERVATION.

"Key Area" Approach ("Enclave" Strategy): an approach to species and habitat protection that concentrates on protecting the areas with the highest or most pivotal concentrations of biota to be protected. The key area approach, which has been traditionally applied by most developed nations, contrasts with the increasingly dominant view within national and international conservation organizations that conservation must focus on wider areas because of geographical interconnections such as migration patterns that exceed the boundaries of presumably "key" areas.

Monuments of Nature: outstanding elements of landscape (usually small, natural objects or areas) that are preserved and protected in their natural condition in many countries for aesthetic, cultural, scientific, and educational reasons. They serve both to enrich the landscape and to convey knowledge about human and natural history to the public.

National Parks: areas designated by national governments to come under strict limitations on land use and economic activity in order to conserve their natural characteristics. The first national park per se was Yellowstone National Park, established in the United States in 1872; yet the practice of reserving areas for parks dates back many centuries. CONSERVATION through the establishment of national parks is typically justified by some combination of aesthetic, recreational, or environmental objectives, including the conservation of biodiversity existing within the park. Recently, economic motives have also come to the fore with the rise of ECOTOURISM bringing in visitors willing to pay for their enjoyment of the park and incidentally to pay for goods and services while engaged in their tours. The challenges to national parks lie in the restrictions put on the resource rights of people living in or near the park, and the fact that many officially designated national parks lack effective enforcement of use limitations, making them "paper parks."

See Michael Wells and Katrina Brandon, *People and Parks: Linking Protected Area Management with Local Communities* (Washington, DC: World Bank, World Wildlife Fund, and U.S. Agency for International Development, 1992); Katrina Brandon, Kent H. Redford, and Steven E. Sanderson, eds., *Parks in Peril: People, Politics, and Protected Areas* (Washington, DC: Island Press, 1998).

Protected Areas: geographically defined areas of land or sea, which are specifically designated for the protection of biological diversity or other conservation objectives and are secured from human use and encroachment. In practice, the purposes for which protected areas are managed vary from scientific research and wilderness protection to maintenance of particular environmental services and protection of specific natural or cultural features. According to the WORLD RESOURCES INSTITUTE, in 1994 there were nearly 10,000 protected areas covering nearly 1,000,000 hectares of lands for a total of 7.1 percent of the earth's surface. The international classification of protected areas includes five categories: a) scientific reserve/strict nature reserve; b) national parks; c) natural monuments/natural landmarks; d) nature conservation reserve/managed nature reserve/wildlife sanctuary areas; and e) protected landscape or seascape areas. This classification scheme reflects different objectives of management and various degrees of protection. Most countries also have their own classifications of protected areas. The results of establishing protected areas are often controversial: in many developing countries in Africa and Asia, for instance, the creation of protected areas resulted in the dislocation of local communities. Today environmentalists agree that the best management of protected areas includes partnership with local communities.

See International Union for the Conservation of Nature (IUCN), *United Nations List of National Parks and Protected Areas* (Gland, Switzerland: IUCN, 1985). *T. S.*

Rainforest Action Network (RAN): a nonprofit, volunteer organization founded in 1985 to protect the native BIODIVERSITY of the rainforests and the human rights of native tribes through public education, communication, and direct action. RAN is headquartered in San Francisco and has over 30,000 U.S. members. It works with environmental and human rights groups in over sixty countries by disseminating information and coordinating worldwide campaigns against specific actions of corporations and government agencies. RAN's campaigns have included a highly publicized 1986 blockade of the entrance to the WORLD BANK to protest their environmentally destructive policies, and a nationwide boycott of Burger King that led to the cancellation of $35 million in beef contracts in Central America.

Red Book: a set of data on RARE, EXTINCT, AND ENDANGERED SPECIES of plants and animals designed to encourage legislation to protect and recuperate the populations of rare and endangered species. The International Red Book is maintained by the WORLD CONSERVATION UNION (IUCN), while many nations have their own Red Books.

Traffic Records Analysis of Flora and Fauna in Commerce (TRAFFIC): a joint program of the WORLD WIDE FUND FOR NATURE and the WORLD CONSERVATION UNION (IUCN), that monitors global trade in wildlife and wildlife products. Founded in 1976, TRAFFIC aims to ensure that wildlife trade is at sustainable levels and in accordance with domestic and international laws and agreements. Through a decentralized network of national offices in eighteen countries, TRAFFIC actively tracks wildlife trade and reports violations of the CONVENTION ON INTERNATIONAL TRADE IN ENDANGERED SPECIES (CITES) trade provisions to governments and the international community. It is also active in raising public awareness about conservation and wildlife trade and in seeking protection for newly threatened species.

World Conservation Strategy (WCS): a watershed document published in 1980 "to draw the attention of both decisionmakers and the general public to the urgent need for the conservation of the world's land and marine ecosystems as an integral part of economic and social development." It was elaborated by the WORLD CONSERVATION UNION (IUCN) in cooperation with the UN ENVIRONMENT PROGRAMME (UNEP) and the WORLD WILDLIFE FUND (WWF) with its stated objectives to: a) maintain essential ecological processes and life support systems; b) preserve genetic diversity; and c) ensure the sustainable utilization of species and ecosystems. The WCS may be considered the first international attempt to launch the concept of SUSTAINABLE DEVELOPMENT. The document defined conservation as "the management of human use of the biosphere so that it may yield the greatest sustainable benefit to present generations while maintaining its potential to meet the needs and aspirations of future generations" and particularly emphasized practical societal benefits from appropriate forms of environmental management. The WCS specified the objectives of living resource conservation, determined the priority requirements for achieving each of these objectives, proposed national conservation strategies, recommended anticipatory environmental policies, and advocated greater public participation in planning and decision making concerning environmental and resource use. The strategy openly recognized the gross disparities among and within nations as the basis of serious ecological deterioration and proposed arrangements for national and international action that would put the world on the road to sustainable economies. During its first decade, forty-five countries developed national plans incorporating these recommendations. In 1991, an updated version of the WCS entitled *Caring for the Earth: A Strategy for Sustainable Living* developed its ideas further and examined various policy and institutional implications.

See International Union for the Conservation of Nature (IUCN), *World Conservation Strategy* (Gland, Switzerland: IUCN, UNEP, WWF, 1980); *Caring for the Earth: A Strategy for Sustainable Living* (London: Earthscan, 1991).

COMMENTARY ON THE WORLD CONSERVATION STRATEGY

It may be hard for us to imagine, after two decades of "sustainable development," but in 1980 it was considered remarkable that an organization dedicated to the protection of the world's resources and natural systems would call for "sustainable utilization" of species and ecosystems. *Robert Healy*

Zakazniks: special purpose reserves used to safeguard certain flora or fauna populations (rather than for the protection of entire ecosystems; see ZAPOVEDNIKS). In contrast to most Western protected areas, in Russia the zakazniks are usually so designated for a limited period, such as five to ten years, on the assumption that the protected species would recover after being given a respite from hunting and development pressures. Currently in Russia there are over 1,500 zoological, botanical, landscape, and hydrological zakazniks, occupying approximately 4 percent of Russia's territory.

See V. F. Protasov and A. V. Molchanov, *Ecology, Health and Nature Use in Russia* (Moscow: Finansy I Statistika, 1995).

Zapovedniks: reserves of pristine natural systems maintained in a state of total inviolability. They are used as benchmarks against which human impact on nature might be assessed. This method of nature preservation was first advocated by Russian environmentalists in the nineteenth century on the assumption that scientific progress and economic activity that depend on the land's carrying capacity required prior study of how nature works. The first zapovedniks were created in Russia shortly after the 1917 Bolshevik Revolution—the first protected areas anywhere to be created by a government exclusively in the interests of the scientific study of nature. Their sites were selected by a master plan as areas representative of historically formed natural ecological communities capable of self-regulation and self-reproduction (see BIOGEOCENOSIS) and have been a basis of environmental monitoring for over seventy years. Russian zapovedniks are of global importance because of their broad expanse, unique assemblages of species, and concentration of native species surpassing that found in other temperate forests. This method of nature preservation provided the basis for developing a comprehensive understanding of the natural environment of the former Soviet Union and also contributed to ecological studies in phytosociology and the ECOLOGICAL ENERGETICS paradigm. In Russia there are currently eighty-nine zapovedniks with a total area of 29.5 million hectares, comprising more than 40 percent of the world's scientific reserves (IUCN Category I). A 1995 presidential decree declared the government's intention to expand this system of protected extensive wilderness to 150 reserves covering 3 percent of the territory. However, at least half of the national zapovedniks are in jeopardy of encroachment due to the economic and political difficulties of the transition period.

See Douglas Weiner, *Models of Nature* (Bloomington: Indiana University Press, 1988); Ministry of Environment of the Russian Federation, *State of the Environment* (Moscow: Gidrometizdat, 1997).

6.3.3 INTERNATIONAL AGREEMENTS AND ORGANIZATIONS RELATED TO BIODIVERSITY CONSERVATION

African Convention on the Conservation of Nature and Natural Resources (1968 Algiers African Convention): an international agreement signed in Algiers in 1968 by representatives of thirty-eight member states of the Organization of African Unity. It came into force in 1969 to "ensure the conservation, utilization, and development of soil, water, flora, and the fauna resources in accordance with scientific principles and with due regard for the best interest of the people." The explicit goals of the Algiers African Convention reflect the transition from the preservationist orientation of the 1933 African Convention to a broader conceptual extent of environmental policy. It is the first international agreement to address the need for conservation education and the integration of conservation into development plans. However, the African Convention failed to establish the necessary institutional mechanism: it lacked a permanent body to oversee and promote implementation. While the African Convention served as a framework for national legislation in some member countries, it never achieved the intended impact and is generally considered as a "sleeping convention" of little practical value.

See Simon Lyster, "Effectiveness of International Regimes Dealing with Biological Diversity from the Perspective of the North," in *Global Environmental Change and International Governance,* eds. Oran Young, George Demko, and Kilaparti Ramakrishna (Hanover, NH: University Press of New England, 1996), 190–91.

African Convention Relative to the Preservation of Flora and Fauna in their Natural State (1933 London Convention): one of the earliest international agreements on wildlife protection, signed in London in 1933. Its members were European states governing colonial Africa. The London Convention included specific hunting restrictions for threatened species, regulations of trade in wildlife, and special measures on the protection of habitat. It was motivated by the then traditional concern for the preservation of colonial big-game hunting grounds and revenues; some of its provisions aimed at the eradication of harmful or "useless" animal species, for instance, would be considered outrageous from today's conservation perspective. However, the regime set up an important precedent of using international trade measures for the wildlife protection that was eventually developed by the CONVENTION ON INTERNATIONAL TRADE IN ENDANGERED SPECIES (CITES). After decolonization, the 1933 London Convention was succeeded by the 1968 AFRICAN CONVENTION ON THE CONSERVATION OF NATURE AND NATURAL RESOURCES (ALGIERS AFRICAN CONVENTION).

Biosafety Protocol. See under SECTION 6.3.2.

Convention on Biological Diversity (CBD): an international agreement that addresses the rapidly increasing rate of species extinction and ecosystem destruction. The CBD was negotiated under the auspices of the UNITED NATIONS ENVIRONMENT PROGRAMME. It was opened for signature at the 1992 United Nations Conference on Environment and Development and entered into force on 29 December 1993. As of February 2000, 177 countries had become parties. The goals of the convention are to promote the conservation of BIODIVERSITY, the sustainable use of its components, and the fair and equitable sharing of benefits arising out of utilization of genetic resources. The central conservation commitments under the convention include the identification and monitoring of components of biological diversity; the establishment of a system of protected areas; the adoption of measures for both IN-SITU and EX-SITU CONSERVATION; the integration of genetic resource conservation considerations into national decision making; and the adoption of incentives for the conservation of biological resources. In distinction from other agreements on nature conservation, the CBD takes a comprehensive rather than a sectoral approach. It also formally proclaims the intrinsic value of biodiversity. Negotiations of the first protocol to the convention, which deals with the "biosafety" of genetically modified organisms, were concluded in January 2000 (see CARTAGENA PROTOCOL). The secretariat of the convention is located in Montreal, Canada.

Convention Concerning the Protection of the World Cultural and Natural Heritage (World Heritage Convention; WHC): an international agreement adopted in 1972 to designate and protect cultural and natural properties of "outstanding universal value." World Heritage sites are selected by an international committee following their nomination by the member state in which the site is located. Sites may qualify if they represent: a) major stages of the earth's history, e.g., the American Everglades; b) ongoing ecological and biological processes in the evolutionary sense, e.g., the Galapagos Islands of Ecuador; c) exceptional natural phenomena, e.g., the Great Barrier Reef, Australia; or d) the most important and significant natural habitats for IN-SITU CONSERVATION of biodiversity, e.g., Manu (Peru). There are also special criteria for cultural sites. By 2000, 147 nations were parties to the convention and 630 World Heritage sites were listed, three-fourths of which were cultural heritage areas and one-fourth natural heritage areas. The WORLD HERITAGE FUND, an administrative body of the WHC operating within the UN EDUCATIONAL, SCIENTIFIC AND CULTURAL ORGANIZATION (UNESCO) structure, has the responsibility of providing technical support and financial assistance to member states in the management of World Heritage sites. Member states are required to ensure the effective management of a heritage site so that it retains the features for which it was nominated.

Convention on the Conservation of European Wildlife and Natural Habitats (Bern Convention): an international agreement signed in Bern, Switzerland, in 1979 and in effect since 1982. Twenty-one European countries, the European Union, Senegal, and Burkina Faso are parties to the convention, which is administered by the Council of Europe. The primary focus of the Bern Convention is on the conservation of endangered natural habitats and the habitats of fauna and flora listed in the annexes to the convention. The effectiveness of the convention's implementation is mixed. Most member states adopted laws prohibiting the killing and exploitation of species protected by the convention. However, most members do not provide legal protection for habitats outside of national parks or nature reserves, even though it is recommended by the treaty.

See Simon Lyster, "Effectiveness of International Regimes Dealing with Biological Diversity from the Perspective of the North," in *Global Environmental Change and International Governance,* eds. Oran Young, George Demko, and Kilaparti Ramakrishna (Hanover, NH: University Press of New England, 1996), 190–91.

Convention on the Conservation of Migratory Species of Wild Animals (Bonn Convention): an international agreement signed in Bonn in 1979 that came into effect in 1983. The convention designated roughly fifty species as requiring direct protection (so-called appendix I species) and a larger set of species requiring less stringent protection measures (so-called appendix II species). Forty-four countries (most of them developed countries) were signatories to the Bonn Convention.

COMMENTARY ON THE BONN CONVENTION

The Bonn Convention was initially proclaimed as the "ultimate modern conservation convention" and characterized by practically ideal institutional arrangements (a permanent secretariat and scientific council, flexible and transparent procedural activities). However, in practice the Bonn Convention was very ineffective. Its failure can be partly explained by the fact that most of the species listed in appendix I either are covered by other international agreements or migrate through territories of nonsignatory countries. In addition, the treaty provisions do not include any material assistance to developing countries, which subsequently are reluctant to join it.

See Simon Lyster, "Effectiveness of International Regimes Dealing with Biological Diversity from the Perspective of the North," in *Global Environmental Change and International Governance,* eds. Oran Young, George Demko, and Kilaparti Ramakrishna (Hanover, NH: University Press of New England, 1996), 190–91. *Natalia Mirovitskaya*

Convention on International Trade in Endangered Species (CITES): a treaty adopted in 1973 that seeks to preserve endangered species of plants and animals by prohibiting international trade in live or dead specimens or in body parts or products derived from them. The controls apply to species listed in appendices to the treaty. Species that are threatened with extinction are listed in appendix I. Trade in them is prohibited except under exceptional circumstances. Species that are not currently endangered, but could become so unless trade in them is regulated, are listed in appendix II. Trade in appendix II species requires an export permit to be granted only if it can be demonstrated that such exports will not be detrimental to the survival of the species. Species can be added or deleted from the appendices or moved from one appendix to another at meetings of the parties that are normally held every two years, the most recent one being in Nairobi, Kenya, in April 2000.

See Ginette Hemley, ed., *International Wildlife Trade: A CITES Sourcebook* (Washington, DC: Island Press, 1994). M. S. S.

Convention on Nature Protection and Wildlife Preservation in the Western Hemisphere (Western Hemisphere Convention): an international agreement signed in 1940 by most Western Hemisphere countries. It includes a comprehensive set of measures to protect the wildlife and natural environment of the region. However, it has been largely ineffective since its adoption and many American conservationists are not even aware of its existence. Limited institutional arrangements (absence of a secretariat and clear procedures of implementation and monitoring) are the main reason for the treaty's relative ineffectiveness. However, the Western Hemisphere Convention has provided a legal basis for a number of cooperative conservation programs between the United States and Latin American nations.

See Simon Lyster, "Effectiveness of International Regimes Dealing with Biological Diversity from the Perspective of the North," in *Global Environmental Change and International Governance*, eds. Oran Young, George Demko, and Kilaparti Ramakrishna (Hanover, NH: University Press of New England, 1996), 191–92.

Convention on Wetlands of International Importance Especially as Waterfowl Habitat (Ramsar Convention): one of the early agreements on wildlife conservation, adopted in 1971 following a conference in Ramsar, Iran, and in force since December 1975. By early 2000, it had 118 contracting parties to the convention, each agreeing to conserve and make "wise use" of their wetlands as regulators of water regimes and as habitats of distinctive ecosystems. Each member nation is required to designate at least one wetland of international importance under the criteria established by Ramsar. Over one thousand wetland sites, encompassing 72.7 million hectares, were so designated for inclusion in the Ramsar List of Wetlands of International Importance. Originally, the Ramsar Convention lacked an implementation mechanism and therefore was very weak. However, changes in the implementation structure of the convention (e.g., the establishment of the Ramsar Convention Bureau as its permanent secretariat and the Wetland Conservation Fund to provide aid to developing countries) gave new impetus to its activities. The number of designated wetlands has therefore increased dramatically over the past few years.

International Board for Plant Genetic Resources (IBPGR): an international agency developed under the auspices of the CONSULTATIVE GROUP ON INTERNATIONAL AGRICULTURE RESEARCH (CGIAR). The IBPGR coordinates an international network of genetic resource centers that foster the collection, conservation, documentation, evaluation, and use of plant germplasm. The saving of genetic wild plant resources as well as crop cultivars safeguards basic genetic material that can be instrumental in gaining or restoring crucial plant characteristics, such as resistance to disease, harsh climatic conditions, poor soil quality, or adverse water conditions. An example of the IBPGR's work is an international effort to obtain, replicate, and replenish the Somali seed stock.

See Stephen Henry Schneider, *Laboratory Earth: The Planetary Gamble We Can't Afford to Lose* (New York: Basic Books, 1994).

International Undertaking on Plant Genetic Resources (IUPGR): an international agreement established by a 1983 resolution of the UN FOOD AND AGRICULTURE ORGANIZATION (FAO) to address the concern for preserving the genetic diversity of seed varieties. The IUPGR encourages worldwide preservation, evolution, and exchange of germplasm under FAO auspices and currently has 113 members states. Its basic premise is that "plant genetic resources are a common heritage of mankind and consequently should be available without restrictions." In its current form, the undertaking is aimed to ensure both EX SITU and IN SITU CONSERVATION of plant genetic resources as well as their sustainable use. The signatories agree to: a) provide access to the materials and ensure their unrestrained export for scientific and plant breeding purposes on the basis of mutual exchange or on mutually agreed terms; b) mount exploration missions to identify plant genetic resources in danger of extinction and put into place legislation protecting plants and their habitats; c) intensify plant breeding and germplasm maintenance activities; d) establish an international GENE BANK; e) build an internationally coordinated network of collections; and f) create an international data system. The IUPGR is SOFT LAW and does not provide any sanctions for noncompliance or dispute settlement mechanisms. Many developing countries and adherents of indigenous knowledge rights have been calling for the IUPGR to serve as a defender of local and national property rights to genetic materials, while some developed country governments have been trying to modify the IUGPR so that it ensures the unrestricted exchange of plant genetic materials. Currently governments are negotiating revision of the undertaking that eventually would become a protocol of the CONVENTION ON BIOLOGICAL DIVERSITY (CBD). See also FAO GLOBAL SYSTEM OF PLANT GENETIC RESOURCES.

1900 London Convention. See CONVENTION DESIGNED TO ENSURE THE CONSERVATION OF VARIOUS SPECIES OF WILD ANIMALS IN AFRICA, WHICH ARE USEFUL TO MAN OR INOFFENSIVE.

Rainforest Action Network (RAN). See under SECTION 6.3.2.

World Conservation Union (IUCN). See under SECTION 5.5.6.

World Wildlife Fund for Nature. See under SECTION 5.5.6.

6.4.1 CLIMATE CONCEPTS, TRENDS, AND CONDITIONS

Carbon Dioxide Equivalent: a quantitative criterion used to compare the emissions from different greenhouse gases on the basis of their GLOBAL WARMING POTENTIAL (GWP). The impact of emissions of a particular gas is determined by multiplying the volume of emissions in metric tons by the GWP, usually expressed in terms of metric tons of carbon dioxide equivalents (MTCDE). Sometimes carbon dioxide equivalents are converted to carbon equivalents (CE). Carbon dioxide equivalents are often used in international climate negotiations, while the carbon equivalents are more often applied in the United States.

Climate Change: variability in weather conditions, both averages and ranges, over extended periods of time. Over the past two million years, naturally occurring swings in global temperatures of 5 to 10 degrees C. have triggered many periods of expanded glaciation known as ice ages, which have been interspersed with warmer, interglacial periods. The last major ice age peaked about 18,000 years ago, but a marked period of cooling known as the Little Ice Age occurred from 1400 to 1850 when temperatures averaged about 1 degree C. below the longer-term norm. Shorter climatic variations of a few years are associated with the El Niño phenomenon and major volcanic eruptions. Scientific evidence suggests that human additions to atmospheric concentrations of carbon dioxide and methane, as well as large-scale changes in land use have brought about an increase in global mean temperatures of 0.3 to 0.6 degrees C. over the past century. A human-induced rise in temperatures in the range of 1–4 degrees C. by 2100 is projected if the world remains on the expected course of greenhouse emissions. Climate changes associated with global warming, including the direction and amount of temperature changes, patterns of precipitation, and frequency and intensity of storms, are expected to vary considerably by region.

See J. T. Houghton, L. G. Meira Filho, B. A. Callander, N. Harris, A. Kattenberg, and K. Maskell, *Climate Change 1995: The Science of Climate Change* (New York: Cambridge University Press, 1996). *M. S. S.*

COMMENTARY ON CLIMATE CHANGE

The challenge for scientists has been to detect whether human-induced climate change is occurring against a backdrop of natural variability in climate, which is readily apparent in repeated swings between periods of expanded glaciation, known as ice ages, and warmer interglacial periods that have been taking place over the past two million years. Significant uncertainties remain as to the amount of climate changes that will take place due to human influences and the impacts such changes will have on the natural environment and human societies. Nevertheless, there is widespread agreement among scientists and policymakers that the seriousness of the threat of climate change warrants precautionary efforts to reduce emissions of greenhouse gases. *Marvin S. Soroos*

Climate Index. See GREENHOUSE CLIMATE RESPONSE INDEX.

Global Warming. See GREENHOUSE EFFECT.

Global Warming Potential (GWP): the index used to translate the level of emission of various gases into one common measure in order to compare their relative role in the global climate change. GWPs are based on the heat-trapping or radiation-forcing properties of different greenhouse gases and the residence time of different gases in the atmosphere. Conventionally, global warming potentials are expressed in terms of carbon dioxide equivalent units, i.e., the impact of a specific greenhouse gas is expressed in terms of the weight of carbon dioxide with the same effect. The International Governmental Panel on Climate Change has developed a table of GWPs for all greenhouse gases and these data are used in calculating necessary reductions in emissions.

COMMENTARY ON GLOBAL WARMING POTENTIAL (GWP)

The global warming potential concept was based on the ozone-depleting potential unit, which was designed to compare the effects of different chlorofluorocarbons, the substances that cause a thinning in the earth's stratosphere. A criticism of the GWP is that it seeks to find a simple index of comparison for greenhouse gases that have dissimilar cycles in the atmosphere and that may affect global warming indirectly through a range of chemical reactions that occur nonuniformly and unpredictably. The point that international agreements tend to spawn simplifying indices is brought home by the recollection that the concept of the Blue Whale Unit (BWU) was used in discussions surrounding the negotiations to ban commercial whaling. *Susan E. Subak*

Greenhouse Climate Response Index: an indicator of climate change or variability that includes such variables as above-normal temperature, above-normal precipitation in the cool months, extreme drought in the warm months, greater-than-normal proportion of annual precipitation during days of high precipitation, and reduced day-to-day temperature swings. The index was developed by the U.S. National Climatic Center and is used to estimate global climate trends.

Greenhouse Effect: the warming effect of the atmosphere on the earth's climate, which has been compared, albeit erroneously, to heat retention by the transparent glass roof of a greenhouse. The atmospheric greenhouse effect occurs as solar energy in the form of high-frequency ultraviolet radiation passes through the atmosphere and warms the earth's surface. The heat is radiated back into space in the form of lower-frequency infrared (IR) rays. A proportion of these outgoing IR rays are absorbed by greenhouse gases in the atmosphere, in particular carbon dioxide, methane, and water vapor, and reradiated back to earth. Human activities, such as the burning of fossil fuels, livestock production, rice cultivation, and land clearing have added substantially to atmospheric concentrations of greenhouse gases, causing more of the outgoing IR rays to be absorbed.

See Melvin A. Benarde, *Global Warning . . . Global Warming* (New York: John Wiley, 1992). *M. S. S.*

While the greenhouse analogy is popularly used to explain global warming, it is somewhat misleading. Temperatures within the greenhouse are higher than outside because the warmed air within it cannot escape the glass enclosure and cooler outside air cannot penetrate it. In contrast, the air near the earth's surface is warm not because it is enclosed, but because so-called greenhouse gases absorb infrared rays radiated from the earth, which releases heat. *Marvin S. Soroos*

Greenhouse Index: an early measure based on the emissions of three major greenhouse gases (carbon dioxide, methane, and CHLOROFLUOROCARBONS), with each gas weighted in accordance with its heat-trapping quality (in carbon dioxide equivalents) and expressed in metric tons of carbon per capita.

Though based on scientific data and seemingly neutral, this index, as used by the WORLD RESOURCES INSTITUTE and the UN ENVIRONMENT PROGRAMME, was nevertheless criticized by scientists from developing countries as an example of "politically motivated mathematics." In the opinion of its critics, emissions of greenhouse gases are misleading unless presented in combination with the population and comparative sink capacity (i.e., the capacity to absorb or "sequester" carbon or other components of greenhouse gases so that they do not enter the atmosphere) of a particular country. Thus, according to the Greenhouse Index, India is among the top five countries responsible for the accumulation of these gases in the atmosphere. However, if its population and absorptive capacity are taken into account, India appears to be the world's lowest per capita net emitter of greenhouse gases.

See Anil Agarwal and Sunita Narain, "Global Warming in an Unequal World: A Case of Environmental Colonialism," *Earth Island Journal* (Spring 1991): 39–40. *Natalia Mirovitskaya*

Sea-Level Rise: the elevation of ocean or sea levels, which in turn may threaten flooding of surrounding land masses. Sea-level rise is one of the anticipated impacts of global climate change, as warmer mean temperatures trigger thermal expansion of ocean waters and accelerate the melting of glaciers and ice sheets. Sea levels have risen an average of 10–25 cm over the past century and with a continuation of current trends are anticipated to rise an additional 15–95 cm by 2100. Sea-level rises vary somewhat by region depending upon whether land is subsiding due to geological forces or human withdrawals of groundwater or is rising as a result of reduced pressure when heavy ice sheets of glacial periods recede ("isostatic rebound"). Levels of inland seas may also rise because of increased precipitation. Rising sea levels and increasingly strong storm surges also associated with a warming climate are likely to alter coastal ecosystems significantly and to jeopardize low-lying coastal cities and agricultural lands. Small island states are especially vulnerable to sea-level rise and storm surges as are states with extensive deltas, such as Bangladesh and Egypt.

See Jodi Jacobson, "Holding Back the Sea," in *State of the World 1990*, eds. Lester Brown et al. (New York: Norton, 1990), 79–97. *M. S. S.*

COMMENTARY ON SEA-LEVEL RISE

Climatic change is affecting the ability of native Alaskans to engage in subsistence activities. The polar ice cap is estimated to be 40 percent thinner than it was in the 1970s, there is accelerated thawing of the permafrost, and storm surges are already eroding village sites and destroying infrastructure. Inuit hunters fear that they will not reach their food resources, as marine species like the Pacific walrus, seals, and polar bears are forced to move further away to feed and bear their young. Current policies to deal with impending risk and crisis due to environmental instability concentrate on mitigation and adaptation rather than the survival of indigenous populations. How these uncertainties are addressed within each of these systems may help us predict the chances of survival for northern cultures and their environment.

See Gunter Weller and Patricia A. Anderson, eds., *Assessing the Consequences of Climate Change for Alaska and the Bering Sea Region* (Fairbanks: University of Alaska Center for Global Change and Arctic System Research, 1999); http://www.besis.uaf.edu. *Susanne Swibold and Helen Corbett*

Shapley Value: a notion of relative contribution or burden taken from game theory. The Shapley value (named after the mathematical economist L. S. Shapley) represents the potential contribution or power of each actor over the entire range of possible interactions; the average value can then be used to assess the rewards or penalties for that actor. Thus the Shapley value can be used to estimate the responsibility for pollution or greenhouse gas emissions, and serve as the basis for assigning charges or obligations to reduce the offending behavior, when the actual contribution or responsibility cannot be directly determined. *R. P.*

6.4.2 CONVENTIONS AND INSTITUTIONS RELATED TO CLIMATE ISSUES

Alliance of Small Island States (AOSIS): an organization of small island states and several small continental countries, which grew out of the Small States Conference on Sea Level Rise held in Malé in 1989. It was formed to give greater voice to the concerns of these countries over the potentially catastrophic threats posed to them by climate change, especially the anticipated rising sea levels and increased incidence and intensity of tropical storms. The unique concerns of these countries were also the subject of the United Nations Global Conference on Small Island Developing States held in Barbados in 1994 and a special session of the United Nations General Assembly in 1999. AOSIS, whose membership now includes forty-two states from all regions of the world, had a strong presence in international negotiations on climate change during the 1990s, with its repeated calls for developed countries to reduce their CO_2 emissions by 20 percent from 1990 levels by 2005.

See Anjali Acharya, "Small Islands Awash in a Sea of Troubles," *World Watch* (November/December 1995): 24–33. *M. S. S.*

COMMENTARY ON ASSOCIATION OF SMALL ISLAND STATES (AOSIS)

While small in combined territory, population, and economic size, AOSIS members are a sizable bloc that comprise approximately 20 percent of the voting membership of United Nations bodies such as the General Assembly. AOSIS has raised discomforting ethical issues in the climate change debate, in particular whether other countries, especially the industrial ones, are morally obliged to reduce emissions of greenhouse gases that would otherwise trigger sea-level rises that jeopardize the very existence of many of the low-lying island states. *Marvin S. Soroos*

Annex Parties: three general categories of countries that are parties to the FRAMEWORK CONVENTION ON CLIMATE CHANGE (FCCC). Annex I includes industrialized countries (both OECD members and transitional-economy [formerly Eastern Bloc] countries); Annex II includes only the wealthier industrialized countries, while Annex III lists developing country parties. Under the convention (in its present form), Annex I parties were supposed to reduce their greenhouse gas emissions by the year 2000 to 1990 levels; Annex II countries are committed to provide financial support for the implementation of the convention.

Berlin Mandate: a ruling adopted by the first conference (March 1995, in Berlin) of the parties to the FRAMEWORK CONVENTION ON CLIMATE CHANGE (FCCC) that concluded that initial commitments undertaken under the convention were not adequate. The Berlin mandate strengthened the commitments of the ANNEX I PARTIES. A powerful coalition of developing nations, however, was able to avoid taking on any substantial commitments of its own.

Clean Development Mechanism (CDM): a key feature of the KYOTO PROTOCOL designed to encourage ANNEX I (industrialized nations) and non-Annex I countries (industrializing and developing nations) to reduce GREENHOUSE GAS (GHG) emissions. The CDM represents an effort to reduce GHG emission efficiently and equitably and extends the JOINT IMPLEMENTATION concept to non-Annex I countries. In doing so, the CDM creates an incentive for private investment in clean technologies in the non-Annex I countries in return for certified emission reductions in Annex I countries. The CDM is anticipated to bring together the private, nonprofit, and public sectors in an effort to curb greenhouse gas emissions. *T. S.*

Conference of the Parties (COP): a governing body of the FRAMEWORK CONVENTION ON CLIMATE CHANGE (FCCC) that is comprised of nation-states that have ratified the convention. The COP now has over 150 members and about 50 observer states. The main objective of the COP is to monitor the implementation of the convention and to make decisions necessary to make it effective.

Framework Convention on Climate Change (FCCC): the first treaty to address the problem of human-induced climate change. Adopted at the 1992 UNITED NATIONS CONFER-

ENCE ON ENVIRONMENT AND DEVELOPMENT (UNCED; RIO CONFERENCE; EARTH SUM-MIT) in Rio de Janeiro, the FCCC establishes a goal of stabilizing concentrations of GREENHOUSE GASES in the atmosphere at levels that will prevent dangerous human interference with the climate system. Toward this end, it calls upon developed countries to accept the goal of limiting their greenhouse gas emissions to 1990 levels by the year 2000. In negotiating the treaty, numerous developed countries pushed for a target date for binding reductions of greenhouse gas emissions, but in the end relented in the face of opposition from the United States, which cited continuing scientific uncertainties about the threat of climate change. The first conference of the parties to the treaty was held in Berlin in 1995, where it was agreed that further measures were needed to limit climate change and that a schedule for mandatory reductions of greenhouse gas emissions should be negotiated for adoption at the third conference of the parties (see KYOTO PROTOCOL).

See Daniel Bodansky, "The United Nations Framework Convention on Climate Change: A Commentary," *Yale Journal of International Law* 18 (1993): 451–558. *M. S. S.*

Intergovernmental Panel on Climate Change (IPCC): an expert body on climate change issues established in 1988 by the UN ENVIRONMENT PROGRAMME (UNEP) and the WORLD METEOROLOGICAL ORGANIZATION (WMO). Consisting of more than 1,400 scientists from around the world, the IPCC was charged with three core tasks undertaken by working groups: to assess scientific information on climate change, to assess its socioeconomic and environmental consequences, and to develop strategies and policy options of human response. The IPCC has accomplished the first and second tasks by providing balanced scientific judgments on links between greenhouse gas emissions and global climate change and by documenting the potentially catastrophic consequences of climate change on freshwater resources, agriculture, and food supplies, forests and species, oceans, settlements, and national security. However, the third IPCC working group, chaired by the United States and dominated by diplomats, has failed to provide clear recommendations for action by nations and the international community as a whole. The IPCC's activities to a large extent affect the dynamics of climate change negotiations.

Joint Implementation (JI; Activities Implemented Jointly). See under SECTION 5.5.5.

Kyoto Protocol (KP): the 1997 protocol to the FRAMEWORK CONVENTION ON CLIMATE CHANGE. If ratified by a sufficient number of nations, it would become the first binding commitment by the world's major polluters to reduce their emissions of GREENHOUSE GASES. The protocol requires the major developed countries to reduce their emissions of six greenhouse gases by 2012. These reductions average 5.2 percent, with 8 percent for the EUROPEAN UNION (EU), 7 percent for the United States, and 6 percent for Japan. A few countries, including Russia, are allowed stabilization of emissions, while some countries are allowed increases (e.g., Australia, 8 percent). The Kyoto Protocol also introduced a few flexibility mechanisms, such as EMISSIONS TRADING, JOINT IMPLEMENTATION, and the CLEAN DEVELOPMENT MECHANISM. The protocol will enter into force once 55 countries have ratified it, providing these countries represent at least 55 percent of 1990 emissions by most industrialized or developed countries.

Carbon dioxide has become the poster child for the global environment. Is this wise? Solving the greenhouse gas problem will not solve, for example, the much more serious problem of maintaining ecosystem services. More to the point, *not* solving the greenhouse gas problem—which is the direction that we are headed with the Kyoto Protocol—means that we have wasted huge amounts of technical and diplomatic capital that could have been used much more fruitfully to address problems that are more scientifically and politically tractable. The Kyoto Protocol is a case history in failed environmental strategy. It depends for its legitimacy on comprehensive scientific prediction of complex, unpredictable systems, and for its effective implementation on concerted political action to forestall environmental consequences that are, and will remain, abstract and contestable. It puts all the scientific and political eggs in one basket, and thus invites both determined opposition and vigorous backlash. *Daniel Sarewitz*

The Kyoto Protocol is one of the most exciting international environmental agreements because many countries committed to more than they can easily achieve. For instance, the wealthiest party to the protocol faces the choice of biting the bullet and accepting more expensive fuel, or implementing ambitious programs for conserving energy, or buying emissions credits from Russia. Purchase of credits from Russia, where emissions have been falling dramatically, is the most likely course. The trading provision in the protocol, which was preserved despite the European Union's mistrust, is a U.S.-Russia initiative and an interesting twist in a post–Cold War world. The Clean Development Mechanism (CDM) also allows industrialized countries to purchase emissions credits, this time from developing countries. This CDM may free developing countries from the host-donor relationship that many objected to with the mechanism's predecessor, Activities Implemented Jointly. Time will tell if the CDM provides a generous flow of funds for innovative and useful programs. *Susan E. Subak*

World Climate Conferences: two major international meetings on the subject of climate change, sponsored by the WORLD METEOROLOGICAL ORGANIZATION (WMO) and the UN ENVIRONMENT PROGRAMME (UNEP). The first of these conferences, held in Geneva in February 1979, drew 350 scientists from 50 countries, who agreed on the need for a major international scientific effort to study the prospects for climate change and as well as its causes and consequences. This took the form of the World Climate Program that was inaugurated in 1980. The second conference was held in November 1990, also in Geneva, at which 747 scientists reviewed the accomplishments of the first decade of the World Climate Program and endorsed the initial report of the Intergovernmental Panel on Climate Change. They further proposed that developed countries reduce their CO_2 emissions by 20 percent by 2005. Government ministers attending the second part of the meeting failed to agree on any schedule for reducing greenhouse gas emissions.

See Robert M. White, "World Climate Conference: Climate at the Millennium," *Environment* 21 (3) (April 1979), 31–33; Jill Jäger and Howard L. Ferguson, eds., *Climate Change: Science, Impacts and Policy: Proceedings of the Second World Climate Conference* (Cambridge: Cambridge University Press, 1991). *M. S. S.*

World Climate Research Programme (wcrp): a major international research program begun in 1980 to determine the human impact on climate and the predictability of climate change. The wcrp is jointly sponsored by the WORLD METEOROLOGICAL ORGANIZATION (WMO), the INTERNATIONAL COUNCIL OF SCIENTIFIC UNIONS (ICSU), and UNESCO'S INTERGOVERNMENTAL OCEANOGRAPHIC COMMISSION. Its secretariat is housed in Geneva with the World Meteorological Organization. The program does not have its own resources: all of its projects are implemented by scientists and staff from participating nations with equipment and facilities that are paid for by these nations. WCRP programs and projects in the 1990s provided a much better understanding of the climate system.

World Meteorological Organization (wmo): a specialized agency of the United Nations, headquartered in Geneva, which coordinates and facilitates monitoring and research on weather and climate throughout the world. The organization was created in 1950 as a successor to the nongovernmental International Meteorological Organization, which dates back to 1873. The WMO created the World Weather Watch in the 1960s, which greatly enhanced the collection, processing, and dissemination of weather data internationally. With the INTERNATIONAL COUNCIL OF SCIENTIFIC UNIONS (ICSU), the WMO undertook the Global Atmospheric Research Program from 1967 to 1982. As concern increased about human-induced climate change, the WMO sponsored two World Climate Conferences in 1979 and 1990 and became a sponsoring partner in the WORLD CLIMATE PROGRAM, inaugurated in 1980, and the Intergovernmental Panel on Climate Change that began work in 1988.

See Arthur Davies, *Forty Years of Progress and Achievement: A Historical Review of the WMO* (Geneva: WMO, 1990). *M. S. S.*

COMMENTARY ON WORLD METEOROLOGICAL ORGANIZATION

The wmo has been widely viewed as a highly effective international agency, whose programs have significantly improved the accuracy and duration of weather forecasts and added much to an understanding of naturally occurring and human-induced variations in climate. Also described as a "temple of rationality" where reason prevailed over ideology, the wmo achieved a high level of East-West scientific cooperation throughout the Cold War era. *Marvin S. Soroos*

6.5 Deforestation and Forestry

6.5.1 FORESTRY PRACTICES, TRENDS, AND PROBLEMS

Clear-Cutting: the practice of cutting a large proportion of trees in a given area. Clear-cutting provides loggers with a high volume of timber from a given site, thus reducing some of the investments in logging roads and other infrastructure to obtain that volume. However, the average quality and value of the trees removed from clear-cutting are likely to be lower than that of other harvesting approaches, such as SELECTIVE LOGGING.

COMMENTARIES ON CLEAR-CUTTING

Clear-cutting entered into controversy in the 1970s when a lawsuit was brought against the U.S. Forest Service over the Monongahela National Forest in West Virginia. Local hunters who had grown weary of the USFS clear-cutting trees sought recourse through the courts in *Izaak Walton League v. Butz*. They claimed that clear-cutting was in violation of the Organic Act of 1897. Judge Maxwell in the Federal District Court ruled that the USFS was indeed in violation of the Organic Act and placed a restraining order on timber harvests from the Monongahela National Forest. The decision was appealed to the Fourth Circuit Court of Appeals, which supported the Maxwell decision. After the "Monongahela Decision" had been handed down, the chief of the Forest Service ordered that timber sales in the nine national forests of the four states within the Fourth District be stopped. The Monongahela decision set a precedent that could be cited in other appellate courts and therefore posed a threat to timbering in the entire national forest system. To avoid losing timber harvests in the rest of the nation, Congress began working on legislation that would better balance local demands with the forest practices employed by the USFS. The subsequent legislation enacted was the National Forest Management Act of 1976. *Toddi Steelman*

Clear-cutting has been heavily criticized by environmentalists because of the danger that regeneration of a large clear-cut area may fail. However, the drawbacks of other logging techniques, such as selective logging with high levels of damage to trees that are not targeted for removal, or HIGH-GRADING, which removes the best trees and leaves the forest with deteriorating genetic stock, puts these criticisms into perspective. Strip-cutting, a variant of clear-cutting that concentrated harvesting into smaller areas or strips, opening up the forest canopy for more rapid regeneration, is gaining favor. *William Ascher*

Deforestation: the reduction in the stock or quality of forests, which may result from natural causes (e.g., forest fires) or human actions such as logging and the conversion of forest to other uses, such as cropland and pasture. Deforestation is linked to many environmental problems and occurs in both temperate and tropical climates. The world's few remaining stands of undisturbed temperate forests, located primarily in Canada, Russia, and the United States, continue to be harvested. Deforestation rates in the tropics have increased steadily from the 1960s to the 1990s. Deforestation contributes to the extinction of species, worsens droughts and floods, releases carbon dioxide into the earth's atmosphere, changes microclimates, and contributes to soil erosion. The basic reasons for deforestation are varied. Lands in tropical countries are often deforested to make room for agriculture. Growing populations mean that more land must be cleared to allow local people to grow crops, build shelter, and provide fuel for cooking and heating. Commercial logging also contributes to a loss of forests in both temperate and tropical climates. *T. S.*

High-Grading: the resource extraction strategy of harvesting the best specimens. The rationale for high-grading is that the value of the timber removed is higher than with other harvesting approaches. Some government regulations encourage high-grading over CLEAR-CUTTING because of the belief that high-grading does not entail as much removal of trees. However, high-grading is often a short-sighted extraction strategy, because it often degrades the forest quality through the removal of the best genetic stock, and collateral damage to trees that are not targeted for removal is often very high. Royalties or taxes on logging that do not discriminate according to quality provoke high-grading, because the extractor has an incentive to extract higher-valued timber taxed at the same rate as lower-valued timber.

Old-Growth Timber: timber stands characterized by a preponderance of trees at or near the maximum age that would prevail under natural conditions (e.g., conditions that present only natural disturbance possibilities, such as forest fires, rather than human-induced disturbances). These may be species found in the final stage of shifts from the early species that thrive in clearings to the successor species (i.e., the "climax of an ecological succession"), or they may be species found at an intermediate stage where natural fire or disease typically prevents attainment of climax vegetation. "Old growth" is often used as a synonym for the popular term "virgin" stands, but old growth may actually have been cut in the past, yet has recovered the requisite qualities over time. Old-growth timber stands are typically characterized by large individual trees, and high wood volumes per unit of area. This makes them very attractive for timber harvest, as both quantity and quality of wood extracted can be very high. Thus old growth is rare in regions with intensively managed timber resources, such as Europe and the eastern United States, and more common where terrain or lack of markets has limited the frequency of harvest, as in Canada, Siberia, and the U.S. Northwest. Environmentalists, particularly in the western United States and Canada, have been active in seeking legal protection of remaining old growth, citing its role as an increasingly rare ecosystem and its relation to specific endangered species such as the northern spotted owl. *R. G. H.*

COMMENTARY ON OLD-GROWTH TIMBER

One of the most celebrated cases of trying to preserve old-growth forests occurred when newly elected U.S. President Bill Clinton brought loggers, conservationists, and scientists together for a two-day conference in Portland, Oregon in 1993. This team was directed to try to come up with a management plan that would balance industry's demand for timber with environmental-ists' demands for science-based resource management to preserve some of the ancient forests in the Pacific Northwest. While expectations for the North-west Forest Plan were high, environmentalists and loggers alike have attacked the plan. Environmentalists say that the plan doesn't adequately protect old-growth trees, while loggers have been disappointed in the amount of timber they have been allowed to cut. *Toddi Steelman*

Plantation Forestry: timber or other tree-product production from planted trees, usu-ally of a very limited number of species for each site, managed intensively to produce the highest volume of output. Some plantations are devoted to nontimber outputs, such as palm oil, rubber, fruits, etc., but the greater focus of forestry policy has been on the prospects for plantation forestry in developing (particularly tropical) countries as an answer to the dual problems of loss of forest cover and the difficulty of meeting the world's growing demand for timber. Plantation forestry permits the scientific applica-tion of silvicultural techniques, and in the temperate zones has enabled commercial timber companies to maintain high timber yields. In tropical countries, however, plan-tation forestry faces greater challenges, including greater susceptibility to pest infes-tation, questionable sustainability of introduced ("exotic") tree species, conflicts with local people displaced by the grant of the land concession to the plantation operators, and often greater costs in processing and in reaching timber markets.

Selective Logging: the practice of harvesting specific varieties and/or sizes of trees, as opposed to CLEAR-CUTTING. Sometimes selective logging is preferred by loggers, be-cause they can concentrate on harvesting trees of greater value, although this runs the risk of HIGH-GRADING that in the long run can lead to the degeneration of the tree stock. More often, though, selective logging practices are required by government regulation, in order to promote regeneration, conserve wildlife, or both. Typical selec-tive logging regulations include the requirement that the trees exceed a minimum size (usually designated in terms of diameter at breast height, or "dbh") and the prohibition on harvesting endangered tree species.

6.5.2 RESPONSES AND STRATEGIES

Declaration of the Development of Forests: an agreement achieved at the UN CONFER-ENCE ON ENVIRONMENT AND DEVELOPMENT (UNCED, OR RIO EARTH SUMMIT) about the use, protection, and SUSTAINABLE DEVELOPMENT of forests. The declaration came out as a compromise between the ambitious desire of the Northern countries to achieve the protection of the forests through a worldwide convention on forests and the reluc-tance on the part of the South to give up its sovereignty over natural resources, the tropical rain forest in particular. The declaration is not binding by international law. However, it might be a step toward negotiating a new regime on the conservation of world forests. See also STATEMENT OF FOREST PRINCIPLES.

Harvesting Plans: plans submitted by loggers to the relevant government agencies in order to obtain permission to proceed with tree harvesting. Harvesting permits are re-quired in a very large proportion of public-lands forestry management regimes, and some governments (e.g., the state of California) require harvesting plans to be filed for logging on private lands. The requirement of harvesting plans is seen by many government officials and environmentalists as crucial for sound management of forests. However, others see the requirement of harvesting plans for logging on private land as an infringement on property rights. A chronic problem with the harvesting-plan system in developing countries is the inadequacy of oversight, which allows many such plans to be empty exercises that bear little relation to good harvesting practices or to the logging that is actually carried out.

Statement of Forest Principles: a document adopted at the UN CONFERENCE ON ENVIRONMENT AND DEVELOPMENT (UNCED, OR RIO EARTH SUMMIT) in lieu of a global forest convention. Negotiations on forest issues at Rio were extremely polarized between the North American "global responsibility" approach and the "sovereign discretion" approach of developing countries. Canada and the United States advocated an agreement on the principles of national responsibility and global concern for forests, as well as targets and timeframes for national forestry plans. Developing countries, led by India and Malaysia, strongly objected to these proposals, which they regarded as an attempt to treat tropical forests as the "common heritage of mankind." The final version of the forest principles ("A Non-Legally Binding Authoritative Statement of Principles for a Global Consensus on the Management, Conservation, and Sustainable Development of All Types of Forests") is simply a series of general guidelines addressing, with no details or commitment, a range of concerns brought to Rio by different parties. It reinforces states' sovereign rights to resources, and emphasizes the importance of international cooperation, fair cost sharing for forest conservation and SUSTAINABLE DEVELOPMENT. It recognizes the impact of poverty and indebtedness on the ability of developing countries to manage their forests and the need for technology transfer.

COMMENTARY ON THE STATEMENT OF FOREST PRINCIPLES

The statement contains a lot of politically correct but practically meaningless phrases. It has been regarded by environmentalists and authorities of most developed countries as worse than no declaration at all because it appears to legitimize unsustainable forest management practices. *Natalia Mirovitskaya*

Sustainable Forestry: forestry management and policies that encourage the preservation of forests as sources of timber and a host of other societal benefits, including important environmental services. An OPTIMAL SUSTAINABLE YIELD management strategy would be applied to each forest stand in order to find the best tradeoff between timber production and the benefits derived from the standing timber stock. In some but not all circumstances, this harvest rate may be the MAXIMUM SUSTAINABLE YIELD rate that maintains a roughly constant timber stock. However, it may be that reconstituting the timber stock to other species would provide more societal benefits and sustainability. Most conflicts over sustainable forestry management revolve around the desire of some groups to maintain diversity in forests, while others wish to provide a sustained yield of fiber for human use. *T. S.*

Tropical Forestry Action Plan (TFAP): a program financed by several international organizations, including the WORLD BANK and the UN FOOD AND AGRICULTURE ORGANI-

ZATION (FAO), to provide technical assistance and financing for tropical countries to develop long-term forestry strategies. In many countries, the TFAPs have been the first efforts to assess forestry resources on a comprehensive basis and to formulate general forestry strategies. However, some critics have castigated the process for leaving local people out of the process, and for catering exclusively to the interests of commercial loggers.

6.5.3 ORGANIZATIONS RELATED TO FORESTRY

Forest Stewardship Council (FSC): an international coalition of organizations and individuals with a stake in forestry issues (representatives from environmental and social groups, the timber trade and the forestry profession, indigenous people's organizations, community forestry groups, and forest product certification organizations from around the world). By early 2000, the FSC had over 350 members from 49 countries. The FSC was founded in 1993 to establish standards for supporting accrediting and monitoring timber certification bodies, which in turn label timber that is produced responsibly. The FSC was established to address the concern about the proliferation of independent timber certification programs, as well as self-labeling initiatives, that have been inappropriately certifying timber operations that do not deserve designation as responsible or sustainable. The FSC strengthens national certification and forest management capacity, as well as promotes forest stewardship by encouraging the development of local forest management standards worldwide: local initiatives are established in twenty-five countries, including Brazil, Bolivia, Belgium, Papua New Guinea, Sweden, Switzerland, and the United Kingdom. The FSC is registered in Mexico and is governed by a general assembly and a board of directors. The composition of the board of directors, and the FSC's decision-making procedures, are designed to ensure a balance between commercial, environmental, and social interests, and between Northern and Southern interests (see also NORTH VS. SOUTH).

Intergovernmental Forum on Forests (IFF): an ad hoc mechanism, successor to the INTERGOVERNMENTAL PANEL ON FORESTS (IPF), established in April 1997 under the UN COMMISSION ON SUSTAINABLE DEVELOPMENT to: a) promote and facilitate the implementation of the Intergovernmental Panel on Forestry's proposals for action; b) monitor progress in implementation toward sustainable forest management; and c) consider matters left pending on the need for financial resources, trade, and environment, and the transfer of environmentally sound technologies to support sustainable forest management. The forum is also mandated to develop a legally binding instrument for the sustainable management of all types of forests, or to develop another permanent international mechanism. The secretariat of the forum is located with the United Nations Secretariat in New York and depends entirely on voluntary financial contributions.

Intergovernmental Panel on Forests (IPF): a mechanism established by the UN COMMISSION ON SUSTAINABLE DEVELOPMENT (CSD) in April 1995 to continue the international forest policy dialogue in the aftermath of the 1992 UN CONFERENCE ON ENVIRONMENT AND DEVELOPMENT (RIO EARTH SUMMIT). The IPF produced over one hundred negotiated proposals for action on issues related to sustainable forest management, including national forest programs, forest assessment, sustainability criteria and indicators, traditional forest-related knowledge, and underlying causes of deforestation. These action proposals are directed to governments, international organizations, and major groups including the business sector. The IPF has also addressed such complex and politically sensitive issues as financial assistance and technology transfer, and trade and environment in relation to forest products and services. Controversies among governments were too deep to reach consensus in that realm. The IPF's mandate expired in 1997 and was succeeded by the INTERGOVERNMENTAL FORUM ON FORESTS (IFF).

International Council for Research in Agroforestry (ICRAF): an autonomous nonprofit international organization established in 1977 and located in Nairobi, Kenya. Its mandate is to initiate, stimulate, and support research leading to more sustainable and productive land use in developing countries through the integration and management of trees in land-use systems. ICRAF is one of the research institutions within the CONSULTATIVE GROUP ON INTERNATIONAL AGRICULTURAL RESEARCH (CGIAR). Since 1986, ICRAF has collaborated with various national and international institutions in research on AGROFORESTRY. Core ICRAF funds are provided by the WORLD BANK.

International Tropical Timber Agreement (ITTA): an international commodity agreement mandated to protect the interests of both producers and consumers of tropical timber. The ITTA arrangements were established in 1983 and 1994, within the framework of the UN CONFERENCE ON TRADE AND DEVELOPMENT (UNCTAD). As of late 2000 the ITTA had 56 members, including 30 producing countries, 25 consuming countries, and the EUROPEAN UNION; member countries account for 95 percent of world tropical timber trade and 75 percent of tropical forests. In distinction to other commodities agreements, ITTA (though certainly far from a conservation treaty) provides for a number of safeguards intended to maintain the ecological balance of tropical forests and encourages members to develop national policies aimed at sustainable utilization and conservation of timber-producing forests and their genetic resources. The agreement is administered by the INTERNATIONAL TROPICAL TIMBER ORGANIZATION (ITTO) with the International Tropical Timber Council as its governing body.

International Tropical Timber Organization (ITTO): an intergovernmental organization consisting of most major tropical timber exporting nations and timber-importing nations, founded through the 1983 INTERNATIONAL TROPICAL TIMBER AGREEMENT. Headquartered in Yokohama, Japan, the ITTO promotes international trade in tropical timber through consultation, research, technical assistance, market intelligence, and encouragement of reforestation, scientific forest management, and improved marketing and more sustainable forestry policies. The ITTO's policy positions have often emphasized conservation. The ITTO's "Year 2000 Objective" was to have all internationally traded tropical timber sourced from sustainably managed forests by that year. The feasibility of these objectives is widely questioned, both because of the slow progress in adopting any serious forest management in many countries, and skepticism about whether sustainable tropical forestry is realistic in principle.

See Duncan Poore, ed., *No Timber without Trees* (London: Earthscan, 1989).

Rainforest Action Network (RAN). See under SECTION 6.3.2.

6.6 Desertification

Club du Sahel: an informal intergovernmental association of nine contiguous African states in the Sahel area (the dry region of the sub-Sahara), eight donor nations, and several multilateral development agencies, created in 1975. The club's member nations are Burkina Faso, Cape Verde, Chad, Gambia, Guinea-Bissau, Mali, Mauritania, Niger, Senegal, as well as Canada, the United States, Austria, Belgium, France, Netherlands, Switzerland, and United Kingdom. The goal of the club was to raise the awareness of the international community on development needs and prospects of the Sahel and to facilitate the mobilization of resources. Current programs include the Network for the Prevention of Food Crisis in the Sahel, a sector of Environmental Resource Management, the Support Unit for Private Sector Development, a sector on Agricultural Transformation, and an Inter-State Commission on the Struggle against Drought.

Convention to Combat Desertification (CCD): an international agreement designed to combat DESERTIFICATION, mitigate the effects of drought in countries affected by serious drought and/or desertification, and secure the basis of existence of the poorest people in these countries. The convention was signed in Paris on October 1994, as the United Nations Convention to Combat Desertification in Those Countries Experiencing Serious Drought and/or Desertification, Particularly in Africa, and entered into force on December 1996. As of the end of 1998, the convention had 145 signatories, but several developed countries, including the United States and Australia, had not yet ratified it. The convention calls for national strategies to integrate arid and semi-arid areas into national economies, land tenure reform, and increased cooperation among donor agencies, national governments, grassroots organizations, NONGOVERNMENTAL ORGANIZATIONS, and local populations. The convention contains four regional implementation annexes: for Africa, Asia, Latin America and the Caribbean, and the northern Mediterranean. Its financial organ, the GLOBAL MECHANISM, is housed within the INTERNATIONAL FUND FOR AGRICULTURAL DEVELOPMENT (IFAD), which is supposed to improve monitoring of aid flows and increase its coordination. This convention is one of only two nearly global international agreements established at the initiative of developing countries despite the opposition of advanced industrialized countries.

Desertification: environmental change that converts land into desert—usually defined as dry regions with less than twenty-five centimeters of rainfall a year, low humidity, and high evaporation rates. These changes may be natural or induced by human action, though most result from a combination of natural and human factors. By the estimates of the UN ENVIRONMENT PROGRAMME (UNEP), over 250 million people are directly affected by desertification; in addition, one billion people in over one hundred developing nations are at risk. Desertification threatens mostly the northern subtropical regions of Africa and Asia and is usually connected with many complex factors, ranging from international trade patterns, poverty, political instability, and DEFORESTATION to the unsustainable land management practices of local communities. Desertification also affects some eighteen developed countries, including the United States, Australia, and Spain.

See United Nations Conference on Environment and Development (UNCED), *Agenda 21* (Rio de Janeiro, Brazil, June 3–June 14, 1992).

COMMENTARY ON DESERTIFICATION

The significance of desertification, and the attempts to address it, have been fraught with uncertainty and contention. Regarding its significance, the term does not really refer to "deserts on the march," as many would believe, insofar as the scientific evidence on the spread of deserts is inconclusive. As an international issue, desertification has been the subject of international concern and cooperation for more than two decades, since the 1977 UN Conference on Desertification, which launched the Plan of Action to Combat Desertification, but the plan has been generally acknowledged to have been a failure because of both the lack of sufficient commitments from the donor community and the inadequate policies of African governments. Some critics also assert that the failure of the plan was the result of erroneous assumptions on the nature of desert ecosystems and insufficiency of reliable data. Despite the opposition of the most industrialized countries and the WORLD BANK, the issue of desertification was brought back to the global environmental agenda in 1992 due to the insistence of African nations and because it became linked with other issues in UNCED. Eventually the UN CONVENTION TO COMBAT DESERTIFICATION entered into force in 1996 with more than 110 signatories. Scientific and political debates around this convention demonstrated that combating desertification could be successful only as a part of a much broader objective: the SUSTAINABLE DEVELOPMENT of countries affected by drought *and* desertification. *Natalia Mirovitskaya*

Global Mechanism (GM): the mechanism for financing the programs emerging in late 1997 from the CONVENTION TO COMBAT DESERTIFICATION (CCD). Housed within the INTERNATIONAL FUND FOR AGRICULTURAL DEVELOPMENT (IFAD), it was created to serve as a "hub" for cooperation and resource transfers for dry-lands poverty alleviation by the UN FOOD AND AGRICULTURAL ORGANIZATION (FAO), the GLOBAL ENVIRONMENT FUND (GEF), the UN DEVELOPMENT PROGRAMME, and the WORLD BANK, as well as bilateral and private-sector initiatives. The basic challenge of the Global Mechanism is to ensure significant amounts of resources so that the CCD avoids the fate of its predecessor—the United Nations Conference on Desertification, of the late 1970s. The significance of the Global Mechanism thus far is not the volume of resources that have been mobilized—rather modest to this point—but rather as an experiment in coordinating the large number of efforts by many international and national organizations to address a specific development and environmental challenge.

Intergovernmental Authority on Drought and Development (IGADD): an agency founded in 1986 by six African nations—Djibouti, Ethiopia, Kenya, Somalia, Sudan, and Uganda—to coordinate their activities in combating drought, desertification, and related environmental disasters. The IGADD is involved in mobilizing financial resources from the United Nations and other international agencies for emergency and preventive relief programs.

Plan of Action to Combat Desertification (PACD): an international program of actions to deal with the problems of DESERTIFICATION that was first adopted by the 1977 United Nations Conference on Desertification and later by the United Nations General Assembly. The plan included twenty guiding principles and twenty-eight recommendations and called for improved land use, measures to improve the living standards of people living in dryland areas, and the integration of efforts to control desertification with other development projects and social programs. The main effort in combating desertification was left to the national governments, in conjunction with aid agencies. The plan accomplished the training of approximately one thousand nationals from developing countries in desertification assessment and prevention, and ensured involvement of NONGOVERNMENTAL ORGANIZATIONS in this effort. However, it failed to generate a strong commitment from the parties, which might be explained by the political weakness of the main recipients (the rural poor) and insufficient involvement of policymakers in the planning process. Officials of the UN ENVIRONMENT PROGRAMME blamed the failure of the PACD on both African governments and the donor community.

See Ramon Lopez-Ocana, "Effectiveness of International Regimes Dealing with Desertification from the Perspective of the South," in *Global Governance: Drawing Insights from the Environmental Experience*, ed. Oran Young (Cambridge, MA: MIT Press, 1997), 125–35.

6.7 Energy Shortages and Pollution

6.7.1 ENERGY TRENDS, PRACTICES, AND PROBLEMS

Energy Crisis: in general, any of a number of acute shortages of energy supply or abrupt increases in energy prices. Specifically, the energy crisis of 1973–74 refers to the dramatic rise of world oil prices following the 1973 Arab-Israeli War and the concerted oil price increases by the ORGANIZATION OF PETROLEUM-EXPORTING COUNTRIES (OPEC) the following year. That energy crisis led initially to fuel shortages around the world, and triggered inflation in many countries. A less dramatic but more severe energy crisis prevails in developing countries that have had increasing difficulty in finding fuel sources, including fuelwood. The search for fuels often results in deforestation and the use of highly polluting energy sources. Whether the future will bring a global energy crisis is very much in debate, depending on predictions about the potential for technology to deliver more efficient energy uses, and theories about the relationships between energy use and environmental impacts.

Energy-GNP Correlation: the striking relationship between economic activity and energy use that prevailed during the period prior to the mid-1970s, reflecting the fact that cheap energy allowed for energy consumption to rise proportionately with economic activities such as industrial production. After the mid-1970s, in industrial countries the energy used per unit of production declined dramatically as energy prices rose, showing that the correlation could be reduced. However, the decline in real energy costs, reflected most visibly by the decline in real oil prices, has relaxed concern about energy economizing.

Energy-Intensive Industry: the industries that require high energy inputs per unit of output. Certain processes, such as the smelting and refining of metals, require very high energy inputs. These industries are often promoted in countries where such raw materials of energy supplies are abundant. These industries are, of course, vulnerable to increases in energy prices. The reliance on energy-intensive industries in energy-abundant countries may sacrifice the higher earnings of exporting energy, if the domestic energy-intensive industries use potentially exportable fuels (such as oil) without high returns.

Fuel Efficiency. See under SECTION 6.15.

6.7.2 ENERGY RESPONSES AND STRATEGIES

Demand-Side Management: the use of programs designed to reduce demand for electricity (or other energy) in lieu of the construction of new generation capacity. This term arises most often in the context of electric utility regulation. In the United States, many state public utility commissions have mandated the use of demand-side management by public utilities to reduce the need for new generating capacity. *D. S.*

COMMENTARY ON DEMAND-SIDE MANAGEMENT

It is quite remarkable that demand-side management, pressed upon private electricity producers by government regulators, results in private firms trying to *reduce* the demand for their product. In one sense this simply reflects how different the world looks to a price-regulated utility than to a conventional private company—so-called public utilities are private enterprises that make their money by pleasing government rate setters who reward good performance, whether in producing a lot of electricity or less electricity, by allowing higher profits. In another sense, it reflects how malleable energy supply and demand can be if a government is serious about controlling energy consumption. *William Ascher*

Ecological Energetics Paradigm. See under SECTION 1.3.2.

Energy Valuation. See under SECTION 5.1.7.2.

Energy-GNP Decoupling: the approach of designing economic activities, particularly in industry and transportation, to reduce the increases in energy consumption as production increases. For many policymakers, restraining the growth of energy use is often attractive for both economic and environmental objectives; the challenge is to do this without sacrificing economic growth. Product and process design is an important element of energy-GNP decoupling. In many contexts in which energy consumption has been subsidized, a crucial economic policy instrument is to charge energy consumers for the full costs of energy consumption, including the environmental costs. Another approach is DEMAND-SIDE MANAGEMENT.

International Atomic Energy Agency (IAEA): an independent intergovernmental organization founded in 1957 to carry out activities related to the development and verification of nuclear energy's peaceful uses. By 2000, the IAEA had 130 member states and was connected by formal agreements with forty governmental and nongovernmental organizations. The agency is associated with the United Nations and is headquartered in Vienna. It serves as the world's central forum for scientific and technical cooperation in the nuclear field, and as an adviser for the application of nuclear safeguards concerning civilian nuclear programs. Under agreements with member states, IAEA inspectors regularly visit nuclear facilities to verify government records on the location and uses of nuclear material, check IAEA-installed instruments and surveillance equipment, and confirm physical inventories of nuclear materials. More than a thousand nuclear installations are under IAEA safeguards, representing 95 percent of the world's nuclear facilities and materials outside of the five declared nuclear powers. Joint projects undertaken by the IAEA range from monitoring nuclear fallout associated with weapons testing in the 1960s to the standardization of reporting on nuclear incidents and accidents in more recent years. The key inducement for governments to comply with IAEA monitoring and restrictions is the financial arrangement whereby governments, especially of developing countries, can obtain technical assistance, equipment, and fissionable material at subsidized prices. The IAEA determines whether nuclear plant design and operations qualify for such subsidies. Many IAEA programs, such as the Isotope Hydrology and Radioactive Waste Management Advisory programs, contribute directly or indirectly to the goals of environmental protection and SUSTAINABLE DEVELOPMENT. However, the most visible activities of the IAEA in the 1990s were the investigations of nuclear programs suspected of being used to develop nuclear weapons.

COMMENTARIES ON THE IAEA

Apart from its regulatory functions, the IAEA played a major role in "development programs" from its inception. These programs were underpinned by the notion that nuclear power represented an appropriate technology for developing nations. Under the guise of technical assistance, the IAEA thus played a prominent role in the attempt to "sell" the nuclear dream of cheap, plentiful electricity around the world. In 1987, US$2,039,600,000 were spent on such schemes in Africa (*IAEA Bulletin*, 30 [2]: 24),a continent where the presence of massive solar gain and an absence of integrated supply systems (national electricity grids) make nuclear power an irrelevance in the eyes of most advocates of sustainable development. The IAEA's acceptance of levels of Material Unaccounted For (MUF) during plutonium audits have formed the basis of fears for the spread of nuclear weapons to sub-state actors (terrorists) since the 1970s. The assignment of personnel to the IAEA from the world's nuclear industry and its dual role as regulator and promoter of nuclear power make it a classic example of the fox left in charge of the henhouse. *Ian Welsh*

The IAEA is a clever mechanism for rich countries to subsidize safer, but more expensive, nuclear reactors in poorer countries. The IAEA does this without making the government of the poorer country appear to need bribing to adopt safer nuclear designs. *William Ascher*

International Energy Agency (IEA): an autonomous body established in 1974 within the framework of the ORGANIZATION FOR ECONOMIC COOPERATION AND DEVELOPMENT (OECD) to foster collaboration on energy issues by its membership, initially comprising twenty-one of the twenty-four OECD (essentially the industrialized) countries. By 2000, the IEA had twenty-five member states, but some new OECD members, most notably oil-exporting Mexico, have not joined. Set up after the 1973–74 dramatic increase in oil prices instigated by the ORGANIZATION OF PETROLEUM-EXPORTING COUNTRIES (OPEC), IEA members were committed by its International Energy Program, which came into force in 1976, to share their oil in emergencies and to strengthen cooperation to reduce their dependence on oil imports. The IEA Long-Term Cooperation Program was designed to secure energy supplies, promote the stability of world energy markets, develop alternative energy sources, and promote cooperation to conserve energy. The IEA thus represents an effort, albeit modest and low-profile, to organize the major oil-importing nations to counter the OPEC cartel.

Organization of Petroleum Exporting Countries (OPEC): the first and so far the most successful intergovernmental commodity cartel of developing countries. Its members are Algeria, Ecuador, Gabon, Indonesia, Iran, Iraq, Libya, Nigeria, Qatar, Saudi Arabia, the United Arab Emirates, and Venezuela. During 1973–74, OPEC members raised the price of crude petroleum in real terms more than sixfold. Another large price increase occurred in 1979–80. In both cases, however, international political crises, such as the 1973 Arab-Israeli War, also contributed to the rise in oil prices because of actual or feared disruptions in petroleum supply, making it difficult to assess the impact of the cartel in itself. While the OPEC secretariat oversees several technical efforts of data collection, research, and resource transfers, its principal activity is to facilitate the negotiation of prices and production ceilings among OPEC members. After 1980, the energy conservation programs of oil-importing countries and the discovery of new oil supplies undercut OPEC's strong position in the energy market. Periodic agreements among OPEC members to significantly restrict their oil production and to resist price cutting often failed because of the enormous temptations to capture windfalls by selling more oil when other OPEC members were adhering to the production restrictions. OPEC has also been a major player in several international environmental negotiations.

See Daniel Yergin, *The Prize: The Epic Quest for Oil, Money and Power* (New York: Simon and Schuster, 1991).

COMMENTARY ON OPEC

The early effectiveness of OPEC seemed quite obvious, but it has come under question, even in terms of whether it was actually a cartel in its heyday of the mid- and late 1970s. This has led to skepticism about the potentials of commodity cartels in general. Were the price increases really due to the concerted effort of several producers to cut back on production in order to force a price increase—the classic definition of a producer cartel—or did they reflect the essentially unilateral production cutbacks by Saudi Arabia, by far the largest oil producer with the greatest reserves? A middle-ground explanation seems most plausible: Saudi Arabia has played the major role in cutting production in order to raise world oil prices, but the incentives for Saudi Arabia to do so are certainly greater if other oil-exporting countries also cut production and try to maintain higher prices, even if each such effort erodes because of cheating. Thus the fact that OPEC provides even weak coordination among most oil-exporting countries has increased Saudi Arabia's willingness and capacity to be a price setter. *William Ascher*

Soft Energy Path: the strategy, most prominently espoused by Amory Lovins, to address energy shortages and high energy prices through decentralized, diversified, small-scale energy sources. A "soft energy path" does not consist only of a collection of technologies and efficiency improvements; it is also defined by the political structure, which avoids the centrism, coercion, vulnerability, and technocracy of a "hard energy path." Soft technologies could be deployed by coercion from the government, but this would contradict the political requisites of a "soft energy path." The "soft path" is in stark contrast with the search for overall national energy solutions that proliferated after the 1973–74 energy crisis.

See Amory Lovins, *Soft Energy Paths: Toward a Durable Peace* (Cambridge, MA: Ballinger, 1977); Ronald D. Brunner and Robin Sandenburgh, eds., *Community Energy Options: Getting Started in Ann Arbor* (Ann Arbor: University of Michigan Press, 1982).

6.8 Fresh Water Scarcity and Pollution

Absolute Territorial Integrity, Doctrine of: one of the international river management models directly opposed to the HARMON DOCTRINE. It asserts the doctrine of water rights that gives the lower riparian state the right to the continued, uninterrupted (or natural) flow of the water from the territory of the upper riparian state. This doctrine, clearly favorable to the lower-basin state, has been criticized for allocating rights without imposing corresponding duties. It has been invoked in situations when the continued flow of water is critical to the survival of the state concerned (for instance, Iraq and the Euphrates River).

See Bonaya A. Godana, *African Shared Water Resources* (Boulder, CO: Lynne Rienner Publications, 1985).

Canada/USA International Joint Commission (IJC): the oldest of U.S.-Canadian intergovernmental organizations, established by the Boundary Waters Treaty of 1909 to deal primarily with the apportionment, conservation, and development of water (including hydropower) resources along the international boundary. According to its charter, the IJC can also serve as a final court of arbitration on any issue between Canada and the United States, but has never been used as such. A unique feature of the agreement is that it gives an equal weight to *preventing* disputes as to *settling* existing ones; therefore the IJC has been one of very few organizations engaged in anticipatory planning. The IJC has carried out a wide range of investigative, legal, administrative, and arbitral functions. Since 1978 the IJC's major emphasis has been on management of the entire Great Lakes ecosystem, where it is believed to have employed some of the most sophisticated uses of public information and participation practices.

See Alfred Duda and David La Roche, "Joint Institutional Arrangements for Addressing Transboundary Water Resource Issues," *Natural Resource Forum*, 21 (2) (1997): 127–37.

Clean Water Act (U.S. Federal Water Pollution Control Act): a series of legislative acts establishing a system for controlling water pollution in the United States. The first Clean Water Act was passed in 1972, in the wake of serious deterioration of many U.S. lakes (e.g., Lake Erie) and rivers. This statute (U.S. Code Title 33, Ch. 26) regulates the discharge of pollutants into U.S. waters through a system of permits issued to dischargers. The statute also establishes a system for financing and regulating municipal sewage treatment plants. The Clean Water Act is administered principally by the U.S. Environmental Protection Agency. *D. S.*

Community of Interests, Doctrine of: one of the international river management models. Essentially, it proposes that state boundaries should be ignored and that the water system ought to be managed as an integrated whole. In that case, all basin states would have a collective right of action and no state could dispose of the water without consultation with and cooperation with the other states. The "community of interests" approach leads to the implementation of basin-wide development programs designed by all riparian states in the basin. The approach, where adopted, is likely to minimize development along the water course, insofar as down-river states would have the power to veto up-river developments that risk either pollution or water depletion.

See Bonaya A. Godana, *African Shared Water Resources* (Boulder, CO: Lynne Rienner Publications, 1985).

Convention on the Law of the Non-Navigational Uses of International Watercourses: an international agreement adopted by the UN General Assembly on May 21, 1997. The convention applies to uses of international watercourses and their waters for purposes other than navigation and to measures of their protection, preservation, and management. The agreement is based on principles of equitable and reasonable utilization and participation, and defines rights and obligations of riparian states in different situations and relating to various types of resource use. Lacking sufficient ratifications, the convention is not yet in force.

Convention on the Protection and Use of Transboundary Watercourses and International Lakes (Helsinki Convention): a regional agreement administered by the UN Economic Commission for Europe to manage transboundary waters, open to member states of the region covered by the UN Economic Commission for Europe, other European states, and the nations of North America. The convention was adopted in March 1992 in Helsinki and entered into force in October 1996. By late 2000, it had been ratified by thirty-two parties, including the EUROPEAN UNION. The convention is aimed at strengthening national and international "protection and ecologically sound management of transboundary waters . . . and related ecosystems." The parties are obligated to undertake comprehensive measures to prevent, control, and reduce water pollution, particularly hazardous substances, from point and diffuse sources. The PRECAUTIONARY PRINCIPLE, the POLLUTER-PAYS PRINCIPLE, and the SUSTAINABILITY PRINCIPLE were recognized as guidelines to protect and conserve not only water resources, but also soil, flora, fauna, air, climate, landscape, and cultural heritage. The Helsinki Convention is one of very few multilateral agreements on transboundary rivers and lakes and can be used as a framework for future international agreements on water resources management.

Equitable Utilization, Doctrine of: one of the international river management models. It has evolved gradually during the discourse on competing theories of international water management (i.e., the HARMON DOCTRINE, models of ABSOLUTE AND LIMITED TERRITORIAL INTEGRITY, and the COMMUNITY OF INTERESTS approach). The doctrine of equitable utilization proposes that each riparian state has a right to utilize the water of the basin and is entitled to a reasonable and equitable share of the basin water. The doctrine reflects three fundamental concerns: a) the socioeconomic needs of the basin states through an objective consideration of various factors and conflicting elements relevant to the use of the basin; b) water distribution among the riparian states to satisfy their needs to the greatest possible extent; and c) maximum benefit for each eco-basin state with the minimum detriment. This doctrine has been used in several schemes of international river management initiated by the UN ENVIRONMENT PROGRAMME (UNEP).

See Bonaya A. Godana, *African Shared Water Resources* (Boulder, CO: Lynne Rienner Publications, 1985).

Harmon Doctrine: a legal doctrine of river management that emphasizes the unlimited sovereignty of each riparian state over its natural resources. Thus the up-river state has no obligation to provide the "natural" volume to the down-river state. The practical implementation of this doctrine in the management of transboundary freshwater resources often results in international conflicts. The principle of EQUITABLE UTILIZATION is suggested as an alternative to the Harmon Doctrine.

Helsinki Rules on the Use of Waters of International Rivers: a document adopted in 1966 by the International Law Association to specify criteria of a reasonable and equitable use of transboundary water resources. Because of complicated and ambiguous definitions of the factors to be considered, the Helsinki Rules have not been formally applied in any actual transboundary water conflicts. However, as one of very few legal documents on water conflict resolution, it has served as the basis for decisions of the INTERNATIONAL COURT.

International Watercourses. See SECTION 3.4.1.

Riparian Rights: rights of use and ownership of streams, rivers, and other flowing water bodies (usually confined to freshwater bodies rather than to oceans and seas). The traditional, so-called natural flow doctrine of riparian rights allows all the owners to use the water so long as the quantity or quality of the flow is not decreased. This doctrine, based in English common law, ensures that the water body remains in its natural state throughout its course and that the owners at the downstream areas have the same opportunities to use it as those upstream. Another, currently much more popular, doctrine of riparian rights is that of "reasonable use," permitting owners to use water resources for their own ends while trying to ensure that it does not interfere with the potential use of others. However, the growing scarcity of fresh water resources, along with diversification of their use, make implementation of the "reasonable use" doctrine more and more conflictual. Discords about riparian rights are expected to become one of the major causes of inter- and intrastate tension in the future, therefore this issue has come to the forefront of environmental legislation and international environmental diplomacy. See also ABSOLUTE TERRITORIAL INTEGRITY, DOCTRINE OF; HARMON DOCTRINE; COMMUNITY OF INTERESTS, DOCTRINE OF; EQUITABLE UTILIZATION, DOCTRINE OF.

Transboundary Waters. See SECTION 3.4.1.

Water Environment Federation (WEF): one of the oldest NONGOVERNMENTAL ORGANIZATIONS, founded in 1928 in the United States to preserve and enhance the global water environment. WEF is a federation of 76 member associations in 27 countries and over 40,000 individual members (primarily water experts) in 85 countries. Through a series of its publications and Internet lists, WEF provides its members with the latest information on waste-water treatment and water quality protection. It also organizes conferences and training programs on water quality and pollution control technologies and reviews environmental legislation and regulations.

Water-Resource Management: a complex system required to match water supply and demand. Sustainable water resource management has been an objective for many countries since the 1980s, while the first proposals for river basin management date back to the eighteenth century. Technically, river basins that are clearly defined biogeophysical systems are particularly suited for monitoring and management purposes. Having started with coordinating the different demands made on water within a basin, river basin management has typically gone through four stages: a) single-purpose management (for instance, flood control or irrigation supply); b) the multipurpose planning approach (dividing water between users for multiple use); c) the integrated river basin planning approach (the attempt to coordinate water use in the basin with other development purposes); and d) comprehensive river basin planning and management (the attempt to coordinate the use of all resources within a basin-wide program). Water resource management becomes particularly challenging in case of INTERNATIONAL WATERCOURSES, because no national government can arbitrate disputes among the localities.

Water Use Charges. See CHARGES.

6.9 Global Environmental Change

Framework for the Development of Environment Statistics (FDES): a conceptual framework developed by the UN Statistics Division in 1984 to coordinate environmental data and related socioeconomic statistics. It is based on STRESS-RESPONSE principles of environmental impact and in its most comprehensive variant reflects the impact of GLOBAL CHANGE.

Global Change: a label for the full range of natural and human-induced changes over time in the earth's environment. U.S. environmental legislation defines it as "changes in the global environment (including alterations in climate, land productivity, oceans or other water resources, atmospheric chemistry and ecological systems) that may alter the capacity of the earth to sustain life." Major international attention focuses on: a) long-term climate change, for instance, caused by GREENHOUSE-EFFECT WARMING; b)

311

OZONE DEPLETION; c) DESERTIFICATION; d) DEFORESTATION; e) BIODIVERSITY LOSS; f) SOIL EROSION AND DEGRADATION; g) SEA-LEVEL CHANGES; and h) increased storm activity and flooding.

See United States National Science and Technology Council, *Our Changing Planet: The 95 Global Change Research Program* (Washington, DC: USNSTC, 1994).

Global Environmental Monitoring System (GEMS). See under SECTION 5.1.5.

International Referral System for Sources of Environmental Information (INFOTERRA). See under SECTION 5.1.5.

Stress-Response Environmental Statistical System. See under SECTION 5.1.5.

6.10 Hazardous Wastes

Convention on the Ban of the Import into Africa and the Control of Transboundary Movements and Management of Hazardous Wastes within Africa (Bamako Convention): an international agreement between the member states of the Organization of African Unity (OAU) adopted on January 30, 1991, and entered into force on April 22, 1998. The Bamako Convention is aimed "to protect human health and the environment from dangers posed by HAZARDOUS WASTES by reducing their generation to a minimum in terms of quantity and/or hazard potential" and generally bans the shipments of hazardous waste from any country. It underlined the determination of African countries to put an end to GARBAGE IMPERIALISM and has created, in effect, the strongest waste trade regime.

Convention on the Control of Transboundary Movements of Hazardous Wastes and Their Disposal (Basel Convention): an international treaty regulating the transnational shipment and disposal of HAZARDOUS WASTES, signed in 1989 in Basel, Switzerland, and entered into force in 1992. Its objectives are to: a) control and reduce transboundary movements of wastes to a level consistent with environmentally sound management; b) minimize the hazardous wastes generated, ensuring their environmentally sound management, including disposal and recovery operations, as close as possible to the source of generation; and c) assist developing countries and countries with economies in transition in environmentally sound management of the wastes they generate. The 1995 Amendment to the Convention, which bans the export of hazardous wastes destined for final disposal, is not yet in force, nor is the Protocol on Liability and Compensation for the Damage Resulting from the Transboundary Movements of Hazardous Wastes.

COMMENTARY ON THE BASEL CONVENTION

The debate on the scope of the convention caused a substantial delay in the negotiating process and resulted in a treaty that is generally regarded as weak. The U.S.-led coalition of VETO STATES advocated an "informed consent agreement" in which signatories to the treaty could export hazardous wastes in case of prior written consent of the importing country. The opposing coalition of developing nations, led by the Organization of African Unity, backed a total ban on hazardous waste trade. Though during formal negotiations the veto coalition succeeded in preventing a total ban, due to the continued pressure of the developing countries and embarrassing media attention, several governments changed their positions. Today, most former veto states have publicly announced a total ban on trade in hazardous wastes. Despite its relatively high participation (134 parties as of late 1999) the Basel Convention is generally regarded as a WEAK TREATY. Although the treaty posits that wastes should be disposed of in an environmentally sound manner, it does not specify what that means. With so much room for interpretation, the issue of liability has not been resolved. Critics of the convention argue that it does not go any further than existing regulations in industrialized countries, which have failed to end legal or illegal dumping. *Natalia Mirovitskaya*

Dirty Dozen. See under SECTION 6.1.1.

Garbage Imperialism. See under SECTION 4.3.1.

Hazardous Substances (and Hazardous Wastes): inputs, products, and byproducts that pose risks to humans and natural systems. A very wide range of substances pose such hazards, including infectious, radioactive, flammable, carcinogenic, mutagenic (mutation-causing), climate-changing, and polluting agents. Hazardous substances can present either short-term acute or long-term environmental hazards. Wastes may arise as byproducts, process residues, spent reaction media, contaminated plant or equipment from either manufacturing operations or the treatment of toxic substances, and from the discarding of manufactured products. Whether a substance is regarded as unacceptably hazardous depends in part on what functions are regarded as worthy of safeguarding. Therefore different countries and international organizations have different classifications as to what constitutes hazardous substances. *R. P.*

International Register of Potentially Toxic Chemicals (IRPTC): a database of hazardous chemicals established in 1976 under the auspices of the UN ENVIRONMENT PROGRAMME'S (UNEP) EARTHWATCH program. The IRPTC is set up to provide: a) information on the production, distribution, release, disposal, and adverse effects of chemicals; b) information about national, regional, and global policies and regulations on potentially toxic chemicals; and c) scientific tools for sharing data to assess the hazards posed by chemicals to human health and the environment.

Nuclear Waste: radioactive materials left over from the production, use, or deployment of weapons, fuel for nuclear reactors, and isotopes used in industry, science, and medicine. Nuclear wastes have the potential to pose serious health risks, ranging from acute radiation sickness to slowly emerging cancers and birth defects. The safe disposal of nuclear wastes, primarily from decommissioned warheads and spent fuel from nuclear power generators, has been an increasing challenge. Schemes for mass underground storage of nuclear wastes, for example in salt domes, have been greatly delayed by uncertainties regarding the long-term viability of storage sites. As a result, the spent nuclear fuels from generating plants are often kept at the sites, which were not designed for long-term waste storage. The costs of the safe disposal of nuclear wastes can be enormous, and have been rarely fully appreciated in the calculations of the relative costs of nuclear energy compared to other energy sources.

COMMENTARY ON NUCLEAR WASTE

While not denying the vast array of incredible technical details one must master to understand problems of nuclear waste disposal, several elementary political considerations often seem to pass unnoticed in unproductive discussions about dealing with nuclear waste. Nowhere in the world where nuclear power or weapons exist have meaningful calculations and allotments been made to dispose of or clean up the waste. "Termination" in decision process terms has not been considered and, if it had been, serious questions would probably have been raised about selecting the nuclear option at all. Scientific judgments about the suitability of geological disposal of nuclear waste materials have misguided the political and popular discussion of what actually to do. As early as 1957 authoritative scientific sources were recommending deep geological disposal of spent nuclear fuel bundles. The need to do so for 10,000 years, more or less, seemed acceptable to the scientists, if done in suitable geological sites. What was completely missing in these calculations and the recommendations that followed was any consideration for the durability of the human institutions required to carry out and then monitor such disposal. No human institution has existed for thousands of years, to state the obvious. Furthermore, most human institutions are fallible, including those entrusted with responsibility for nuclear materials and processes. The THREE MILE ISLAND, CHERNOBYL, and multiple other nuclear disasters (both actual and too close for comfort) all now exist in the collective human consciousness never to be ignored or removed. The current debates about nuclear waste disposal around the world can often be traced back to unreasonable scientific assumptions about the durability of human institutions for the long haul and/or to human discomfort about scientific pretensions in general. Diminished trust and confidence in science and scientists are now a worldwide consequence.
Garry D. Brewer

Prior Informed Consent (PIC): a new procedure in international environmental policy meant to obtain and disseminate the decisions of importing countries as to whether they wish to receive shipments of hazardous chemicals and to ensure compliance to these decisions by exporting countries. It was introduced in 1989 in response to increasing concerns about shipments of chemicals that have been banned or severely restricted in the First World to developing countries and countries in transition (see GARBAGE IMPERIALISM). PIC helps participating countries learn more about the characteristics of potentially hazardous materials that may be shipped to them, initiates a decision-making process on the future import of these chemicals by the importing-country governments, and facilitates the dissemination of this decision to other countries. Managed jointly by the UN FOOD AND AGRICULTURE ORGANIZATION (FAO) and the UN ENVIRONMENT PROGRAMME (UNEP), the PIC procedure is intended to promote shared responsibility between exporting and importing countries in protecting human health and the environment. PIC provisions were included in the amendments of the 1985 International Code of Conduct on the Distribution and Use of Chemicals and the 1987 London Guidelines for the Exchange of Information on Chemicals in International Trade. Currently, 155 countries adhere to it. The CONVENTION ON THE PRIOR INFORMED CONSENT PROCEDURE FOR CERTAIN HAZARDOUS CHEMICALS AND PESTI-CIDES IN INTERNATIONAL TRADE, negotiated in 1996–98, entrenched the PIC requirement as a legally binding principle. The text of the convention was opened for signature in September 1998.

Superfund: the system established by the U.S. Comprehensive Environmental Response, Compensation and Liability Act of 1980 for cleaning up inactive HAZARDOUS WASTE disposal sites in the United States. Under that program, individual potentially responsible parties are held liable for the cost of cleaning up wastes at these sites. In many instances, the costs of thorough cleanup have been staggering. This program has proven controversial, however, and has been criticized for being too costly, too slow, and for imposing liability for cleanup costs in sometimes arbitrary ways. *D. S.*

6.11 Natural and Humanmade Disasters

Chernobyl Nuclear Disaster: the nuclear reactor accident that occurred on 26 April 1986, in Chernobyl, Ukraine (then part of the Soviet Union). Explosions at one of the reactors at the power plant site released 400 times more radioactive material into the atmosphere than the atomic bomb dropped on Hiroshima. The Chernobyl disaster is by far the greatest nonmilitary nuclear catastrophe, with over 350,000 people evacuated from their homes, a large area rendered unfit for agriculture, several thousand cases of thyroid cancer traceable to exposure to iodine isotopes, especially in Ukraine and Belarus, and higher levels of radioactivity experienced throughout the Northern Hemisphere. The long-term health effects on the 200,000 workers who were deployed to contain and clean up the wreckage are still unknown, although the incidence of quickly emerging problems was less than first feared. Ukraine, Belarus, and Russia were most heavily affected, but significant radioactivity traveled through the atmosphere to Scandinavia as well. The cross-boundary impact of the disaster was an important wake-up call for international concern. This led to several conventions (e.g., the 1987 CONVENTION ON ASSISTANCE IN THE CASE OF A NUCLEAR ACCIDENT OR RADIOLOGICAL EMERGENCY) and increased appreciation for the role of the INTERNATIONAL ATOMIC ENERGY AGENCY (IAEA) in pressing for better nuclear reactor design. The Chernobyl reactor design was not among those endorsed by the IAEA.

Convention on Assistance in the Case of a Nuclear Accident or Radiological Emergency: an international agreement adopted in 1986 in Vienna, in force since 26 February 1987. The main objectives of the Assistance Convention are to set out an international framework to provide prompt assistance in the event of a nuclear accident or radiological emergency. This assistance would be provided directly from states, through or from the INTERNATIONAL ATOMIC ENERGY AGENCY (IAEA), and from other international organizations. The convention was one of several initiatives following the CHERNOBYL NUCLEAR REACTOR DISASTER of 1986. By early 1999, 72 states and three intergovernmental organizations had become parties to the convention. The IAEA secretariat is in charge of administering the convention.

Convention on Nuclear Safety: an international agreement adopted in 1994 in Vienna, in force since 24 October 1996. The main goals of the agreement are to achieve and maintain a high level of nuclear safety worldwide through national measures and international cooperation. This involves safety-related technical cooperation and strengthening nuclear reactor safety systems, as well as mitigating the consequences of nuclear accidents. The core of the agreement is the obligation of parties to adhere to nuclear reactor safety standards according to benchmarks established by the INTERNATIONAL ATOMIC ENERGY AGENCY. By early 2000, the convention had fifty-three parties.

Convention on the Transboundary Effects of Industrial Accidents: an international agreement adopted in 1992 in Helsinki, awaiting sufficient ratifications to come into force, probably by 2001. A regional agreement covering Europe and North America, it requires parties to notify other members about any proposed or existing hazardous activity capable of causing transboundary effects, or any industrial accident that causes transboundary effects. The convention also calls for cooperation to prevent such industrial accidents and prepare for prompt responses should such accidents occur. Thus the convention involves research and development as well as exchange of information and technology regarding industrial accidents, and efforts to harmonize relevant national policies and practices. The convention also provides for mutual assistance in the event of industrial accident and for public participation in relevant decision-making processes. European members are particularly concerned about the potential for major industrial accidents in the transitional countries of Eastern Europe and the former Soviet Union; most of the activities within the convention's plan relate to the development and implementation of preventive measures in these countries.

Natural Disasters: events of natural origin that result in human casualties, property damage, disruptions of community life, or negative environmental impact. Forest fires, earthquakes, landslides, hurricanes, droughts, floods, tornadoes, volcanic eruptions, blizzards, and tsunamis are the principal forms of natural disasters. For most of human history, natural disasters were attributed to forces beyond human understanding, control, or intervention. Within the past few decades, however, natural disasters have been reconceived as having, at least in part, human as well as natural origins, whether in causing the disaster (e.g., droughts caused by DESERTIFICATION in turn brought about by DEFORESTATION) or in exposing humans to the damage of natural disasters (e.g., building cities in earthquake-prone areas). Thus certain natural disasters are now seen as the result of the complex interplay between physical agents and social setting that may be mitigated if not completely prevented. Current disaster planning strategies include four main steps: a) prevention or mitigation; b) preparedness; c) response; and d) recovery. Many countries have created special institutions to deal with these tasks (e.g., the U.S. Federal Emergency Management Agency and the Russian Ministry of Emergencies).

When environmental activists talk about natural disasters, they usually invoke them as harbingers of global warming. This rhetorical strategy is both a moral and a political failure. A moral failure because the devastating present-day impacts of natural disasters—felt most acutely in the developing world —has little, if anything, to do with anthropogenic global warming; and a political failure because natural disasters offer a potent, but largely neglected, organizing principle for international environmental action. Poverty, environmental degradation, and vulnerability to disasters are close partners. Upslope deforestation has been implicated in catastrophic landslides triggered in Central America by Hurricane Mitch in 1998. Migration of populations to environmentally fragile coastal zones will increase human and environmental vulnerability to storms and floods regardless of future changes in global climate. Natural disasters should be central to policy debate on sustainability and global environmental justice, not ammunition in the fight to advance the global warming agenda. *Daniel Sarewitz*

Office of the United Nations Disaster Relief Coordinator (UNDRO): a coordinating mechanism for the UN system, established in 1972 as the result of a General Assembly resolution. UNDRO performs three primary functions: a) to mobilize and coordinate the relief assistance of the various organizations and specialized agencies of the UN system, or governments, intergovernmental and NONGOVERNMENTAL ORGANIZATIONS, and of national voluntary agencies, in response to requests from countries stricken by disasters; b) to study and promote the prevention and preparedness aspects of disasters, including the prediction of natural disasters and their control; and c) to serve as the focal point of the UN system for the dissemination of disaster-related information.

To serve these functions, UNDRO gathers and disseminates background information on disaster-prone countries, technological developments in the areas of disaster relief, preparedness, and prevention, and operational information on current disaster emergency situations.

Three Mile Island Nuclear Disaster: the meltdown of a nuclear reactor core at the Three Mile Island nuclear facility near Harrisburg, Pennsylvania, in March 1979, resulting in the release of radioactive materials into the air and the Susquehanna River, and moderate but not fatal exposure of the personnel at the facility. Mass evacuations from the Harrisburg area were accompanied by tremendous apprehension of a major disaster. The accident resulted from a series of design flaws, gauge malfunctions, and human error. The unit suffering the damage was permanently decommissioned and shielded to minimize further radiation leakage; other units resumed operation after five years. Although not nearly as damaging to human life as the later CHERNOBYL DISASTER, the Three Mile Island accident was an enormous blow to the future of the nuclear power industry in the United States. The credibility of the industry's claims of reactor safety, based on multiple fail-safe systems, was severely undermined. One lesson drawn from the Three Mile Island disaster was that the complexity of safety systems, in making the operation of the reactor even more challenging, poses its own risks.

United Nations International Decade on Natural Disaster Reduction: a United Nations designation for the 1990s to encourage all countries to improve their planning for natural disasters. In particular, the UN emphasized the importance of disaster planning strategies, which include a wide range of measures, from stronger land regulations to better effectiveness of emergency response organizations. The goal of the decade was to reduce or at least stabilize human casualties, property losses, and social disruptions associated with natural disasters.

6.12 Ocean's Resource Use and Degradation

6.12.1 JURISDICTIONAL ISSUES

6.12.1.1 Boundary Concepts and Issues

Contiguous Zone: a maritime zone seaward from the TERRITORIAL SEA but not exceeding twenty-four nautical miles, where coastal states are permitted to act in the interests of national security or to enforce national laws pertaining to such matters as customs, health and sanitation, immigration, and environmental protection. Only a few coastal nations have declared contiguous zones.

Continental Shelf: the seabed and subsoil of a continent's landmass extending into the ocean, thus typically extending beyond the limits of its territorial sea. The concept as an element of international law, based on its geological meaning, was established by the 1958 Convention on Continental Shelf and the 1982 UN CONVENTION ON THE LAW OF THE SEA (UNCLOS). According to the UNCLOS definition, the outer limit of the continental shelf is where the continental landmass drops down to the deep ocean floor. However, if a state's physical continental shelf does not extend to 200 nautical miles, the Law of the Sea Convention allows the coastal state to claim that breadth for continental shelf jurisdiction. When the dropoff occurs beyond 200 nautical miles, a coastal state can claim a continental shelf of up to 350 nautical miles in accordance with submarine topography. In this case, the state has to inform the Commission on the Limits of Continental Shelf (CLCS), which provides recommendations on setting the outer limits of the shelf. In accordance with international law, the coastal state has sovereign rights for the purpose of exploring and exploiting the natural resources of its continental shelf.

Creeping Jurisdiction: an informal label for the gradual process of unilateral expansions of exclusive jurisdiction of a coastal state seaward from international boundaries. Among the most recent examples are the Chilean concept of a PRESENTIAL SEA and Canadian attempts to manage stocks beyond its 200-mile zone in the North Atlantic.

Enclosed and Semi-Enclosed Seas: in the context of the 1982 UN CONVENTION ON THE LAW OF THE SEA (UNCLOS), a gulf, basin, or sea surrounded by two or more states and connected to another sea or the ocean by a narrow outlet or consisting entirely or primarily of the territorial seas and EXCLUSIVE ECONOMIC ZONES of two or more coastal states. Most enclosed and semi-enclosed seas exemplify so-called large marine ecosystems; they are also most affected by overexploitation, pollution, and habitat destruction and therefore are in need of uniform environmental management. The Law of the Sea Convention obligates states bordering an enclosed or semi-enclosed sea to cooperate with each other (directly or through an appropriate regional organization) in the exercise of their rights and in the performance of their duties. In some regions, such as parts of the Mediterranean, the environmental cooperation of bordering states has been very effective.

Exclusive Economic Zone (EEZ): the area beyond the twelve-mile territorial waters of coastal states out to 200 nautical miles in which the adjacent coastal state may exercise exclusive rights to explore, exploit, conserve, and manage the living and nonliving natural resources of the waters, the seabed, and the underlying subsoil. All other states retain certain high-sea freedoms such as navigation, overflight, and the laying of pipelines in the EEZs. Coastal states are expected to conserve living resources in the EEZs and promote their optimum exploitation, which may involve granting access to foreign fishing vessels if there is a surplus of fish beyond what the coastal state can harvest. Priority access in EEZs is to be given to neighboring landlocked countries and to countries with short coastlines. EEZs are significant because most of the productive ocean fisheries and exploitable offshore oil and natural gas reserves are located within them, and thus these resources come under the control of coastal states.

See Bernard H. Oxman, David O. Caron, and Charles L. O. Buderi, eds., *The Law of the Sea: U.S. Policy Dilemma* (San Francisco: ICS Press, 1983). *M. S. S.*

Freedom of the Seas Principle: a guiding principle of the HIGH SEAS regime. It provides that the high seas, or any part of them, may not be claimed by any state, but are available to all members of the international community for activities such as navigation, fishing, laying of submarine cables and pipelines, and overflight, if undertaken with reasonable regard to the interests of other states. The principle of the freedom of the seas came into being when the ocean's resources were thought to be inexhaustible and exclusive claims on them did not make economic sense. Since the 1950s, with a dramatic increase in the use of the oceans, the principle has been under attack. One of the most radical proposals was the introduction of COMMON HERITAGE OF MANKIND status for the ocean's seabed.

Geographically Disadvantaged States (GDS): in the context of the 1982 UN CONVENTION ON THE LAW OF THE SEA (UNCLOS), coastal or landlocked states whose geographical position makes them dependent upon the exploitation of the living resources of the EXCLUSIVE ECONOMIC ZONES (EEZS) of other states in the region or subregion, for adequate supplies of fish to meet the nutritional needs of their populations. According to the UNCLOS, geographically disadvantaged and landlocked states have the right to participate on an equitable basis in the exploitation of the surplus living resources in the EEZs of other states in their region or subregion. The limitation is that a coastal state has the discretionary power to decide whether there is a surplus and who gets access to it.

High Seas (International Waters): areas of the oceans lying beyond the jurisdiction of coastal states and not included in the EXCLUSIVE ECONOMIC ZONES (EEZS), TERRITORIAL SEAS, internal waters, or archipelagic waters (i.e., waters of island chains). The fundamental principle governing the high seas is FREEDOM OF THE HIGH SEAS, which includes freedom of navigation, overflight, fishing, and scientific research, as well as freedom to lay submarine cables and pipelines and to construct artificial islands and other installations permitted under international law. The high seas are reserved for peaceful purposes and no state may lay claim to any part of it.

Maritime Boundaries: the limits of the maritime zones established off the shores of coastal states for administration and resource management. The rules governing these zones have been established by the 1958 and 1982 UN CONVENTION ON THE LAW OF THE SEA (UNCLOS). Maritime boundaries separate jurisdictions of neighboring states as well as different zones, such as territorial seas, contiguous zones, and EXCLUSIVE ECONOMIC ZONES (EEZS) within one state's jurisdiction.

Presential Sea, Concept of: a geographically defined maritime area of the HIGH SEAS adjacent to an exclusive economic zone where the coastal state's interests could be directly involved. The concept was adopted by the government of Chile when high-seas fisheries in the southeast Pacific threatened the productivity of its EXCLUSIVE ECONOMIC ZONE (EEZ). In accordance with the 1991 Chilean Fisheries Law, the government of Chile could undertake in its presential sea activities in relation to conservation, pollution control, security of navigation, etc. It has often been presented as an example of CREEPING JURISDICTION, though Chilean legal experts defend it as a legitimate development of the international law of STRADDLING STOCKS and HIGHLY MIGRATORY SPECIES.

See Thomas Clingan, "Mar Presencial (The Presential Sea): Deja-vu All Over Again?" *Ocean Development and International Law* 24 (1993): 93–97; Francisco Orrego Vicuna, *The Changing International Law of High Seas Fisheries* (New York: Cambridge University Press, 1999).

Territorial Sea: a zone within which a coastal state can exercise most prerogatives of sovereignty, including control over the waters, the ocean bed, subsoil, and air space. The limitation is that the coastal state cannot limit the rights of vessels from other countries to navigate the zone under the doctrine of "innocent passage." The breadth of the territorial sea, according to the 1982 UN CONVENTION ON THE LAW OF THE SEA (UNCLOS) should not exceed twelve nautical miles.

6.12.1.2 Maritime Conventions, Programs, and Implementing Institutions

Agreement Relating to the Implementation of Part XI of the 1982 United Nations Convention on the Law of the Sea: an international agreement regarding the deep seabed mining provisions of the 1982 UN CONVENTION ON THE LAW OF THE SEA (UNCLOS). To try to resolve the highly contentious issue of who has rights to mine the deep seabed, the agreement was adopted in 1994 by the UN General Assembly and came into force on 28 July 1996. By the end of 1999, the agreement had 132 parties. The agreement identifies various issues that have impeded the universal acceptance of UNCLOS. These include how costs are distributed among parties, institutional arrangements, the decision-making mechanisms, transfer of technology, production policy, and the financial terms of contracts for deep seabed mining. One of the most important provisions is that the agreement and the deep seabed mining provisions of UNCLOS shall be interpreted and applied together as a single instrument; however, in the event of an inconsistency between the two, the provisions of the agreement shall prevail. This provides a mechanism to bring into UNCLOS the countries previously dissatisfied with the deep seabed mining provisions of the convention. This complicated arrangement illustrates both the potential economic importance of deep seabed mining and the difficulties of finding universally acceptable arrangements for this issue's resolution.

Antarctic Treaty: the foundation agreement for the current international regime governing activities on the Antarctic continent and surrounding oceans south of 60 degrees south latitude. Concluded in 1959, the treaty provides that the area will be used exclusively for peaceful purposes. All military activities are banned, as are the testing of nuclear weapons and the disposal of radioactive waste. All states may conduct scientific research anywhere in the area provided that it is done in a nonsecretive way, with all installations and findings open to inspection by any of the parties. The treaty sidestepped the politically sensitive issue of the status of previous national claims to sectors of Antarctica (e.g., by Argentina and Chile) by neither legitimizing nor invalidating them. Finally, it established the conditions for states to become members of the Antarctic Treaty Consultative Parties, which meets periodically to consider additional agreements pertaining to activities in the region, such as harvesting living marine resources and exploitation of minerals. See also CONVENTION ON THE REGULATION OF ANTARCTIC MINERAL RESOURCE ACTIVITIES (CRAMRA).

See M. J. Peterson, *Managing the Frozen South: the Creation and Evolution of the Antarctic Treaty System* (Berkeley: University of California Press, 1988). *M. S. S.*

Commission on the Limits of the Continental Shelf (CLCS): an international mechanism created in 1997 in accordance with the 1982 UN CONVENTION ON THE LAW OF THE SEA (UNCLOS). Its function is to make recommendations to coastal states on matters related to the establishment of the limits of their CONTINENTAL SHELF in cases where outer limits extend beyond 200 nautical miles. The CLCS is composed of twenty-one members who are experts in the field of geology, geophysics, or hydrography, elected by parties to this convention from among their nationals, reflecting the need to ensure equitable geographical representation. The CLCS considers the data and other material submitted by coastal states concerning the outer limits of the continental shelf, makes recommendations, and provides scientific and technical advice.

International Tribunal for the Law of the Sea: an international dispute-resolution forum established in Hamburg in 1996 in accordance with the 1982 UN CONVENTION ON THE LAW OF THE SEA (UNCLOS). It also has jurisdiction to provide advisory opinions at the request of the Assembly or the Council of the International Sea-Bed Authority on legal questions arising within the scope of their activities. It is composed of twenty-one independent members elected by parties to the convention from among persons with recognized competence in the field of the law of the sea and representing the principal legal systems of the world.

Law of the Sea: the part of public international law that governs the rights and duties of nation-states and other subjects of international law regarding the use of ocean space and resources. See UN CONVENTION ON THE LAW OF THE SEA.

Regional Seas Programme: a set of regional programs within the framework of the UN ENVIRONMENT PROGRAMME (UNEP). It was established in 1974 with the premise that countries of each region share a common interest and have a mutual need for sustainable development of marine resources. Because coastal waters, estuaries, and ENCLOSED AND SEMI-ENCLOSED SEAS suffer the earliest and most severe degradation from human activity, a starting point for combating marine pollution is to link coastal nations together in a common commitment to mitigate and prevent degradation. Each regional program is shaped to address the specific needs of the targeted marine region as perceived by coastal nations. However, the institutional structure of regional sea programs is similar: a) an action plan for cooperation on management, protection, rehabilitation, development, monitoring, and research of coastal and marine resources; b) an intergovernmental FRAMEWORK CONVENTION; and c) a series of detailed PROTOCOLS on protection against particular environmental hazards (e.g., pollution by dumping, oil spills, land-based pollution).

Thus far, the program has resulted in the formation of nine regional ocean protection regimes, covering the Black Sea, the wider Caribbean, the East African Seaboard, the Persian Gulf, the Mediterranean, the Red Sea and the Gulf of Aden, the South Pacific, the South-East Pacific, and the Atlantic coast of West and Central Africa.

United Nations Convention on the Law of the Sea (UNCLOS): the international convention establishing a legal regime for practically all uses of the oceans and their resources. It is the most comprehensive intergovernmental agreement relating to global commons. With 320 articles and 9 annexes, it is probably the most complex environmental agreement ever achieved, and took over ten years to negotiate. A second agreement, modifying the provisions of the convention that address the exploitation of the mineral resources of the deep seabed, was opened for signature in mid-1994. The treaty was signed in 1982 by 119 states and came into effect in November 1994. By the end of 1999, it had 132 parties, including the European Union, but the United States, Canada, and several other developed nations had not yet ratified the convention. Their major objection was the treatment of seabed mining, for which the convention called for the sharing of proceeds. Yet with such comprehensive coverage and membership, the convention has provided an unprecedented degree of order in maritime activities, which previously had been marked by very high levels of conflict over national jurisdictions and rights to exploit living and mineral resources. UNCLOS has two affiliated agreements: the AGREEMENT RELATING TO THE IMPLEMENTATION OF PART XI OF THE UN CONVENTION ON THE LAW OF THE SEA and the FISH STOCKS AGREEMENT.

United Nations Conference on the Law of the Sea III (UNCLOS III): the third and most recent meeting in a series of United Nations efforts to codify the rules by which nations utilize oceans and seas. UNCLOS III took place from 1974 to 1982 and after years of difficult negotiations was able to reach consensus on nearly all points except part XI, relating to the exploitation of deep seabed minerals. Under the provisions of part XI, major developed countries would be required to share their marine technology with other nations, and financial benefits from the mining in the area would be shared with the United Nations supervisory agency (the "Enterprise"), which in addition would acquire functions of control over the activities in the area. The proposed regime exemplified the concept of the COMMON HERITAGE OF MANKIND and was eventually rejected by the United States, the United Kingdom, France, Germany, and Japan. Though at the final session of the UNCLOS III (December 1982), representatives of 117 countries signed the convention, opposition by the most advanced maritime states brought the regime to a deadlock for the next twelve years, until a breakthrough occurred with the AGREEMENT RELATING TO THE IMPLEMENTATION OF PART XI OF THE 1982 UN CONVENTION ON THE LAW OF THE SEA.

6.12.2 COASTAL PROCESSES AND MANAGEMENT

Coastal Zone: the area where sea and land ecological processes interact with human influences and controls. Exact definitions of coastal zone boundaries vary, depending on biogeographical conditions, the pattern of uses, and the legal specifics of particular countries. In some countries, the coastal zone is defined broadly as an area between the top of the watershed (as the landward boundary) and the edge of the exclusive economic zone (as the seaward boundary). In other cases, it encompasses smaller areas on both the land and sea sides. Most often, the coastal zone covers three administrative areas: the offshore waters, beyond low tide and within national jurisdiction; the coastal margin between low and high water tides and including estuaries; and the littoral landward zone, including headlands, beaches, and coastal settlements. These areas are usually administered by different agencies reporting to a variety of departments with potentially conflicting objectives and styles of management. Coastal zones typically contain unique habitats and ecosystems that provide different resources and environmental services to coastal communities; serve as important sources of economic development for these communities and are often a preferred site for urbanization; and are characterized by an ever-increasing competition for space, resources, and environmental services. The combination of extreme environmental importance and fragility, high human impact, and complicated administrative structure has made it extremely challenging for coastal zones to enjoy SUSTAINABLE DEVELOPMENT.

Coastal Zone Management (CZM): the management of ecosystems spanning the entire COASTAL ZONE. The term was initially introduced in the 1972 U.S. Coastal Management Act in reference to the management of all uses in the coastal zone. However, in practice most CZM programs have been primarily focused on the management of shore land resources and much less on water-related issues. Also, the prevalent type of coastal management until recently was sectoral, when different uses of the coastal zone (such as fisheries, oil and gas development, coastal tourism, port development) were managed separately by relevant agencies. See also INTEGRATED COASTAL MANAGEMENT.

Integrated Coastal Management (ICM; **Integrated Marine and Coastal Area Management** [MCAM]; **Integrated Coastal Zone Management** [ICZM]): the management and jurisdictional arrangements that place resource, economic, and environmental management of an entire coastal area within the same planning and regulatory framework. As a means of addressing the specific character of a coastal zone and the complexity of human interests involved, ICM is based on the premise that the pattern of coastal and ocean resources and space should correspond to the natural interconnectedness of these realms. It is designed to harmonize goals and policies among different levels of government and among different coastal and marine sectors. Typically, the objectives of integrated coastal management are: a) encouragement of natural processes of coastal defense and protection; b) adequate legislative protection of natural zones essential to this purpose, such as headlands, dunes, salt marshes, and wetlands, their clearance from settlement and monitoring; c) encouragement of the retention of natural beach through a particular design of coastal defense works; and d) addressing the vulnerability of the area to sea-level rise and increased storminess.

See Keith M. Clayton and Timothy O'Riordan, "Coastal Processes and Management," in *Environmental Science for Environmental Management,* ed. Timothy O'Riordan (London: Longman, 1995).

COMMENTARY ON INTEGRATED COASTAL MANAGEMENT

Good performance in integrated coastal management may be assessed by three criteria: optimization of multiple objectives, maintenance of life-support systems, and responsive management. These objectives and criteria are not easily met, even in cases of ideal administration. In practice, the interplay of different political interests involved in coastal development makes their simultaneous and effective implementation close to infeasible. *Natalia Mirovitskaya*

Land Reclamation: in the context of coastal management, coastal construction activity aimed at gaining land from the sea for agriculture, urban, or industrial development. This can be done by filling in wetlands or semi-enclosed bays, constructing dikes, or building dams and other barriers to exclude coastal waters. Opposition to land reclamation often centers on the concerns for the elimination of wetland habitats and for the vulnerability of settlements established in low-lying areas, which frequently remain at risk of flooding, beach erosion, and storms.

Marine Protected Areas: areas of coastal land or waters that are designated to protect coastal and marine resources, preserve biological diversity, increase public awareness, and provide sites for recreation, research, and monitoring. Marine protected areas often face greater challenges than many types of strictly terrestrial protected areas, because of the difficulties of monitoring and enforcing limitations on extractive activities such as overfishing, waste dumping, and damaging recreational diving.

Police Power: in the context of coastal management, the inherent power of a governmental agency (federal, provincial, or local) to control certain activities (if they are reasonably related to promotion and maintenance of the health, safety, morals, and general welfare of the public) by setting conditions and constraints on the way in which property or resource owners use their assets. Environmental and resource management in most countries relies on this legal principle and it has been advocated as the basis for INTEGRATED COASTAL MANAGEMENT. However, in the United States the use of police power in environmental and resource management has been subject to harsh criticism on behalf of property owners whose property has been taken away by the government without due compensation.

Public Trust Doctrine: a legal doctrine granting the nation or subnational jurisdiction ownership of lands lying under navigable waters, to hold such lands in trust for the benefit of the people of that jurisdiction. These submerged lands may not be sold or otherwise alienated by the country, except in a manner that promotes the public interest. The doctrine presumes that managing resources as a commons should be preferred over privatizing such resources and, if private developments are allowed, the public should receive financial benefits from such developments. The public trust doctrine is prevalent in countries with a common-law tradition. Some scholars argue that in such countries the public trust doctrine should serve as the legal basis for coastal zone management.

See Jack H. Archer et al., *The Public Trust Doctrine and the Management of America's Coasts* (Amherst: University of Massachusetts Press, 1994).

Sectoral Management: a coastal management approach that treats different uses of the COASTAL ZONE (e.g., fishing, coastal navigation, oil and gas development, port development) separately, by different agencies and governments. As recognition of the interactions among ocean and land activities has grown, sectoral management has come under increasing criticism. Nevertheless, in practice most coastal areas are still subject to sectoral management.

Special Area Management: environmental and resource management that designates an entire area in the COASTAL ZONE as a unit and manages that unit regardless of government jurisdictions. This kind of management is usually recommended for bays, estuaries, and other coastal ecosystems that are geographically integral and are best managed as a whole. Plans for special area management typically are integrated into the management program of the larger coastal area of which they are a part.

6.12.3 FISHERIES

6.12.3.1 Trends, Practices, and Problems

Anadromous Species: fish species that spawn in the freshwater streams of coastal states, but spend most or part of their life span in the ocean. They include salmon, shad, steelhead trout, some herring, sturgeons, and smelt. According to the 1982 UN CONVENTION ON THE LAW OF THE SEA (UNCLOS), states in whose rivers these stocks originate should have the primary interest in and responsibility for them. Fisheries for anadromous stocks should be conducted only within EXCLUSIVE ECONOMIC ZONES (EEZS). In cases where anadromous stocks migrate into or through the waters landward of the outer limits of the exclusive economic zone of a state other than the state of origin, UNCLOS requires this state to cooperate with the state of origin in the conservation and management of such stocks, preferably through regional organizations. One of the serious potential problems of the international management of anadromous species is related to the practice of salmon "farming"—artificial enhancement of stocks by several nations in one region (such as the northern Pacific, for instance) may exceed the capacity of the habitat, which would in turn detrimentally affect native stocks and thereby cause international conflict.

Aquaculture: intensive farming and husbandry of aquatic species (such as fish, mollusks, shrimps, and algae). Global production from aquaculture is currently estimated at 15 million tons per year (compared to a marine catch of 80–85 million tons) and has been rising continually as a source of food, a means of economic diversification, and a tool in fishery management. From a legal perspective, coastal states have exclusive right to practice, permit, and regulate aquaculture (including the right to ban navigation at the sites of marine plantations) within their internal waters and territorial sea. In accordance with the principle of FREEDOM OF THE SEAS, any state can practice aquaculture in the HIGH SEAS as long as the rights of other states and norms of international law are not violated. Currently, aquaculture is limited to fresh water and (to a much lesser extent) shallow coastal seawater.

COMMENTARY ON AQUACULTURE

The possibility is slim that in the near future the oceans' fish resources will be largely "farmed" rather than hunted. However, the environmental consequences of aquaculture are already evident in a number of regions, where the composition of fish stocks has been changed. By analogy with the GREEN REVOLUTION, one may expect that the introduction of species may have high adverse impacts, changing the receiving ecosystems through interbreeding, predation, competition for food and habitats, and genetic pollution of indigenous stocks. Therefore, carefully drafted international rules for aquaculture are necessary. *Natalia Mirovitskaya*

Catadromous Species: fish and related species that are born in the oceans, spend most or part of their life span in fresh water, but eventually return to the open seas to spawn and die. Eels are a typical example of catadromous species. According to the 1982 UN CONVENTION ON THE LAW OF THE SEA (UNCLOS), responsibility for the management of these species and their ingress and egress during migration belongs to the coastal state in whose waters catadromous species spend the greater part of their life cycle. These species can be harvested only in waters landward of the outer limits of EXCLUSIVE ECONOMIC ZONES (EEZS).

Driftnets: fishing gear that is made of a single or several rectangular panels of net webbing linked together and suspended vertically in the water by floats on the top of the panel and sinkers at the bottom. They drift with the winds and currents, creating a webbing curtain in which the fish are enmeshed. Modern industrial fleets maintain driftnets of 5–50 km in length. Driftnets have increased yields dramatically in many places. They also make considerable impact both on targeted and nontargeted species, or bycatch, including marine mammals and birds. In the international arena, the issue of driftnet fishing was first raised in a meeting of the South Pacific Forum (Kiribati, 1989) when the member states expressed grave concern over the damage inflicted to the economy and environment of the South Pacific region by this technology, usually used by LONG-DISTANCE FISHING fleets. Following this meeting, the members of the forum signed the so-called Wellington Convention, which effectively restricted driftnet fishing in the South Pacific. The Organization of Eastern Caribbean Seas and EUROPEAN COMMUNITY immediately followed by restricting driftnet fishing in the waters under their control. Beginning on 31 December 1992, the UN General Assembly imposed a global moratorium on all large-scale pelagic driftnet fishing on the HIGH SEAS, including ENCLOSED AND SEMI-ENCLOSED SEAS. The moratorium has been extended on an annual basis since then.

Exotic Species Introduction: the introduction of non-native fish to enhance fish food yields or the attractiveness of sport fishing. The ecological impact of the introduction of exotic fish species is difficult to anticipate; in some circumstances, such as the introduction of sport fish in Africa's Lake Victoria, it has resulted in the dramatic decline of native species and the reduction of opportunities for traditional fishers to pursue their livelihoods. The introduction of exotic species sometimes occurs inadvertently when bodies of water are connected by canals.

Fish Aggregation Devices (FADS): platforms of bamboo, logs, or other materials that are anchored or placed in the water to attract fish, particularly tuna, making them easier to harvest. FADS are widely used in the southern and western Pacific and are the subject of intense legal and environmental debate. Their opponents argue that such devices divert fish stocks, though no serious transnational effects have been documented thus far.

Fisheries Population Crash: the abrupt decline of fish populations, often caused by OVERFISHING or changes in environmental conditions. Given the great difficulties of gauging fish populations, and the difficulty of regulating fish catches in both domestic and international waters, overfishing periodically brings fish populations below the levels of sustained regeneration. Several fish species that were mainstays of the fishing industry (such as sole) have disappeared for all practical purposes. Macroclimatic changes, such as El Niño, can lead to temporary crashes of fish populations, as occurred with the anchovetas off the Pacific coast of South America in the 1960s.

COMMENTARIES ON FISHERIES POPULATION CRASH

One of the true environmental tragedies of the late twentieth century is the appalling degradation of the world's fisheries, one of the truly renewable and sustainable resources on the planet. The tragedy is mainly the product of greed and the stupidity of both those entrusted with stewardship of these resources and those who labor to harvest and profit from them. Alarms about the precarious status of selected fisheries around the world began sounding as early as the 1960s with the advent of highly efficient trawler and refrigerator operations in the North Sea, on Georges Bank, and elsewhere in the North Atlantic. The substitution of technology for manpower allowed order of magnitude improvements in catch per unit of fishing effort, however measured. These capital-driven improvements diffused around the world and the technology itself continued to improve relentlessly, e.g., navigation and communication aids, weather reporting satellites, improved on-shore processing, and marketing. Warnings about the increasing number of endangered fisheries mounted during the 1970s and early 1980s in a growing number of studies, conferences, books, and the like, although decisions to curtail or otherwise change matters to lessen fishing pressure either were not made or were woefully inadequate to the task. Currently, some of the world's greatest fishing grounds are by all accounts in nothing less than a shambles. The Georges Bank, shared by Canada and the United States, is in such a serious state of depletion that wholesale job loss and related economic displacements are now the norm in most of the maritime provinces of Canada. Similar hardship in New England seaports exists, although the general economic devastation is moderated by the more robust U.S. economy, as compared to sparse Newfoundland, for instance. The root problems of greed and stupidity are never to be underestimated in coming to understand what happened, why, and who is to blame. *Garry D. Brewer*

The Economist stated in 1998 that approximately 60 percent of the world's fish stocks were being harvested near or at unsustainable levels. Throughout the world's oceans, too many fishers use too much equipment to chase too few fish. Each fisher has an individual incentive to increase the catch, while the fisheries are limited. OVERFISHING, habitat degradation, and bycatch have severely depleted many U.S. fisheries. Like many coastal countries, the United States became concerned about the status of its fisheries in the 1970s. The 1972 Magnuson-Stevens Fisheries Conservation and Management Act (FCMA) protected U.S. fisheries from other countries by establishing authority over the 200 nautical-mile EXCLUSIVE ECONOMIC ZONE (EEZ) and the fisheries therein. The FMCA also articulated standards for the management of U.S. fisheries and the process for maintaining them. Since that time, foreign fishing in American waters has been eliminated, however U.S. fisheries have continued to suffer. For instance, of the 157 known fishery stocks managed by the National Marine Fisheries Service (NMFS), 73 or 46 percent are "overutilized," thereby compromising their ability to contribute to the maximum long-term potential yield. In 1995 the maximum long-term potential yield from U.S. fisheries resources was estimated at 8.14 million metric tons, however the recent average yield was only 5.06 million metric tons, a 38 percent shortfall. Until American fish stocks recover to a more sustainable ecological state, the fishing industry will continue to lose $3–8 billion dollars a year, as estimated by the Center for Marine Conservation.

See *The Economist*, "Going Deep," May 23, 1998. *Toddi Steelman*

Highly Migratory Species: fish species that can migrate long distances across the oceans, in and out of EXCLUSIVE ECONOMIC ZONES (EEZS). The 1982 UN CONVENTION ON THE LAW OF THE SEA (UNCLOS), which listed such species in its annex, included nine tuna and twelve bullfish species, two tuna-like species, four species each of sauries, pomfrets, dolphin fish, oceanic sharks, and cetaceans. Some of these species, particularly tuna, are of great economic importance. According to UNCLOS, the coastal state and other states whose nationals fish for the highly migratory species shall cooperate directly or through appropriate international organizations to ensure conservation and to promote the objective of optimum utilization of such species throughout the region, both within and beyond the exclusive economic zone. In case no appropriate international organization exists, they shall cooperate to establish such an organization and participate in its work.

Long-Distance Fishing: a type of fishing performed far away from the fishers' home base. Such operations usually require the use of refrigerated factory trawlers or "mother ships" that allow fleets to travel vast distances from the home country and to stay at sea for longer periods without having to return to shore. Also, long-distance fleets often use nonselective fishing equipment, which results in high volume of "bycatch" (undersized target species, nontarget species, and other marine life such as mollusks, jellyfish, turtles, and porpoises). These practices are a point of conflict between long-distance fishing nations on one side and local fishermen and environmentalists on the other. A major part of the world's long-distance fleet belongs to Russia, China, South Korea, Japan, Spain, Poland, and Taiwan.

COMMENTARY ON LONG-DISTANCE FISHERIES

The Canadian Broadcasting Corporation produced an excellent documentary on "pockets of hope," examples of people working to change government policies to preserve the oceans and the livelihood of the communities that depend on them. Included in the program is a fascinating glimpse at how the fisherwomen of southern India successfully mobilized the fishworkers of India into a powerful political movement that challenged the national government, the IMF, and the WORLD BANK, and expelled the foreign factory fleet from a twenty-two-kilometer zone in the Indian Ocean (a singular example in this century of women revolutionizing a fishery and exerting their considerable influence). This story is compared to Canada's Bay of Fundy, after the collapse of the Atlantic cod fishery, where coastal communities rejected INDIVIDUAL TRANSFERABLE QUOTAS (ITQS) and won the right to manage their fisheries on an integrated, sustainable basis.

See Canadian Broadcasting Corporation, "*The Nature of Things*—Fisheries: Beyond the Crisis," Show 10, 1997–1998. *Susanne Swibold and Helen Corbett*

Open-Access Fisheries: fishing areas where no authority has the legal or effective means to enforce restrictions on fishing. The open seas beyond a nation's territorial jurisdiction are particularly prone to this problem, but there are often weak fishing regulations and unclear property rights in national rivers, lakes, and territorial waters as well. Open-access fisheries are therefore vulnerable to OVERFISHING, sometimes leading to a FISHERIES POPULATION CRASH. Once the population crash has occurred, it is often easier to impose a ban on fishing (a FISHING MORATORIUM) followed by a more stringent regulatory regime to prevent that pattern from reoccurring. See also OPEN-ACCESS RESOURCES.

Overcapitalization: one of the most serious problems faced by the global fisheries industry, which can be defined simply as "too many vessels chasing too few fish." According to the UN FOOD AND AGRICULTURE ORGANIZATION (FAO), the size of the world's fishing fleet increased at twice the rate of the increase in the global marine catch between 1970 and 1990. Overinvestment in previous years, when fishing stocks were believed to be inexhaustible, produced an explosion in the number of fishing vessels, many of which are now aging, economically inefficient, and operate only with the support of government subsidies. This undermines the sustainability of fisheries and the viability of the fishing industry itself.

Overfishing: fish harvesting at rates that exceed the MAXIMUM SUSTAINABLE YIELD (MSY), which leads to a net decline in the fish populations. According to the UN FOOD AND AGRICULTURE ORGANIZATION (FAO), currently almost 70 percent of all fish stocks are either fully to heavily exploited (44 percent), overexploited (16 percent), depleted (6 percent), or very slowly recovering from overfishing (3 percent); in a third of the world's major fishing regions, the annual catch is down 20 percent or more from peak years. Explanations of overfishing are manifold: difficulties in reaching international agreements, reluctance of the fishing industry suffering from overinvestment, and excess fleet capacity to limit catches on a voluntary basis, use of nonspecific technology, etc. Overfishing can lead to a FISHERIES POPULATION CRASH, and the economic dislocations that this crash may trigger. However, there may be instances in which overfishing and even the commercial disappearance of a particular fish species may be economically desirable, if the populations of other fish take the place of the depleted species, or if the immediate income gain of overfishing is particularly valuable. The issue of overfishing in the high seas was addressed by the 1995 FISH STOCKS AGREEMENT.

COMMENTARY ON OVERFISHING

Human greed and shortsightedness are the root causes of overfishing. These are rooted in the human fear of scarcity that fuels fantasies of wealth and power. The population crashes of Pacific sardines, Peruvian anchovetas, and North Atlantic cod have all underscored the fallacy of depending on maximum sustainable yield in fisheries management. There are too many variables to accurately predict fish populations: we cannot physically see the resource, we do not consider interactions among species, and we can not predict ecological conditions that contribute to population fluctuations. Quotas are set on peak population levels. The industry responds by bringing more boats and capital into the fishery. When populations decline, as in the Atlantic cod, governments are pressured to subsidize the fishery and maintain quota levels. In the long term, the fishery is overharvested and the industry moves on to another species to exploit. The collapse of the cod fishery put 20,000 people out of work in Canada. The government spent enormous amounts of money on job retraining, which, they admit, failed. The Bering Sea pollock fishery is currently following the same destructive course.

See D. Ludwig, R. Hilborn, and C. Walters, "Uncertainty, Resource Exploitation, and Conservation: Lessons from History," *Science* 260 (2) (April, 1993).
Susanne Swibold and Helen Corbett

Straddling Stocks: stocks of fish species that occur both within the EXCLUSIVE ECONOMIC ZONE (EEZ) of a coastal state and in an area of the high seas adjacent to the zone. Most of the species occurring in the high seas are straddling stocks. Therefore depleting the stock on the high seas will have a negative impact on the stock within the EEZ. The regulation of fishing of straddling stocks became an issue of particular importance to the international fisheries community, with growing reports of violence between fishing vessels from coastal and LONG-DISTANCE FISHING states. Several countries, including Britain and Norway, sent naval ships to protect fishing fleets on the high seas, while some coastal states fired on foreign fleets. The coastal states most concerned about the impact of high-seas fishing on their domestic harvest include Argentina, Australia, Canada, Chile, Iceland, and New Zealand. Six countries are responsible for 90 percent of long-distance fishing: Russia, Japan, Spain, Poland, the Republic of Korea, and Taiwan. Recently, the United States and China substantially expanded their high-seas operations as well. A regulatory regime for the exploitation of straddling stocks was finally defined by the UN CONFERENCE ON STRADDLING FISH STOCKS AND HIGHLY MIGRATORY FISH STOCKS.

COMMENTARY ON HIGHLY MIGRATORY SPECIES OR STRADDLING STOCKS

Once upon a time there was a fish, the walleye pollock (*Theragra chalcogramma*) that humans didn't fancy much because it tasted so bland. Huge numbers of these fish spawned, grew, and matured across boundaries shared by Russia and the United States in the cool, northern waters of the Bering Sea. Then the Japanese discovered a way to make this fish tasty. Pollock was suddenly the object of *the chase* for the world's surimi market. Americans knew their share of this now-valuable resource was being hammered in the open waters of the central Bering Sea (the "Doughnut Hole") by large-capacity foreign ships. The United States and Russia signed an agreement, closing the Doughnut Hole to the foreign fleet. Where did the boats turn next? To Russia, reeling from economic collapse and desperate for foreign currency. Soon Russia's pollock fishery was decimated in the open waters of the Sea of Okhotsk (the "Peanut Hole"). Where next? The western Bering Sea zone of Russia where the foreign fleet can fish pollock without regulation. Some 40 to 60 percent of their nets will be filled with juvenile, undersized pollock on their migration back to American waters. This story has not played itself out, but its ending is as predictable as it is tragic. Foresight, prudence, and comprehension of the interconnections between species were abandoned in the rush to catch pollock. And what of the other species of animals, birds, and fish connected to pollock in the Bering Sea food chain? The fishing industry is the first to declaim responsibility for the concurrent population declines of other species. The Sufi have a saying, "How would you like it if the man from whom you bought fruit consumed it before your eyes, leaving you only the skins?"
Susanne Swibold and Helen Corbett

6.12.3.2 Fisheries: Responses, Strategies, and Institutions

Agreement for the Implementation of the Provisions of the UN Convention on the Law of the Sea of 10 December 1982 Relating to the Conservation and Management of Straddling Fish Stocks and Highly Migratory Fish Stocks (1995 Fish Stocks Agreement): an international agreement affiliated with the UN CONVENTION ON THE LAW OF THE SEA (UNCLOS). It was adopted on 4 August 1996 and will become legally binding after ratification by thirty countries. The agreement would legally bind countries to conserve and sustainably manage fish stocks and to settle peacefully any disputes that arise over fishing on the high seas. The treaty sets out principles for the conservation and management of STRADDLING and highly migratory fish stocks and establishes that such management should be based on the PRECAUTIONARY PRINCIPLE and the best available scientific information. The agreement calls for: a) establishing detailed minimum international standards for the stocks' conservation and management; b) ensuring compatibility and coherence between measures taken within national jurisdictions and in the adjacent high seas; c) creating effective mechanisms of compliance and enforcement of these measures in the high seas; and d) establishing effective mechanisms of dispute settlement.

The agreement also calls for regional fishing organizations (where none currently exist) and devolves to them responsibility for regulating and enforcing sustainable fishing practices, including rights to take action against any boat that undermines the agreed conservation regime. Whether such an arrangement is feasible remains to be seen. Some of the world's largest fishing nations (Chile, Mexico, Peru, Poland, and Thailand) have not yet signed the agreement.

Allowable Catch (Total Allowable Catch): the maximum fish that can legally be harvested, as determined by the government of a coastal state. The allowable catch is a basic concept in the international fisheries regime, initially designed to restore or maintain a stock at particular levels as determined by biological, environmental, and economic factors. According to the 1982 UN CONVENTION ON THE LAW OF THE SEA (UNCLOS), a coastal state (at its discretion) determines the allowable catch, its capacity to harvest it, and if a surplus exists, provides access to other states to harvest the fish. Though it is usually assumed that allowable catch is based on objective, scientific criteria, in practice the ambiguity of the term and the level of discretion granted to the coastal state provide it with an opportunity to set practically any size of allowable catch, as long as it does not lead to OVERFISHING.

Convention on the Conservation of Antarctic Marine Living Resources (CCAMLR): an international agreement adopted at Canberra in 1980, in force since 7 April 1982. This agreement, part of the Antarctic Treaty System, was motivated by the rapid depletion of Antarctic fish stocks due to unrestrained fishing. It obligates each party to develop conservation regulations for its fishing fleets, and to contribute to the study and management of the Antarctic's total ecosystem. By the end of 1999 the convention had twenty-nine parties and is coordinated by the intergovernmental Commission for the Conservation of Antarctic Marine Living Resources.

Fishing Moratorium: a ban on fishing designed to allow the fish population to regenerate. A fishing moratorium should come in advance of a FISHERIES POPULATION CRASH, but often the support for a moratorium is inadequate until the crash has occurred. A fishing moratorium can cause significant economic hardship to the communities dependent on fishing.

Individual Transferable Quotas (ITQS). See under SECTION 5.2.2.2.

International Code of Responsible Fishing: a set of international principles and regulations for the implementation of responsible fishing practices. The objectives are to ensure efficient conservation, management, and development of living aquatic resources, with due respect to ecosystems and biodiversity. The elaboration of the code was initiated at the 1992 International Conference on Responsible Fishing (Cancun, Mexico), organized by the UN FOOD AND AGRICULTURE ORGANIZATION (FAO). The FAO Assembly approved the code in 1995. It is not legally binding, although most parts of it are based on relevant provisions of international law. One of the main goals of the code is "to deter reflagging of vessels as a means of avoiding compliance with applicable conservation and management rules for fishing activities in the high seas."

International Commission for the Regulation of Whaling (International Whaling Commission [IWC]): an intergovernmental commission established by the 1946 International Convention for the Regulation of Whaling to conserve whale stocks and facilitate the orderly development of the whaling industry. Its headquarters are in Cambridge, England. The members of the IWC, which by the end of 1999 numbered forty states, meet annually to consider and adopt revisions to rules pertaining to the harvesting of whales, such as bans on killing certain highly endangered whale species, limits on the numbers and sizes of whales that may be taken, and the designation of specified regions as whale sanctuaries. The IWC adopted progressively more restrictive quotas and rules up through 1980, but these management schemes failed to reverse the decline in whale stocks. In 1982, the IWC adopted a moratorium on all commercial harvesting of whales that became effective in 1986 and remains in effect. The moratorium does not, however, affect subsistence whaling by aboriginal peoples and permits limited killing of whales for scientific purposes. The commission also sponsors and promotes international research on whales that can be used to monitor stocks and determine sustainable catch levels.

See Peter J. Stoett, *The International Politics of Whaling* (Vancouver: University of British Columbia Press, 1997). *M. S. S.*

International Fisheries Commissions: international regulatory bodies that coordinate management of ocean fisheries. These are voluntary associations of states (from two to thirty) that have a common interest in harvesting a fishery and keeping it productive. Most of the fishing commissions were created under the auspices of the UN FOOD AND AGRICULTURE ORGANIZATION (FAO), which facilitates their work through its Committee on Fisheries. Fisheries commissions usually concentrate their activities on specific species and certain regions, such as the Inter-American Tropical Tuna Commission. Other commissions, such as the North-Atlantic Fisheries Commission and South Pacific Forum, deal with a variety of commercial stocks in a geographically defined area. The prerogatives of these institutions are defined by their charters and range from conducting research and making recommendations to adopting fishing regulations. The introduction of EXCLUSIVE ECONOMIC ZONES (EEZS) has decreased the fishery commissions' domain rather substantially and since the 1980s most of these commissions are in a semi-dormant status. However, the adoption of the 1995 FISH STOCKS AGREEMENT was supposed to give a new impetus to their revival.

Maximum Sustainable Yield (MSY): the largest harvest or catch of a fish, marine mammal, or any other marine or freshwater species that maintains the population (or "stock") of that species. Although the MSY may be sustainable, it does not necessarily maximize sustained benefits, because recompositions of the stock may have higher overall benefits. See also MAXIMUM SUSTAINABLE YIELD (section 1.2.6).

Optimal Sustainable Yield: the harvesting principle that calls for a catch rate that optimizes the full set of objectives of resource management, rather than relying on a simpler rule such as MAXIMUM SUSTAINABLE YIELD. The optimal sustainable yield depends on the value of catch, the growth rate of the remaining stock, and the other goods and services that the stock provides. For example, a lower catch rate may allow a fish population to reach a higher level that encourages sport fishing, or a desired expansion of species, such as seals, that feed off of that stock.

6.12.4 MARINE ENVIRONMENT

6.12.4.1 Marine Pollution

Bilge Dumping: the practice of pumping ship waste into the water, which is of growing concern, particularly in high-traffic river and lake waterways. Several international agreements regulate bilge dumping in these circumstances.

Contamination: the presence of elevated concentrations of noxious or harmful substances in water, sediments, or marine organisms.

Dumping: in the context of the UN CONVENTION ON THE LAW OF THE SEA (UNCLOS), any deliberate disposal of wastes or other matter from vessels, aircraft, platforms, or other humanmade structures. This includes the deliberate disposal of the vessels, aircraft, and platforms themselves.

Eutrophication: a change in the state of aquatic ecosystems resulting from an abundance of nutrients that first stimulate plant and algae growth and then result in oxygen depletion and the collapse of other species. In some circumstances, eutrophication occurs naturally as a slow process that may take thousands of years. However human-induced eutrophication occurs much more rapidly. A large increase of nutrient supply (particularly phosphorous and nitrogen) to the marine food chain originating from agricultural runoff, sewage, and industrial wastes triggers rapid growth of algae and plants, sometimes including those that are toxic to people and animals. The growth of the algae (often in the form of "algae blooms") and plants exhausts the oxygen in the water, reducing the abundance of fish and other aquatic animals, eventually transforming the entire aquatic ecosystem. Pollution prevention policies are the most effective remedy for human-induced eutrophication.

Marine Pollution: the introduction by humans, directly or indirectly, of substances into the marine environment (including estuaries) resulting in deleterious effects such as harm to living resources, hazards to human health, hindrance of marine activities (including fishing and other legitimate uses of the sea), impairment of quality for use of seawater, and reduction of amenities. Marine pollution includes, but is not limited to, CONTAMINATION. Thus marine pollution could involve activities that increase sedimentation, oil production that creates oil slicks, the disposal of plastic objects that harm marine mammals, etc.

See Joint Group of Experts on the Scientific Aspects of Marine Environmental Protection, *The Review of the Health of the Oceans* (Paris: UNESCO, 1982).

Oil Pollution: discharges of petroleum and petroleum products into the marine environment. It is usually classified into two categories: operational discharges (i.e., rather common oil slicks) and large-scale occasional discharges (i.e., major OIL SPILLS), which have substantial but different ecological and environmental impacts. Most of the oil discharged into oceans comes from operational sources: the continuous low-level discharges resulting from production, processing, and disposal of oil products and from transportation. Occasional discharges result from accidents at sea, accidents at offshore oil production facilities and pipelines or from ecological terrorism. Occasional discharges attract much more public attention than operational discharges, which might explain the fact that legislation and policy in that realm are better developed.

Oil Spills: uncontrolled discharges of petroleum that result from "blowouts" of drilling platforms and wells (particularly offshore), structural collapses of oil transport and refining facilities, and shipping accidents. Oil spills are particularly damaging to the environment when they occur offshore and the oil drifts onto beaches.

6.12.4.2 Marine Pollution Conventions and Their Implementing Institutions

Baltic Sea Conventions: the 1974 and 1992 Helsinki Conventions on the protection of the marine environment of the Baltic Sea area. These agreements commit parties to controlling pollution from land-based sources and from dumping that affects the Baltic Sea. The 1992 agreement, which came into force in January 2000 with the ratification of the Baltic littoral countries and the European Union, calls on parties to apply the PRECAUTIONARY PRINCIPLE to reduce emissions that might pose a hazard, as well as POLLUTER PAYS PRINCIPLE; and also to use the best environmental practices and the best available technologies.

Barcelona Convention. See CONVENTION FOR THE PROTECTION OF THE MARINE ENVIRONMENT AND THE COASTAL REGION OF THE MEDITERRANEAN.

Convention for the Prevention of Marine Pollution by Dumping from Ships and Aircraft (Oslo Convention): an international agreement among European states adopted in 1972 and in effect from April 1974 until March 1998. It prohibited the dumping of harmful substances into the sea by ships and aircraft, and established a system of permits for dumping other materials. The convention, which covered parts of the North Atlantic and the Arctic, had fourteen parties. In March 1998 it was superceded by the Oslo and Paris Convention (OSPAR CONVENTION). The provisions of the Oslo Convention remain in force, however, as part of the OSPAR Convention, until such time as OSPAR may change them.

Convention on the Prevention of Marine Pollution by Dumping of Wastes and Other Matter (London Dumping Convention [LDC]): an international agreement adopted in 1972, in force since 1975 prohibiting the disposal at sea of certain wastes likely to threaten human health, harm living resources and marine life, or interfere with other legitimate uses of the sea. Permits are required to dump other wastes. The treaty covers all the world's oceans: high seas as well as territorial water of the coastal states. By the end of 1999, seventy-seven states were parties to the LDC, whose secretariat is headquartered with the INTERNATIONAL MARITIME ORGANIZATION (IMO). Parties of the convention have clashed over the dumping of nuclear waste and incineration of waste at sea. The scope of the LDC has been expanded several times and currently includes a moratorium on dumping decommissioned nuclear submarines at sea and the disposal of all industrial waste. The LDC has been also working on establishing a mechanism of control over land-based emissions.

Convention for the Prevention of Marine Pollution from Land-Based Sources (Paris Convention): an international agreement adopted among European nations in June 1974, which went into effect in May 1978, seeking to protect the oceans and seas from land-based sources of pollution, including atmospheric pollutants as well as direct discharges into the water. The Paris Convention, which protected the waters of the northeast Atlantic, from Greenland to Gibraltar, had fourteen parties. In 1998, it was superceded by the OSPAR CONVENTION, which merged the OSLO COMMISSION and the PARIS COMMISSION into a broader organization to address pollution in the North Atlantic. The Paris Convention served as a model for other regional agreements.

Convention for the Protection of the Marine Environment of the North-East Atlantic (OSPAR Convention): a convention to amalgamate and strengthen the functions of the 1972 CONVENTION FOR THE PREVENTION OF MARINE POLLUTION BY DUMPING FROM SHIPS AND AIRCRAFT (OSLO CONVENTION) and the 1974 CONVENTION FOR THE PREVENTION OF MARINE POLLUTION FROM LAND-BASED SOURCES (PARIS CONVENTION), by a 1992 decision of the ministers governing the commissions established by the earlier conventions. The OSPAR Convention went into force in March 1998 with the ratification by the fourteen parties of the previous agreements. In addition to continuing the provisions of the earlier agreements (which are, however, subject to change), the OSPAR Convention required the application of a) the PRECAUTIONARY PRINCIPLE; b) THE POLLUTER PAYS PRINCIPLE; c) BEST AVAILABLE TECHNIQUES (BAT); and d) BEST ENVIRONMENTAL PRACTICE (BEP), including clean technology. The OSPAR Commission, headquartered in London, is a rare example of an amalgamated international organization.

Convention for the Protection of the Marine Environment and the Coastal Region of the Mediterranean (Barcelona Convention): an agreement among Mediterranean coastal states and the EUROPEAN UNION to monitor the environmental condition of the Mediterranean Sea and to strengthen and harmonize national environmental legislation relating to the Mediterranean. The convention, originally called the Convention for the Protection of the Mediterranean Sea against Pollution, was adopted in Barcelona in 1976, growing out of the 1975 MEDITERRANEAN ACTION PLAN. It entered into force in 1978, and by 2000 had twenty-one parties. The convention provides the framework for the action plan, and has been followed by protocols proposing to combat OIL POLLUTION, prohibit ship and aircraft waste dumping, regulate the transport of HAZARDOUS WASTES and seabed mineral exploitation, protect BIOLOGICAL DIVERSITY, and establish special protected areas. The change in its title reflects the tendency of framework agreements to expand their scope, in this case from pollution to a broader range of environmental and sustainability issues.

International Convention for the Prevention of Pollution from Ships (MARPOL): a major international agreement adopted in 1973 by the Intergovernmental Maritime Consultative Organization (now the INTERNATIONAL MARITIME ORGANIZATION [IMO]), in force since 1978. As of the end of 1999, the basic MARPOL Convention had 108 parties. MARPOL reaffirmed the controls on oily discharges contained in the 1954 International Convention for Prevention of Pollution of the Sea by Oil and its amendments, while adding numerous additional rules pertaining to the construction and operation of marine vessels. Under MARPOL all new large oil tankers must be constructed with segregated ballast tanks, which keep ballast water from mixing with oil residues in the cargo tanks. Coastal states are required to provide shoreside facilities for treatment of ballast and tank washings containing noxious liquid substances. MARPOL is also significant for strengthening the enforcement mechanisms involving flag, coastal, and port states. A 1978 protocol added to MARPOL requires all existing tankers to be equipped with either segregated ballast tanks or crude-oil washing systems, and all new tankers are required to have both types of systems.

See Ronald B. Mitchell, *Intentional Oil Pollution of the Sea: Environmental Policy and Treaty Compliance* (Cambridge: MIT Press, 1994), 93–103. M. S. S.

International Conventions on the Civil Liability for Oil Pollution Damage (CLC 1969; CLC 1992): agreements adopted first in 1969, in force since 1975, to ensure that adequate compensation is available to persons affected by pollution damage through oil escaping during maritime transport by ship and to standardize international rules and procedures for determining questions of liability and adequate compensation in such areas. The 1992 convention extends the scope of agreement into the territorial sea and the EXCLUSIVE ECONOMIC ZONES (EEZS) of the parties. By mid-1999 the new convention had forty-five parties, leaving seventy-seven parties adhering to the 1969 convention until such time as they ratify the new one. For the time being, the two conventions coexist. The CLC is supplemented by the International Convention on the Establishment of an International Fund for Compensation for Oil Pollution Damage ("Fund Convention"), which entered into force in 1978 and seeks to ensure full compensation to victims of oil pollution damage and to distribute the economic burden between the shipping industry and oil cargo interests. The Fund Convention has sixty-six parties and is also headquartered with the INTERNATIONAL MARITIME ORGANIZATION (IMO).

International Convention on Oil Pollution Preparedness, Response, and Cooperation (OPRC): an international agreement adopted in London in 1990. Its objective is to prevent marine pollution by oil, advance the adoption of adequate response measures in the event that oil pollution does occur, and provide for mutual assistance and cooperation between states. By mid-1999, the convention had fifty parties; its secretariat is headquartered with the INTERNATIONAL MARITIME ORGANIZATION (IMO).

International Maritime Organization (IMO): a specialized agency of the United Nations responsible for measures to improve the safety of international shipping and to prevent marine pollution from ships. As of early 2000, the IMO had 157 member states and two associate members. Its budget is comprised of individual contributions made by member states, the size of which depends primarily on the tonnage of their merchant fleet. During the initial phase (1950–1970s), its mandate was to develop international legislation relating to shipping safety and pollution prevention. Since the 1980s, however, the focus has been on ensuring that the countries that have accepted international agreements properly implement them. The IMO might have the most successful record of activities among the UN agencies: agreements elaborated under its auspices apply to over 98 percent of world merchant shipping tonnage. *M. S. S.*

Mediterranean Action Plan (MAP): a framework for environmental cooperation among Mediterranean coastal states and other interested parties adopted in 1975. Its members—twenty regional nations and the EUROPEAN UNION—pledge "to take all appropriate measures to prevent, abate and combat pollution in the Mediterranean sea area and to protect and improve the marine environment in that area." MAP has concentrated on: a) integrated planning for the development and management of the resources of the Mediterranean; b) development of a framework convention and related protocols; and c) establishment of institutional and financial arrangements to execute this plan. Since 1979, funding for these purposes has been provided by the Mediterranean Trust Fund created by the signatories, while coordination is provided by a Working Group headquartered in Athens. The adoption of the MAP was followed by the 1976 Convention for the Protection of the Mediterranean Sea against Pollution (Barcelona Convention) and a series of protocols addressing different types of pollution.

See Lynton Caldwell, *International Environmental Policy*, 2d ed. (Durham: Duke University Press, 1996), 181–84.

6.12.5 MARINE SCIENTIFIC RESEARCH

Intergovernmental Oceanographic Commission (IOC): an agency established in 1960 in Paris to promote multilateral scientific research on the nature and resources of the oceans. The IOC is a specialized agency of the UN EDUCATIONAL, SCIENTIFIC AND CULTURAL ORGANIZATION (UNESCO). As of early 2000, the IOC had a membership of 126 governments that have voting rights to determine objectives for the commission. The IOC is currently focused on four distinct project areas: international oceanographic research programs, a global ocean observation system, international education and training programs, and dissemination of research findings.

International Council for the Exploration of the Seas (ICES): one of the oldest international organizations. It was established in 1902 in Copenhagen to implement international marine research programs. The ICES Convention, renegotiated in 1964, has in its current membership Belgium, Great Britain, Germany, Denmark, Ireland, Iceland, Spain, Canada, the Netherlands, Norway, Poland, Portugal, Russia, the United States, Finland, and France. ICES objectives are to assist in conducting marine scientific research (especially on fish stocks), develop scientific programs, disseminate information, and provide nonbinding recommendations on fisheries and fish stocks conservation. ICES activities are focused on the Atlantic and adjacent seas. Activities of ICES and its members are coordinated by its bureau and twelve permanent committees focusing on different fields of marine scientific research.

6.12.6 MINERAL RESOURCES OF THE OCEANS

Convention on the Regulation of Antarctic Mineral Resource Activities (CRAMRA): an international agreement adopted by the ANTARCTIC TREATY CONSULTATIVE PARTIES (ATCP) in Wellington, New Zealand, in 1988, after a decade of sometimes contentious negotiations. The agreement would have permitted exploration and extraction of minerals, petroleum, and other natural resources in the Antarctic region, but only in accordance with strict conditions designed to ensure that such activities would not cause unacceptable environmental damage. CRAMRA never came into force, however, as environmental NONGOVERNMENTAL ORGANIZATIONS strongly opposed it on grounds that any extraction of minerals would pose a serious threat to the fragile Antarctic environment. CRAMRA was superseded by the Protocol on Environmental Protection to the Antarctic Treaty, which was adopted by the ATCP in Madrid in 1991. The Madrid Protocol prohibits all commercial mining in the region and in effect designates Antarctica as a World Park in which human activity is restricted to scientific research. For the next fifty years, any modification of the prohibition on mining activities must have the approval of all ATCP members.

See Christopher C. Joyner, "Antarctica," in *A Global Agenda: Issues Before the 48th General Assembly of the United Nations,* eds. John Tessitore and Susan Woolfson (Lanham, MD: University Press of America, 1993), 206–11. M. S. S.

Enterprise: a United Nations body that should be created under the provisions of the UN CONVENTION ON THE LAW OF THE SEA (UNCLOS) as "the operating arm" of the International Seabed Authority. Under the initial 1982 agreement, mines must turn half their claims over to the Enterprise, transfer mining technology, and pay a $500,000 fee. The Enterprise would then share the proceeds with all countries. Such a regime, based on a principle of "common heritage of mankind" was not acceptable to the most advanced states, mainly the United States, whose opposition stalled the UNCLOS negotiations for over twelve years. Under the renegotiated agreement, the Enterprise will operate like a mining company owned by the International Seabed Authority.

International Seabed Area ("The Area"): the seabed, ocean floor, and subsoil located beyond the limits of national jurisdiction. In the mid-1960s the mining industries of the United States and the Soviet Union recognized the potential importance of metallic minerals like cobalt, nickel, and manganese in the form of nodules scattered on the ocean bed at a depth of some 1,000 meters. An expectation that deep seabed mining on a commercial scale would be possible in the near future gave a powerful impetus to the negotiations over the UN CONVENTION ON THE LAW OF THE SEA III. The provisions on mining activities in the Area are described in the convention's Part XI—the most elaborate and controversial part of the convention. The original part was shaped along the principles of the NEW INTERNATIONAL ECONOMIC ORDER. The new agreement (renegotiated in 1994) is market-oriented, emphasizes cost-effectiveness, and keeps institutional machinery to a minimum. Developing countries' involvement in the mining operations in the Area is currently in the form of joint ventures.

International Seabed Authority: the institution through which parties to the UN CONVENTION ON THE LAW OF THE SEA (UNCLOS) organize and control activities in the INTERNATIONAL SEABED AREA beyond the limits of national jurisdiction, particularly with a view to administering the resources of that area. The functions, membership, and management of the authority are modified by the AGREEMENT ON THE IMPLEMENTATION OF PART XI of the convention. The authority functions through three main organs: the assembly (the supreme organ comprised of all members of the authority); the council (the executive organ with thirty-six members); and the twenty-one-member

Legal and Technical Commission, which assists the council in making recommendations.

6.13 Space and the Electromagnetic Spectrum

6.13.1 TRENDS, PRACTICES, AND PROBLEMS

Electromagnetic Spectrum: the range of waves of radiant energy of different frequencies and wavelengths (very short or ultraviolet, visible, and very long or infrared) that
envelop the earth. The electromagnetic spectrum may be considered the "fifth environment" coexisting with the atmosphere, biosphere, inorganic terrestrial environment,
and space. Various components of the electromagnetic spectrum have different properties and they affect the earth, its people, and their environments in different ways. Currently, the electromagnetic environment is used mainly for communication purposes.
Some scholars argue that space and the electromagnetic spectrum should be considered together as the Orbit Spectrum Resource, because the control and use of satellites
depend upon the availability and width of radio frequencies.

Geostationary Orbit (GEO): the orbit 22,300 miles above earth, at which a satellite com-
pletes one orbit in 24 hours and therefore remains over one spot on earth. The fact
that the satellite in geostationary orbit can maintain permanent contact with earth-
bound transmitters and receivers makes it a unique resource for telecommunications
and meteorology. The International Telecommunications Convention has defined the
GEO as a "limited natural resource (which) must be used efficiently and economically
so that countries or groups of countries may have equitable access." In practice, a growing demand for "orbital slots" provokes discontent on the part of developing nations
over the alleged monopolization of this "last frontier." A group of equatorial countries
(Colombia, Brazil, Zaire [now the Democratic Republic of Congo], Indonesia, Kenya,
Congo, Uganda, and Ecuador) in their 1976 Bogota Declaration attempted to claim sovereignty over the GEO as an overhead extension of their national territory. Though
this step did not receive wide international support, the political controversy between
North and South over the "common heritage" status of space and equitable access to
orbits remains heated. The current allocation of GEO slots is negotiated through the
INTERNATIONAL TELECOMMUNICATIONS UNION (ITU).

Low Earth Orbit (LEO): an orbit within a few hundred miles of earth, starting at an altitude of about 100 miles. Low earth orbits are usually polar or inclined to provide extensive coverage of the earth's surface. They can be used for remote sensing, radar, navigation, meteorology, communication systems, and a wide range of other purposes. The
U.S. government maintains a detailed catalog of more than 8,000 objects at low earth
orbits. The growing abundance of space vehicles and satellites in low earth orbit has
stimulated calls for international cooperation in creating a "traffic system" for near-
earth space.

Military Uses of Space: the placement of objects in space for either offensive or defensive military purposes. Military space activities can be roughly divided into six categories: communications, meteorology, surveillance and reconnaissance, navigation,

geodesy, and space-borne weapons and defensive systems. In communications and meteorology, military satellites have always outnumbered civilian systems: the NORTH ATLANTIC TREATY ORGANIZATION (NATO) has long had its own system of communication satellites, as does the United States, the United Kingdom, France, and Russia. The initiatives to place "defense" systems into space, such as the so-called Star Wars system proposed by the Reagan administration, have drawn intense criticism because, according to the logic of nuclear deterrence, the first foolproof defense system would free its owner to deploy offensive weapons with impunity. Despite numerous attempts to establish the demilitarization of the space commons, military uses of space are not likely to be prohibited.

Remote Sensing: a technique of gathering information on earth and its resources from SATELLITES. Remote sensing has enormous potential in providing precise data about geography, geology, forests, crops, pollution, urban growth, and other aspects of the global environment. It was initially developed by the U.S. National Aeronautic and Space Administration (NASA) through its LANDSAT program, whose information was provided to cooperating nations at minimal cost. However, in 1985 the U.S. government transferred much of its remote sensing operation to the private sector; most valuable information is currently priced at commercial rates. Increasingly, remote sensing systems are used for intelligence and by multinational companies surveying for natural resources. There is concern that commercially significant data relating to a specific country will not be available to the government or businesses of that country. Many governments of developing nations consider the current international system of remote sensing a threat to their national economic sovereignty. After prolonged debate in the United Nations, a set of PRINCIPLES RELATING TO REMOTE SENSING OF EARTH FROM SPACE was adopted.

Satellites, Artificial: various scientific and technological objects launched into orbit around the earth or other celestial bodies. Satellites are used to observe, transmit images, and obtain information about earth and other space bodies as well as the environment around them. They are widely used in military intelligence, meteorological forecasting, remote sensing, navigation, and communications. Some satellite systems are created as joint ventures of the international community. For instance, the COSPAS-SARSAT system, used in ground and sea search-and-rescue missions, was established in 1979 by Canada, the United States, France, and the Soviet Union. The system receives distress signals, determines a position of the sender, and transmits data to rescue authorities. COSPAS-SARSAT can be credited with saving thousands of lives. Meteorological satellites have greatly improved weather forecasting, enabling much earlier warnings of severe weather conditions such as hurricanes and typhoons.

Satellite Power Stations (SPSS): orbiting facilities that would collect solar energy and transmit it to earth as microwaves. The SPS potential is one of the most prominent ideas in the discussions over the future use of space. As a result of research performed in several countries during the 1970s, it was established that, provided the costs of space transportation fell substantially, satellite power stations could supply large quantities of power. Such an undertaking would involve the development of an extensive infrastructure in LOW EARTH ORBIT, including building some thousands of square kilometers of orbiting solar arrays. This idea has enormous environmental ramifications and can be realized only subject to international agreements on several different matters.

Space Commercialization: business applications involving objects in space, travel through space, or activities on celestial bodies. Such activities are currently largely confined to the use of satellites, especially for communications and scientific applications, but the long-term potentials are far broader, including manufacturing, mining, power generation (e.g., SATELLITE POWER STATIONS), transport, and even tourism.

Space Debris: humanmade trash in orbit (broken pieces of rocket boosters, pieces of old satellites, lost tools, and other debris). The volume of large space debris is roughly 20,000 objects, with an estimated weight of 2 million kilograms. Some of these items travel at a speed of 22,000 miles per hour and may cause catastrophic damage to operational satellites and spacecraft. More than 40 nuclear-powered devices are also in space, carrying an estimated ton of radioactive materials. Space trash in LOW EARTH ORBIT can stay up for months or years but eventually falls and is burned in the atmosphere. Space debris in GEOSYNCHRONOUS ORBIT stays there for centuries and presents a real hazard for the continuing functioning of satellite systems. The amount, location, and ownership of this debris are documented by the U.S. Space Command, which advises space users on "launch windows" and safe orbits. The impact of space debris on the earth's climate may also become a serious concern. The space nations are now trying to reduce the amount of space debris and to develop sophisticated methods of cleaning the cosmos. This is now widely accepted as an international problem requiring coordinated solutions.

Space Pollution: contamination of the space environment by the byproducts of human activities in a variety of sizes and composition. Two types of pollution can occur in space: back pollution, which, though arising in space, adversely affects the atmosphere or the surface of the earth; and forward pollution, which arising from earth or its atmosphere, affects the quality of the space environment. Solid waste, hazardous waste, and nuclear waste are of immediate concern. Solid wastes include all kinds of nontoxic SPACE DEBRIS found in orbit, which pose collision hazards. Radioactive wastes are the residues of nuclear-powered space objects that can release damaging radiation. More esoteric forms of space pollution include electromagnetic pollution coming from abandoned satellites and biological contamination of space and celestial bodies by earthborne organisms. Despite the apparently remote likelihood of harm from these types of pollution, they cannot be disregarded. Space pollution will become a matter of growing concern over the next decades as commercial exploitation of space increases.

Space Race: the twenty-year long contest for superiority in space travel that paralleled the Cold War between the United States and the Soviet Union. It began in 1957 when the Soviet Union launched the first artificial satellite, *Sputnik 1*, followed in 1958 by the U.S. *Explorer*. Further programs, the Soviet Luna program and the U.S. Apollo program, accomplished unprecedented breakthroughs in space exploration, including landing on the moon and the development of space stations. Despite scientific interest, the programs were resource-intensive and have faced declining funding since the late 1970s. The space race took divergent paths in the two countries: the United States focused on the development of reusable space shuttles, while the Soviets chose the development of space stations. Each path has substantial ramifications for the biosphere and the space environment, though little has been published about it.

6.13.2 RESPONSES AND STRATEGIES

Agreement Governing the Activities of States on the Moon and Other Celestial Bodies (Moon Agreement): an international agreement adopted by the UN General Assembly in 1979 (Resolution 34/68) and entered into force in July 1984. The agreement reaffirms and elaborates many of the provisions of the Outer Space Treaty as applied to the moon and other celestial bodies. In particular, it provides that those bodies should be used exclusively for peaceful purposes, that their environments should not be disrupted, and that the United Nations should be informed of the location and purpose of any station established on those bodies. In addition, the agreement provides that the moon and its natural resources are the COMMON HERITAGE OF MANKIND and that an international regime should be established to govern the exploitation of such resources when it becomes feasible. The agreement also calls for "equitable sharing" of the benefits for all state parties. According to the Moon Agreement, exploration of the surface or subsurface of celestial bodies must not disrupt the equilibrium of the space environment or contaminate it. Also, precautions should be taken to prevent samples taken from celestial bodies from posing a hazard to inhabitants of earth and its environment. The Moon Agreement has been ratified by only nine states, only one of which (India) is a space power; France, which signed the agreement in 1980, still has not ratified it. Most governments believe that it does not provide a good legal framework for business activity on the moon.

Agreement on the Rescue of Astronauts, the Return of Astronauts, and the Return of Objects Launched into Outer Space of 1968 (Rescue Agreement): an international agreement adopted by the UN General Assembly in 1967 and entered into force in December 1968. Parties to the agreement are to provide assistance to the crews of spacecraft in the event of accident or emergency and promptly return them to the launching state. According to the Rescue Agreement, astronauts are to be treated as "envoys of mankind." Following the commercialization of space, this statement is considered to be outdated and the business community has been pushing for its renegotiation. The agreement also establishes the procedure of providing assistance to launching states to recover space objects that return to earth outside of that state's territory. It also specifies that if a hazardous object is discovered in an area of space where another nation is conducting space operations, the finder can demand that the launching state must do whatever is necessary to eliminate the problem. By mid-1999, the Rescue Agreement had been ratified by eighty-five states.

Convention for Registration of Objects Launched into Space of 1976 (Registration Convention): a treaty that establishes a regime for notifying the United Nations about objects launched into outer space, including their general function as well as location. The convention also requires reporting of any launch by a private firm or intergovernmental organization. The convention, which entered into force in 1976 as an elaboration of the OUTER SPACE TREATY, is one of the major agreements that comprise the body of international space law. Parties to the Registration Convention have to declare what they have put into orbit, including any hazardous or potentially polluting materials, in order to lessen the chance of harming another party to the treaty. Insofar as parties abide by the convention, it reduces the secrecy of space activity. The international register of launching is maintained on the basis of information provided by member states and intergovernmental organizations that are parties to the convention; the secretariat also provides information about launchings by states that are not parties to the convention. By 2000, forty nations had ratified the Registration Convention.

Convention on International Liability for Damage Caused by Space Objects (Liability Convention): an agreement that requires a launching state to be absolutely liable to pay compensation for damage caused by its space objects on the surface of the earth or to aircraft, and liable for damage due to its faults in space. It also provides for procedures for the settlement of claims for damages. The Liability Convention, which elaborates on the OUTER SPACE TREATY and came into force in 1972, is one of the major treaties that comprise the body of international space law. By mid-1999, it had been ratified by eighty states. The most notable application of this agreement occurred in 1978, when the Soviet satellite *Cosmos 954* crashed into the North-Western Territories of Canada. The satellite carried a nuclear-power reactor that was scattered over a large area. Canada's claim was eventually settled by the USSR in the amount of $3 million Canadian.

New World Information and Communication Order (NWICO): the concept of a global regime that would govern transnational collection and dissemination of information, including data flows, broadcasting, and remote sensing activities. From 1975 to 1989, developing countries dominated the debates in the UN EDUCATIONAL, SCIENTIFIC AND CULTURAL ORGANIZATION (UNESCO) in their advocacy of the NWICO. One of the major points was a notion of "prior consent" to accept direct satellite transmissions originating beyond national boundaries. The UNESCO Mass Media Declaration proposed principles legitimizing control over transnational media in order to protect national cultural sovereignty against the "electronic colonialism" of the North. The only practical result of this declaration was the temporary withdrawal of the United States and Great Britain from UNESCO. However, NWICO debates were important in recognizing the commercial value of space-based global information flows and its increasing threat to the economic, political, and cultural sovereignty of developing nations.

Principles Relating to Remote Sensing of Earth from Space: a set of principles adopted by the UN General Assembly in 1986 (Resolution 41/65). The principles provide that remote sensing activities should: a) be carried out for the benefit and in the interests of all countries, taking into particular consideration the needs of the developing countries; b) include international cooperation and technical assistance; and c) promote the protection of the environment and protection of humankind from natural disasters. One of the principles specifies that when one country acquires data over another country, the sensed country should have access to the data on a nondiscriminatory basis and on reasonable cost terms. In order to maximize the availability of benefits from remote sensing, the principles encourage regional cooperation in the establishment and operation of collection, storage, processing, and interpretation facilities.

Space Law: principles of law that govern the exploration and use of space (essentially areas beyond the earth's atmosphere) by states, businesses, and other actors. Space law was born with the first launch of a humanmade satellite in 1957 (Soviet *Sputnik 1*) and was developed under the auspices of the United Nations, through its Committee on the Peaceful Uses of Outer Space. The major international agreements on outer space are the 1967 OUTER SPACE TREATY, the 1968 RESCUE AGREEMENT, the 1972 LIABILITY CONVENTION, the 1976 REGISTRATION CONVENTION, the MOON TREATY, and the 1963 TEST BAN TREATY. Multilateral and bilateral cooperation agreements, such as the Intergovernmental Agreement on the International Space Station, are also basic parts of the contemporary space law. Western politicians and experts now argue that most of the space law created during the Cold War is outdated and requires further development, particularly in relation to space commercialization. Proponents of this trend argue that without the introduction of private-property rights in space, businesses would be reluctant to invest in such economically desirable projects as satellite power stations, mines, space hotels, etc. Another idea currently under discussion is a law of "Space Salvage" that might provide an incentive for businesses to collect space debris and therefore initiate recycling as a major orbital business.

Treaty on Principles Governing the Activities of States in the Exploration and Use of Outer Space, Including the Moon and Other Celestial Bodies of 1967 (Outer Space Treaty): a cornerstone international agreement regulating the national activities of outer space in its exploration and exploitation. It was adopted by the UN General Assembly in 1966 and entered into force in October 1967. The Outer Space Treaty provides the basic framework for international space law, including the following principles: a) the exploration and use of outer space shall be carried out for the benefit and in the interests of all countries and shall be the province of all humankind; b) space shall be free for exploration and use by all states; c) space is not subject to national appropriation by claim of sovereignty, by means of use or occupation, or by any other means; d) states shall not place nuclear weapons or other weapons of mass destruction in orbit or on celestial bodies or station them in space in any other manner; e) the moon and other celestial bodies shall be used exclusively for peaceful purposes; f) astronauts shall be regarded as the envoys of humankind; g) states shall be responsible for national space activities whether carried out by governmental or nongovernmental activities; h) states shall be liable for damage caused by their space objects; and i) states shall avoid harmful contamination of space and celestial bodies. By mid-1999, the Outer Space Treaty had been ratified by ninety-five countries, including the major space nations (i.e., nations capable of regularly launching space vehicles and placing satellites into orbit).

6.13.3 ORGANIZATIONS AND INITIATIVES RELATED TO OUTER SPACE

Committee on the Peaceful Uses of Outer Space (COPUOS): the United Nations body set up in 1959 to develop space law, review the scope of international cooperation in outer space, devise programs to be undertaken under United Nations auspices, and encourage continued research and dissemination of information on space matters. As of mid-1999, the committee had sixty-one member states. It is comprised of the Scientific and Technical Subcommittee and the Legal Subcommittee, while the UN Office for Outer Space Affairs acts as its secretariat. The COPUOS has been primarily responsible for the development of the first generation of space law, however over the past decade or so its activities have decreased. Although the current agenda of the committee covers all the major fields of space activities (remote sensing, telecommunications, space transportation, life and basic space sciences), these agenda items are used mainly for exchange of information. Many government leaders stress the need for structural reform of the COPUOS.

European Space Agency (ESA): an organization of fourteen European countries that cooperate in all aspects of the development of satellites for scientific and commercial purposes. It was founded in 1975, is currently headquartered in Paris and has an operating budget of $2 billion a year. ESA also has centers in Noordwijk in Holland, Damstadt and Cologne in Germany, Frascati in Italy, and a launching site in French Guinea. The agency's mission includes the development and launch of communications and weather satellites, scientific spacecraft, and space transportation systems. ESA member nations (Austria, Belgium, Denmark, Finland, France, Germany, Ireland, Italy, the Netherlands, Norway, Spain, Sweden, Switzerland, and the United Kingdom) jointly fund and decide upon projects for the agency. One of the major accomplishments of the ESA has been the Ariane liquid-propellant rocket that launched European scientific and communications satellites into space. In the late 1990s the ESA suffered difficult economic times and had to cancel several projects.

International Maritime Satellite Organization (Inmarsat): a consortium of eighty-six member countries and a major international satellite provider. Inmarsat was established in 1979 through an international convention and became operational in 1982. Its function is to coordinate the use of artificial satellites for surveillance and communications over the oceans. The Inmarsat system allows ships and offshore drilling rigs to communicate by radio with each other and with radio stations at land. Each member country has a signatory (normally the national telecommunications provider) that owns a share in the organization proportional to its usage of the system and also distributes the Inmarsat services to users. In 1998, the Inmarsat General Assembly decided to convert the entity from an intergovernmental organization to a private public-stock company over a two-year period, although an intergovernmental body would continue to ensure that Inmarsat meets its obligation to maintain the Global Maritime Distress and Safety System.

International Space Station: a joint venture of the United States, Russia, Canada, Japan, and the fourteen member nations of the EUROPEAN SPACE AGENCY (ESA), designed as a permanent international laboratory in space. Six astronauts will be able to spend from three to five months on board. This opportunity is extremely promising for research in medicine, industrial materials, and communications technology. In addition to its obvious scientific merits, the International Space Station is presented as a model of international cooperation.

International Telecommunications Satellite Organization (Intelsat): an international satellite system formed in 1964 by eleven original state members, which had 143 members as of mid-1999. This nonprofit organization, which combines elements of both an international agency and a private corporation, currently handles most of the world's international telephone and television communications. The Intelsat system consists of satellites that circle the earth in GEOSYNCHRONOUS ORBIT. Thus, each satellite stays over a particular point on the earth's surface and can receive and transmit signals from ground-based relay stations. Contemporary satellites handle up to 22,500 telephone calls and three television channels simultaneously.

International Telecommunications Union (ITU): the world's oldest official international organization, a direct descendant of the International Telegraphic Union of 1865. In 1947 the ITU became a United Nations specialized agency with the main function to maintain and expand international cooperation in telecommunications of all kinds. The highest authority belongs to the Plenipotentiary Conference of all members that meets every four years to decide on policy issues. The Administrative Council, composed of forty-one members, meets annually and coordinates the activities of four permanent bodies at the ITU headquarters in Geneva. Since the 1960s, the ITU has coordinated the efforts of governments and the private sector in using the ELECTROMAGNETIC SPECTRUM for communications so as to avoid mutual interference and ensure common technical standards and operability. Every four years the ITU convenes the World Administrative Radio Conference to grapple with the allocation of frequencies for the growing number of different services offered through satellites. The resulting agreements have the legal status of treaties among its members, which numbered 189 by late 1999. Though managing the allocation of the spectrum and orbital slots is more a technical than a political issue, recently the ITU has given particular emphasis to considerations of jurisdiction and equity among users.

United Nations Programme on Space Applications: a priority work program of the UN Office for Outer Space Affairs, established by the General Assembly in 1971 with the initial mandate to create awareness among policymakers of the benefits that could be derived from the applications of space technology and to encourage training and education programs that would provide people from developing countries with the necessary knowledge, skills, and experience in these applications. This mandate was fulfilled through the organization of training courses, workshops, and other meetings on various aspects of space technology. Currently, the major objectives of the program are the promotion of cooperation in space science and technology between industrialized and developing countries, and stimulation of the growth of indigenous space technology development in developing countries. Most recent accomplishments of the program include establishments of regional centers for space science and technology education in Brazil, India, Mexico, Morocco, Nigeria, and countries of eastern Europe; establishment of the Asia-Pacific Satellite Communication Council; development of a satellite-based cooperative information exchange network in Africa; and development of the Integrated Coastal Area Management project in the Caribbean.

COMMENTARY ON SPACE COMMERCIALIZATION AND COLONIZATION

Colonizing the moon or other planets will also require careful regulatory attention. Regardless of whether there is life to be found in space, our actions in space can affect life here. A UK Friends of the Earth spokesperson asserted that as there was no animal or plant life on the moon, the organization did not have any concerns over any future colonization attempts. The only caveat to this was any situation where competition over colonization resulted in open conflicts or environmental degradation on earth. This position neglects the cost to the earth's biosphere (in terms of resources and pollution) of colonizing space. *Ian Welsh*

World Meteorological Organization (WMO). See under SECTION 6.4.2.

6.14 Ozone Depletion

Chlorofluorocarbons and Related Compounds: the family of artificial compounds considered environmentally benign for a long time before the MOLINA-ROWLAND HYPOTHESIS that they react with atmospheric compounds to reduce the OZONE LAYER. This group includes chlorofluorocarbons (CFCs), bromofluorocarbons (halons), methyl chloroform, carbon tetrachloride, methyl bromide, and hydrochlorofluorocarbons (HCFCs). These chemicals are usually referred to as ozone-depleting substances. Some of them also play a role in global climate change. Severe limitations on the production and use of these compounds are required under the MONTREAL PROTOCOL.

Molina-Rowland Hypothesis: a groundbreaking discovery of the role of CHLOROFLUOROCARBONS (CFCs) in ozone chemistry. In 1974, Mario Molina and Sherwood Rowland published their widely noted article in *Nature* on the threat to the ozone layer from CFCs. Due to their chemical stability and nontoxicity, CFC gases were widely used by industry and households, for instance, in spray bottles, as the cooling medium in refrigerators, and in plastic foams. Molina and Rowland determined that the chemically inert CFC could gradually be transported up to the ozone layer and there, under intensive ultraviolet light, be separated into their constituents, notably chlorine atoms. They calculated that unabated use of CFC gases would severely deplete the OZONE LAYER within a few decades. The Molina-Rowland Hypothesis was followed by the discovery of the ozone hole over Antarctica; this discovery gave impetus to the creation of an international ozone regime, particularly the MONTREAL PROTOCOL.

See Mario J. Molina and F. Sherwood Rowland, "Stratospheric Sink for Chlorofluoromethanes: Chlorine Atom-Catalysed Destruction of Ozone," *Nature* 249 (June 1974): 810–12.

Montreal Protocol on Substances that Deplete the Ozone Layer: an international agreement concluded in September 1987 as a supplement to the VIENNA CONVENTION ON THE PROTECTION OF THE OZONE LAYER. The original protocol requires the parties to reduce production and use of CHLOROFLUOROCARBONS (CFCs) by 20 percent by 1993 and by 50 percent by 1999, except for developing countries, which are allowed a ten-year grace period. Production and consumption of halons, another class of ozone-depleting substances, were not to exceed 1986 levels after 1993. Mounting scientific evidence of the immediacy of the OZONE DEPLETION problem led to the adoption of amendments and adjustments to the Montreal Protocol in London (1990), Copenhagen (1990), and Vienna (1995) that apply to additional ozone-depleting substances and provide for a complete phase-out of their production. Under the terms of these agreements, halons were to be phased out by 1994; CFCs, carbon tetrachloride, and methyl chloride by 1996;

and HCFCs used as a substitute for CFCs by 2020. Negotiations continue on restricting methyl bromide, a chemical widely used in agriculture. The agreements also provide for a multilateral fund to assist developing countries' shift to substitutes for the banned substances.

See Karen T. Litfin, *Ozone Discourses: Science and Politics in Global Environmental Protection* (New York: Columbia University Press, 1994). *M. S. S.*

COMMENTARY ON MONTREAL PROTOCOL

The Montreal Protocol as amended is widely considered to be one of the most significant international environmental agreements. It has led to a drastic reduction in the flow of CFCs and most other ozone-depleting substances into the atmosphere, such that the ozone levels in the stratosphere may recover to preindustrial levels during the twenty-first century. This effective international response to ozone layer depletion is often cited as a reason for optimism that the problem of global climate change can be similarly managed, even though the latter problem poses much more complex challenges to the international community. *Marvin S. Soroos*

Ozone Layer Depletion: diminished concentrations of naturally occurring ozone in the stratosphere at altitudes of 10–40 kilometers, due to the presence of human pollutants, such as CHLOROFLUOROCARBONS (CFCs), halons, carbon tetrachloride, methyl chloride, and methyl bromide in the atmosphere. The greatest amount of thinning has taken place over Antarctica during the southern spring season, where by the late 1980s ozone concentrations had dropped by up to 60 percent over an area the size of the North American continent, a phenomenon known as the "Antarctic ozone hole." Lesser, but nevertheless significant, reductions in ozone are also observed at other latitudes. The thinning of the ozone layer is significant because ozone molecules shield the planet from intense ultraviolet (UV) radiation from the sun. Increased doses of UV radiation can cause human health problems such as skin cancer, disruption of the immune system, and cataracts. UV radiation also damages many other species of flora and fauna. Especially vulnerable are microscopic organisms such as phytoplankton and zooplankton, which are at the bottom of the food chain and thus are a key source of nutrition for other species.

See Sharon L. Roan, *Ozone Crisis: The Fifteen-Year Evolution of a Sudden Global Emergency* (New York: John Wiley, 1989). *M. S. S.*

Vienna Convention on Protection of the Ozone Layer: a treaty adopted in March 1985 that provided the framework for the international regime addressing the problem of OZONE LAYER DEPLETION. The treaty calls upon the parties to take measures to "control, limit, reduce, or prevent human activities under their jurisdiction should it be found that these activities have or are likely to have adverse effects resulting from modification or likely modification of the ozone layer." The convention did not set a timetable for binding reductions of CHLOROFLUOROCARBONS (CFCs) or other ozone-depleting substances, as was advocated by the United States and several other countries during the negotiations that began in 1982. Nevertheless it is considered significant for its application of the PRECAUTIONARY PRINCIPLE to situations of scientific uncertainty. The agreement also calls for international cooperation in monitoring, research, and scientific assessments and the exchange of information on OZONE DEPLETION and its potential impacts. Follow-up negotiations led to the 1987 MONTREAL PROTOCOL ON SUBSTANCES THAT DEPLETE THE OZONE LAYER.

See Peter H. Sand, "The Vienna Convention is Adopted," *Environment* 27 (5) (June 1985): 19–20, 40–43. *M. S. S.*

COMMENTARY ON VIENNA CONVENTION ON PROTECTION OF THE OZONE LAYER

The Vienna Convention is a typical framework in that it "lacks teeth" in the sense of mandating any reductions of pollution emissions responsible for ozone depletion. Thus, the significance of this convention lays in the institutional mechanisms it established for negotiating supplemental agreements, namely the MONTREAL PROTOCOL ON SUBSTANCES THAT DEPLETE THE OZONE LAYER and subsequent amendments and adjustments. *Marvin S. Soroos*

6.15 Transportation Energy Consumption and Pollution

Alternative Fuel Vehicles (AFVs): automobiles and other vehicles operating in part or in whole on fuels other than gasoline or diesel. These alternative fuels may offer the potential for lower pollution, greater efficiency, lower cost, and, in some instances, lower reliance on imports. Increasing the number of AFVs is one approach to meeting government requirements to lower pollution or to increase fuel efficiency; in the United States the introduction of AFVs in state vehicle fleets has been required by the federal government. To meet FLEET FUEL EFFICIENCY TARGETS, automobile manufacturers have developed AVFs, predominantly using natural gas. Other promising fuel sources include direct electricity-powered vehicles (requiring improved battery technology) and hydrogen fuel cells. The major obstacles to the widespread use of AVFs are the high costs of the vehicles themselves and the need to develop the infrastructure for fueling AVFs on a massive scale. Nonetheless, significant progress has been made, for example, in gas-powered taxi fleets in Tokyo and state vehicles in California.

Car-Pooling Incentives: measures to encourage commuters to join with others rather than drive more vehicles. Car-pooling incentives include providing express lanes accessible only to vehicles with a minimum number of passengers, subsidizing car-pool vans, and charging more for parking at commuting destinations.

Catalytic Converters: pollution-control devices that remove carbon monoxide, hydrocarbons, and nitrous oxide pollutants from exhaust emissions through the chemical interaction between these pollutants and metals (usually platinum and rhodium) that serve as catalysts (i.e., stimulants of chemical reactions) for the oxidation processes that convert these pollutants into carbon dioxide, water, nitrogen gas, and other less noxious emissions. Catalytic converters were required in automobiles and other vehicles in the 1980s in developed countries; the requirement has gradually been extended into developing countries. The inability of catalytic converters to operate with leaded gasoline was one reason for the phase-out of leaded gasoline during this period. The challenge of introducing catalytic converters into developing countries is heightened by the additional costs they require of each vehicle, the need to retrofit older vehicles, and the difficulty of enforcing regulations requiring catalytic converters.

Corporate Average Fuel Efficiency Standards (CAFÉ Standards). See FLEET FUEL EFFICIENCY TARGETS.

Fleet Fuel Efficiency Targets: targets for improving the average fuel efficiency (expressed as miles per gallon or kilometers per liter) of automobiles or other vehicles sold by a manufacturer in a given target year. In the United States, these requirements were introduced with the 1975 Corporate Average Fuel Efficiency (CAFÉ) standards, which called for an average 27.5 miles per gallon for automobiles and 20.7 miles per gallon for light trucks. While these standards were met by the 1980s, environmental groups have been pressing for much more ambitious standards, arguing that automobile gasoline mileage ought to be increased to 45 miles per gallon for cars and 34 miles per gallon for light trucks. They contend that the federal government's failure to increase the standards has permitted manufacturers to devote the improvements in engine and drivetrain design to allow for heavier vehicles rather than to reduce fuel consumption and pollution. Oil consumption is a major consideration; automobiles and trucks account for roughly 40 percent of the oil consumed in the United States. Another consideration is the threat of greenhouse gas emissions, inasmuch as carbon dioxide emissions are essentially proportional to hydrocarbon fuel consumption.

COMMENTARY ON CORPORATE AVERAGE FUEL ECONOMY
(CAFÉ) AND FUEL EFFICIENCY

CAFÉ standards were found to have produced major improvements in transportation energy use through the 1980s. Alas, much of this ground has been lost due to the exception of the category of vans and light trucks (which also includes increasingly popular urban assault vehicles). *Steve Rayner*

Fuel Efficiency: 1) the usable energy that can be derived from a given fuel. Fuel efficiency is affected by both the type and purity of the fuel itself, and by the devices that convert fuel into work.

2) the efficiency of fuel consumption by vehicles, usually expressed in terms of miles per gallon or kilometers per liter for automobiles and trucks. Vehicle fuel efficiency depends largely on the efficiency of the engines and the weight of the vehicles. The 1973–74 energy crisis, and subsequent high oil prices, created intense demand for more fuel-efficient vehicles, but the decline in the real price of gasoline, diesel, and other transportation fuels dampened this demand in the 1980s and 1990s. Heavier vehicles, reflecting safety concerns as well as consumer tastes, made a resurgence, particularly

in the 1990s in countries in which fuel prices declined in real terms. Therefore the major impetus for improvements in fuel efficiency has come from governments rather than from consumers. See also FLEET FUEL EFFICIENCY TARGETS.

Intermodal Split: the mix of different forms of transportation serving a particular area, and the proportions of use or facilities of these different transportation modes. For example, a corridor between two cities may be served by highways handling 50 percent of the traffic; a railroad handling 30 percent; and aircraft handling 20 percent. The intermodal split is one indication of the diversity and efficiency of transportation: the split between automobile and mass transit modes is often a good indicator of the success of efforts to reduce energy consumption and congestion.

Mass Transit: transportation modes that have a high density of passengers per vehicle, especially for travel within metropolitan areas. Rail, bus, and subway transport are the primary modes of mass transit. Mass transit is widely viewed as superior to the predominance of private automobiles, which often cause congestion, higher fuel consumption, and greater pollution. However, the neglect of mass transit systems in many localities arises from the greater inconvenience of passengers to get to and from mass transit stations, and the large investments that mass transit systems entail. Mass transit systems often deteriorate unless users are charged sufficiently to cover operating, maintenance, and modernization costs.

Non-Point-Source Pollution: pollution caused by multiple sources of emissions, each contributing a small portion of the total pollution, as contrasted with point sources such as factories or refineries. Motor vehicles account for a large proportion of non-point-source pollution. Non-point sources of pollution are typically more difficult to monitor or regulate than are point sources.

Transportation Infrastructure: the facilities—roads, railbeds and railroad stations, ports, airports, regulatory institutions—that permit the transport of people and goods. The weakness of a transportation infrastructure is an enormous obstacle to economic development in many developing countries, because of the high costs of transporting agricultural produce, industrial goods, and raw materials. It can even be argued that national boundaries have been shaped by transportation limitations, in that inaccessible areas often develop autonomously (e.g., the nineteenth century breakup of the Central American Federation and many contemporary de facto secessionist movements in Africa). In particular, the costs of building roads are often enormous, especially in areas of difficult terrain. Because of the magnitude of investments, transportation infrastructure development is often subject to corruption and RENT-SEEKING. The construction of transportation infrastructure often has environmental effects that can only be anticipated through careful ENVIRONMENTAL IMPACT ASSESSMENTS.

Transportation-System Maintenance: the upkeep of transportation infrastructure such as roads and railbeds. The sustained usefulness of transportation systems depends heavily on their maintenance, yet maintenance is neglected in many developing countries, often because there is little political gain in funding maintenance compared to building new roads and other transportation facilities. In many circumstances, the sensible strategy of charging transportation users to provide sustained maintenance is difficult to implement because of its unpopularity.

References

Acharya, Anjali. "Small Islands Awash in a Sea of Troubles." *World Watch* (November/December 1995): 24–33.

Adriannse, Albert, et al. *Resource Flows: The Material Basis of Industrial Economies.* Washington, DC: World Resources Institute, 1997.

Agarwal, Anil, and Sunita Narain. "Global Warming in an Unequal World: A Case of Environmental Colonialism." *Earth Island Journal* (Spring 1991): 39–40.

Aghion, Philippe, and Peter Howitt. *Endogenous Growth Theory.* Cambridge, MA: MIT Press, 1998.

Ahmad, Yusuf J., Salah El Serafy, and Ernst Lutz, eds. *Environmental Accounting for Sustainable Development.* Washington, DC: World Bank, 1989.

Allenby, Braden R., and Deanna J. Richards, eds. *The Greening of Industrial Ecosystems.* Washington, DC: National Academy Press, 1994.

Anderson, Philip E. "More is Different." *Science* 177 (4 August 1972): 393–96.

Anderson, Terry, and Donald Leal. *Enviro-Capitalists. Doing Good While Doing Well.* Lanham, MD: Rowman and Littlefield, 1997.

Andraca, R. de, and K. McCready. *Internalizing Environmental Costs to Promote Eco-Efficiency.* Business Council for Sustainable Development, 1994.

Archer, Jack H., et al. *The Public Trust Doctrine and the Management of America's Coasts.* Amherst: University of Massachusetts Press, 1994.

Arrow, Kenneth. *Economic Growth, Carrying Capacity, and the Environment.* Cambridge: Cambridge University Press, 1996.

Art, R., and R. Jervis, eds. *International Politics: Enduring Concepts and Contemporary Issues.* New York: HarperCollins, 1996.

Ascher, William, and Robert G. Healy. *Natural Resource Policymaking in Developing Countries.* Durham, NC: Duke University Press, 1990.

Asian Development Bank. *Economic Policies for Sustainable Development.* Manila: Asian Development Bank, 1990.

Atkinson, Giles, et al. *Measuring Sustainable Development: Macroeconomics and the Environment.* Lyme, NH: Edward Elgar, 1997.

Ausubel, Jesse H., and Hedy E. Sladorich, eds. *Technology and Environment.* Washington, DC: National Academy Press, 1989.

Auty, R. M. *Sustaining Development in Mineral Economies: The Resource Curse Thesis.* Oxford: Clarendon Press, 1993.

Ayres, Robert U. "Industrial Metabolism." In *Technology and Environment,* edited by Jesse H. Ausubel and Hedy E. Sladorich. Washington, DC: National Academy Press, 1989.

Babcock, Richard. *The Zoning Game.* Madison: University of Wisconsin Press, 1956.

Baker, Susan, Maria Kousis, and Dick Richardson, eds. *The Politics of Sustainable Development: Theory, Policy and Practice within the European Union.* New York: Routledge, 1997.

Banfield, Edward. *The Moral Basis of a Backward Society.* Chicago: Free Press, 1958.

Barbier, E., J. Burgess, and C. Folke. *Paradise Lost? The Ecological Economics of Biodiversity.* London: Earthscan, 1994.

Barker, Terry, and Jonathan Kohler, eds. *International Competitiveness and Environmental Policies.* Cheltenham, UK: Edward Elgar, 1998.

Barlett, Scott. *Economic Development and Environmental Policy.* Rome: UN Food and Agriculture Organization, 1996.

Barnett, Richard, and John Cavanagh. *Global Dreams, Imperial Corporations and the New World Order.* New York: Simon and Schuster, 1994.

Barney, Gerald O., ed. *The Global 2000 Report to the President: Entering the Twenty-First Century.* New York: Pergamon, 1980.

Bartelmus, Peter. *Environment, Growth and Development: The Concept and Strategies of Sustainability.* London: Routledge, 1994.

Batchelor, Martine, and Kerry Brown, eds. *Buddhism and Ecology.* London: Cassell, 1992.

Beck, Ulrich. *Risk Society: Towards a New Modernity.* London: Sage, 1992.

———, Anthony Giddens, and Scott Lash. *Reflexive Modernization.* Oxford: Polity, 1994.

Bellamy, Ian. *The Environment in World Politics: Exploring the Limits.* Cheltenham, UK: Edward Elgar, 1997.

Beltratti, Andrea. *Sustainable Growth and the Green Golden Rule.* Cambridge, MA: National Bureau of Economic Research, 1993.

Benton, Ted. *Natural Relations: Ecology, Animal Rights and Social Justice.* London: Verso, 1993.

———, ed. *The Greening of Marxism.* New York: Guilford Press, 1996.

Bergh, Jeroen. *Ecological Economics and Sustainable Development: Theory, Methods, and Applications.* Cheltenham, UK: Edward Elgar, 1996.

Berkes, Fikret. *Sacred Ecology.* Philadelphia: Taylor and Francis, 1999.

———, ed. *Common Property Resources: Ecology and Community-Based Sustainable Development.* London: Belhaven, 1989.

———, and Carl Folke. "A Systems Perspective on the Interrelations between Natural, Human-made and Cultural Capital." *Ecological Economics* 5 (1) (1992): 1–8.

———, eds. *Linking Social and Ecological Systems Management Practices and Social Mechanisms.* Cambridge: Cambridge University Press, 1998.

Benarde, Melvin A. *Global Warning . . . Global Warming.* New York: John Wiley, 1992.

Berry, Michael A., and Dennis A. Rondinelli. "Proactive Environmental Management: A New Industrial Revolution." *Academy of Management Executives* 12(2) (1998): 38–50.

Bertalanfi, Ludwig von. *General System Theory.* New York: George Braziller, 1968.

Birnie, Patricia W., and Alan E. Boyle. *International Law and the Environment.* Oxford: Clarendon Press, 1992.

Bodansky, Daniel. "The United Nations Framework Convention on Climate Change: A Commentary." *Yale Journal of International Law* 18 (1993): 451–558.

Body, Richard. *Agriculture: The Triumph and the Shame.* London: Maurice Temple Smith, 1982.

Bogdanov, A. *Tektology: The Universal Organizational Science.* St. Petersburg, 1993.

Bojo, Jan. *Environment and Development: An Economic Approach.* Dordrecht, Netherlands: Kluwer, 1990.

Bookchin, Murray. *The Philosophy of Social Ecology: Essays of Dialectical Naturalism.* New York: Black Rose Books, 1995.

Booth, N. *How Soon is Now? The Truth about the Ozone Layer.* Hemstead, UK: Simon and Schuster, 1994.

Boserup, Ester. *Women's Role in Economic Development.* London: Allen and Unwin, 1970.

———. *Economic and Demographic Relationships in Development.* Baltimore, MD: Johns Hopkins University Press, 1990.

Bourdieu, Pierre. *Distinction: A Social Critique of the Judgment of Taste.* Cambridge, MA: Harvard University Press, 1984.

Boyle Alan E., ed. *Environmental Regulation and Economic Growth.* Oxford: Clarendon Press, 1994.

Braidotti, Rosi, Ewa Charkiewicz, Sabine Hausler, and Saskia Wieringa. *Women, the Environment and Sustainable Development: Towards a Theoretical Synthesis.* London: Zed Books, 1994.

Brandon, Katrina, Kent H. Redford, and Steven E. Sanderson, eds. *Parks in Peril: People, Politics, and Protected Areas.* Washington, DC: Island Press, 1998.

Brekke, Kjell Arne. *Economic Growth and the Environment: On the Measurement of Income and Welfare.* Cheltenham, UK: Edward Elgar, 1997.

Brunner, Ronald D., and Garry D. Brewer. *Organized Complexity: Empirical Theories of Political Development.* New York: Free Press, 1971.

Brunner, Ronald D., and Robin Sandenburgh, eds. *Community Energy Options: Getting Started in Ann Arbor.* Ann Arbor: University of Michigan Press, 1982.

Brush, Stephen B., and Doreen Stabinsky, eds. *Valuing Local Knowledge: Indigenous People and Intellectual Property Rights.* Washington, DC: Island Press, 1996.

Bryner, Gary. *From Promises to Performance: Achieving Global Environmental Goals.* New York: Norton, 1997.

Buell, John, and Tom DeLuca. *Sustainable Democracy: Individuality and the Politics of the Environment.* Thousand Oaks, CA: Sage, 1996.

Burton, John, ed. *Conflict: Human Needs Theory.* London: Macmillan, 1990.

Caincross, Frances. *Green, Inc.: A Guide to Business and Environment.* Washington, DC: Island Press, 1995.

Caldwell, Lynton. *International Environmental Policy.* 2d ed. Durham, NC: Duke University Press, 1996.

Campbell, Dennis, ed. *International Environmental Law and Regulation.* New York: John Wiley, 1997.

Canadian Broadcasting Corporation. "*The Nature of Things*—Fisheries: Beyond the Crisis." Show 10, 1997–1998, Toronto, Ontario.

Cardoso, Fernando Henrique, and Enzo Faletto. *Dependency and Development in Latin America.* Berkeley: University of California Press, 1979.

Carraro, Carlo, and Jerzy A. Filar, eds. *Control and Game—Theoretic Models of the Environment.* Boston: Birkhäuser, 1995.

Ceballos-Lascurain, Hector. *Tourism, Ecotourism and Protected Areas.* Gland, Switzerland: International Union for Conservation of Nature, 1996.

Chertov, Marian, and Daniel Esty, eds. *Thinking Ecologically.* New Haven: Yale University Press, 1977.

Chichilnisky, Graciela, Geoffrey Heal, and David A. Starrett. *International Emission Permits: Equity and Efficiency.* San Francisco: Annual Congress of ASSA, 1996.

Chichilnisky, Graciela, Geoffrey Heal, and Alessandro Vercelli, eds. *Sustainability: Dynamics and Uncertainty.* Dordrecht, Netherlands: Kluwer, 1998.

Cicin-Sain, Biliana, and Robert Knecht. *Integrated Coastal and Ocean Management: Concepts and Practice.* Washington, DC: Island Press, 1998.

Clark, Tim W. *Averting Extinction: Reconstructing Endangered Species Recovery.* New Haven: Yale University Press, 1997.

———. "Conservation Biologists in the Policy Process: Learning How to Be Practical and Effective." In *Principles of Conservation Biology*, 2d ed., edited by G. K. Meffe and C. R. Carroll, 575–97. Sunderland, MA: Sinauer Associates, 1997.

———, Richard P. Reading, and Alice L. Clarke, eds. *Endangered Species Recovery: Finding the Lessons, Improving the Process*. Washington, DC: Island Press, 1994.

Clarke, John, and Léon Tabah, eds. 1995. *Population, Environment, Development—Interactions*. Paris: CICRED, 1995.

Clayton, Anthony, and Nicholas J. Radcliffe. *Sustainability: A Systems Approach*. Boulder, CO: Westview Press, 1996.

Clayton, Keith M., and Timothy O'Riordan. "Coastal Processes and Management." In *Environmental Science for Environmental Management*, edited by Timothy O'Riordan. London: Longman, 1995.

Clifford, Mary, ed. *Environmental Crime: Enforcement, Policy and Social Responsibility*. Gaithersburg, MD: Aspen Publications, 1998.

Club of Rome. *The First Global Revolution. Report of the Council of the Club of Rome*. New York: Club of Rome, 1991.

Cohen, Joel E. *How Many People Can the Earth Support?* New York: W. W. Norton, 1995.

Convention on the Protection of the Marine Environment of the Baltic Sea Area. Helsinki, 1992.

Cosbey, Aaron, and David Runnalls. *Trade and Sustainable Development. A Survey of the Issues and a New Research Agenda*. Winnipeg: IISD, 1992.

Costanza, Robert. "Embodied Energy, Energy Analysis and Economics." In *Energy, Economics and Environment*, edited by Herman Daly and Alvaro Umaña. Boulder, CO: Westview Press, 1981.

Costanza, Robert, John Cumberland, Herman Daly, Robert Goodland, and Richard Norgaard. *An Introduction to Ecological Economics*. Boca Raton, FL: St. Lucie Press, 1997.

Cottrell, William Frederick. *Energy and Society: The Relation Between Energy, Social Changes, and Economic Development*. New York: McGraw-Hill, 1955.

Crane, Barbara. "International Population Institutions: Adaptation to a Changing World Order." In *Institutions for the Earth: Sources of Effective International Environmental Protection*, edited by Peter M. Haas, Robert O. Keohane, and Marc A. Levy, 351–92. Cambridge, MA: MIT Press, 1993.

Cropper, M., and C. Griffiths. "The Interaction of Population Growth and Environmental Quality." *American Economic Review: Papers and Proceedings* 84(2) (May 1994): 250–54.

Crosby, A. W. *Ecological Imperialism 900–1900*. Cambridge: Cambridge University Press, 1986.

Dalal-Clayton, Barry. *Getting to Grips with Green Plans*. London: Earthscan, 1996.

Daly, Herman E., ed. *Towards a Steady State Economy*. San Francisco: Freeman, 1973.

———. *Steady-State Economics: The Economics of Biophysical Equilibrium and Moral Growth*. San Francisco: W. H. Freeman, 1977.

———. *Beyond Growth: the Economics of Sustainable Development*. Boston: Beacon Press, 1996.

———, and John B. Cobb. *For the Common Good: Redirecting the Economy towards Community, the Environment and a Sustainable Future*. Boston: Beacon Press, 1989.

Dasgupta, Partha, and Karl-Göran Mäler, eds. *The Environment and Emerging Development Issues*. Oxford: Clarendon Press, 1997.

Dasmann, Raymond F. *Environmental Conservation*. 5th ed. New York: John Wiley, 1984.

David, Wilfried. *The Conversation of Economic Development. Historical Voices, Interpretations, and Reality*. Armonk, NY: M. E. Sharpe, 1997.

Davies, Arthur. *Forty Years of Progress and Achievement: A Historical Review of the WMO*. Geneva: World Meteorological Organization, 1990.

Davis, Shelton H., and Katrinka Ebbe. *Traditional Knowledge and Sustainable Development*. Washington, DC: World Bank, 1995.

Deelstra, Tjeerd. "The European Sustainability Index Project." In *A Sustainable World*, edited by Thaddeus Trzyna. Sacramento: International Center for Environment and Public Policy, 1995.

DeSimone, Livio, and Frank Popoff. *Eco-efficiency: The Business Link to Sustainable Development.* Cambridge, MA: MIT Press, 1997.

De Souza, Anthony, and J. Brady Foust. *World Space-Economy.* Columbus, OH: Charles Merrill, 1979.

Dewan, M. L., ed. *Towards a Sustainable Society: Perceptions.* Delhi: Clarion Books, 1995.

Dieren, Wouter van, ed. *Taking Nature into Account.* New York: Springer-Verlag, 1995.

Dobson, Andrew. *Green Political Thought.* London: Unwin Hyman, 1990.

Dobson, A., and Lucardie, P., eds. *The Politics of Nature: Explorations in Green Political Theory.* London: Routledge, 1993.

Dogse, P., and B. von Droste. "Debt-for-Nature Exchanges and Biosphere Reserves." *MAB Digest* 6. Paris: UNESCO, 1990.

Douglas, Mary, Des Gasper, Steven Ney, and Michael Thompson. "Human Needs and Wants." In *Human Choice and Climate Change.* Vol. 1, *The Societal Framework*, edited by Steve Rayner and Elizabeth L. Malone. Columbus, OH: Battelle Press, 1998.

Dovers, S. "A Framework for Scaling and Framing Policy Problems in Sustainability." *Ecological Economics* 12(2) (1995).

Doyle, John, and Gabrielle J. Persley, eds. *Enabling the Safe Use of Biotechnology: Principles and Practice.* Washington, DC: World Bank, 1996.

Dragun, Andrew, and Kristin Jakobsson, eds. *Sustainability and Global Environmental Policy: New Perspectives.* Cheltenham, UK: Edward Elgar, 1997.

Dryzek, John. *Rational Ecology.* Oxford: Blackwell, 1987.

Duda, Alfred, and David La Roche. "Joint Institutional Arrangements for Addressing Transboundary Water Resource Issues." *Natural Resource Forum* 21(2) (1997): 127–37.

Dunlap, Riley E., and Angela G. Mertig, eds. *American Environmentalism.* New York: Taylor and Francis, 1992.

Eckersley, Robyn. *Environmentalism and Political Theory: Toward an Ecocentric Approach.* Albany: SUNY Press, 1992.

———, ed. *Markets, the State, and the Environment: Towards Integration.* South Melbourne: Macmillan Education Australia, 1995.

Economist. May 30, 1992. Survey, 7.

Ehrlich, Paul. *The Population Bomb.* London: Pan, 1972.

El Serafy, Salah. "The Proper Calculation of Income from Depletable Natural Resources." In *Environmental Accounting for Sustainable Development*, edited by Yusuf Ahmad, Salah El Serafy, and Ernst Lutz. Washington, DC: World Bank, 1989.

Elliott, E. D. 1997. "Toward Ecological Law and Policy." In *Thinking Ecologically*, edited by Marian R. Chertow and Daniel C. Esty. New Haven: Yale University Press, 1997.

Endberg-Pedersen, Poul, et al. *Assessment of NDP: Developing Capacity for Sustainable Human Development.* Copenhagen: Centre for Development Research, 1996.

Environmental Fund. *Behind the "Food Crisis."* Washington, DC: Environmental Fund, 1977.

Faucheux, Sylvie, Martin O'Connor, and Jan van der Straaten, eds. *Sustainable Development: Concepts, Rationalities and Strategies.* Dordrecht, Netherlands: Kluwer, 1998.

Fisher, Kurt, and Johan Schot, eds. *Environmental Strategies of Industry.* Washington, DC: Island Press, 1993.

Foreman, David. *Confessions of an Eco-Warrior.* New York: Crown Books, 1991.

Foster, John, ed. *Valuing Nature: Economics, Ethics and Environment.* London: Routledge, 1997.

Fox, Warwick. *Toward a Transpersonal Ecology: Developing New Foundations for Environmentalism.* Toronto: Shambhala Press, 1992.

Freese, Curtis, ed. *Harvesting Wild Species: Implications for Biodiversity Conservation.* Baltimore: Johns Hopkins University Press, 1997.

Fromm, Erich. 1976. *To Have or to Be.* New York: Harper and Row, 1976.

Gandhi, Mohandas Karamchand. *Hind Swaraj and Other Writings.* Cambridge: Cambridge University Press, 1997.

Garling, Tommy L., and Gary W. Evans, eds. *Environment, Cognition and Action: An Integrated Approach.* New York: Oxford University Press, 1991.

Garner, Robert. *Environmental Politics.* London: Prentice Hall/Harvester Wheatsheaf, 1996.

Georgescu-Roegen, Nicholas. *Energy and Economic Myths.* New York: Pergamon Press, 1976.

GESAMP. *The Review of the Health of the Oceans.* Paris: UNESCO, 1982.

Giddens, Anthony. *The Consequences of Modernity.* Oxford: Polity, 1990.

———. *Modernity and Self-Identity.* Oxford: Polity, 1991.

Gillespie, Alexander. *International Environmental Law, Policy and Ethics.* Oxford: Clarendon Press, 1997.

Gleditch, Nils Peter, ed. *Conflict and the Environment.* Dordrecht, Netherlands: Kluwer, 1997.

Global Environmental Management Initiative. *Total Quality Environmental Management: A Primer.* Washington, DC: GEMI, 1992.

Godana, Bonaya A. *African Shared Water Resources.* Boulder, CO: Lynne Rienner Publications, 1985.

Gold, Mary. *Sustainable Agriculture: Definitions and Terms.* Beltsville, MA: National Agricultural Library, 1994.

Goldsmith, Edward. *Blueprint for Survival.* Harmondsworth: Penguin, 1972.

Gore, Albert. *Earth in the Balance: Ecology and the Human Spirit.* Boston: Hougton Mifflin, 1992.

Gorz, Andre. *Ecology and Politics.* Boston: South End Press, 1980.

———. *Farewell to the Working Class.* London: Pluto Press, 1982.

———. *Paths to Paradise: On the Liberation from Work.* London: Pluto Press, 1985.

Gottlieb, Robert. *Forcing the Spring: The Transformation of the American Environmental Movement.* Washington, DC: Island Press, 1993.

Graham, John D., and Jonathan B. Wiener, eds. *Risk versus Risk: Tradeoffs in Protecting Health and Environment.* Cambridge: Harvard University Press, 1995.

Grossman, Gene, and Alan Krueger. *Environmental Impacts of a North American Free Trade Agreement.* Working paper no. 3914. Cambridge, MA: National Bureau of Economic Research, 1991.

Gunderson, Lance H., C. S. Holling, and Stephen S. Light, eds. *Barriers and Bridges to the Renewal of Ecosystems and Institutions.* New York: Columbia University Press, 1995.

Gupta, Joyeeta. *The Climate Change Convention and Developing Countries: From Conflict to Consensus?* Dordrecht, Netherlands: Kluwer, 1997.

Haas, Peter. "Do Regimes Matter? Epistemic Communities and Mediterranean Pollution Control." *International Organization* 43 (1989): 378–403.

———, Robert O. Keohane, and Marc A. Levy, eds. *Institutions for the Earth: Sources of Effective International Environmental Protection.* Cambridge, MA: MIT Press, 1993.

Hammond, Allen, Albert Adriaanse, Dirk Bryant, and R. Woodward. *Environmental Indicators: A Systematic Approach to Measuring and Reporting on Environmental Policy Performance in the Context of Sustainable Development.* Washington, DC: World Resources Institute, 1995.

Hanly, N. *Environmental Economics in Theory and Practice.* Oxford University Press, 1997.

Hanna, Susan, and Mohan Munasinghe, eds. *Property Rights and the Environment: Social and Ecological Issues.* Stockholm: Beijer International Institute of Ecological Economics, 1995.

Hardin, Garrett. "The Tragedy of the Commons." *Science* 162 (1968): 1241–48.

———. "Living on a Lifeboat." *BioScience* 24 (1974): 10.

———, and John Baden, eds. *Managing the Commons.* San Francisco: Freeman, 1977.

Harrison, D. "Whales under Stress as Man Crowds Sea," *The Observer,* October 18, 1998, 7.

Hartwick, John M., and Nancy D. Olewiler. *The Economics of Natural Resource Use.* Reading, MA: Addison-Wesley, 1998.

Harvey, David. *Justice, Nature and the Geography of Difference.* Oxford: Blackwell, 1996.

Hawken, Paul. *The Ecology of Commerce: A Declaration of Sustainability.* New York: Harper Business, 1993.

Heal, Geoffrey M. *Interpreting Sustainability.* Enrico Mattei Discussion paper 1.95. Milan: Fondazione ENI, 1995.

Healey, Michael J., and Brian W. Illbery. *Location and Change: Perspectives on Economic Geography.* Oxford: Oxford University Press, 1990.

Heidenheimer, Arnold J., Michael Johnston, and Victor T. LeVine, eds. *Political Corruption: A Handbook.* New Brunswick, NJ: Transaction Publishers, 1989.

Heinsbergen, P. van. *International Legal Protection of Wild Fauna and Flora.* Washington, DC: IOS Press, 1997.

Hemley, Ginette, ed. *International Wildlife Trade: A CITES Sourcebook.* Washington, DC: Island Press, 1994.

Henderson, Hazel. *Creating Alternative Futures: The End of Economics.* West Hartford, CT: Kumarian Press, 1996.

Hirschman, Albert O. *The Strategy of Economic Development.* New Haven, CT: Yale University Press, 1958.

Holland, John H. "Complex Adaptive Systems." *Daedalus* 121 (Winter 1992): 17–30.

Holling, C. S. "Resilience and Stability of Ecological Systems." *Annual Review of Ecology and Systematics* 4 (1973): 1–23.

———. *Adaptive Environmental Assessment and Management.* New York: John Wiley, 1978.

Hopkins, Terence K., and Immanuel Wallerstein, eds. *Processes of the World-System.* Beverly Hills, CA: Sage, 1980.

Horgan, John. "From Complexity to Perplexity." *Scientific American* (June 1995): 104–9.

Hotelling, Harold. "The Economics of Exhaustible Resources." *Journal of Political Economy* 39 (1931): 137–75.

Houghton, J. T., L. G. Meira Filho, B. A. Callander, N. Harris, A. Kattenberg, and K. Maskell. *Climate Change 1995: The Science of Climate Change.* New York: Cambridge University Press, 1996.

Hurrell, Andrew, and Benedict Kingsbury, eds. *The International Politics of the Environment.* Oxford: Clarendon Press, 1992.

Independent Commission on Population and Quality of Life. *Caring for the Future: Making the Next Decades Provide a Life Worth Living.* Oxford: Oxford University Press, 1996.

International Union for the Conservation of Nature (IUCN). *United Nations List of National Parks and Protected Areas.* Gland, Switzerland: IUCN, 1985.

———, with the UN Environment Programme and World Wildlife Fund. *World Conservation Strategy.* Gland, Switzerland: IUCN, 1980.

———. *Caring for the Earth. A Strategy for Sustainable Living.* Gland, Switzerland: IUCN, 1991.

Jacobson, Jodi. *Environmental Refugees: A Yardstick of Habitability.* Washington, DC: Worldwatch, 1988.

———. "Holding Back the Sea." In *State of the World 1990*, edited by Lester Brown et al., 79–97. New York: Norton, 1990.

Jäger, Jill, and Howard L. Ferguson, eds. *Climate Change: Science, Impacts and Policy: Proceedings of the Second World Climate Conference.* Cambridge: Cambridge University Press, 1991.

Johnson, Perry. *ISO 14000: The Business Manager's Guide to Environmental Management.* New York: John Wiley, 1997.

Johnson, Stanley, and Guy Corcell. *The Environmental Policy of the European Communities.* 2d ed. London: Kluwer, 1995.

Joyner, Christopher. "Antarctica." In *A Global Agenda: Issues Before the 48th General Assembly of the United Nations*, edited by John Tessitore and Susan Woolfson, 206–11. Lanham, MD: University Press of America, 1993.

Kahn, James R. *The Economic Approach to Environmental and Natural Resources.* New York: Dryden Press, 1995.

Karlsson, Gail V. "Habitat II." In *A Global Agenda: Issues Before the 51st General Assembly of the United Nations,* edited by John Tessitore and Susan Woolfson, 253–59. Lanham, MD: Rowman and Littlefield, 1996.

Kasperson, Jeanne, Roger Kasperson, and B. Turner II. "Regions at Risk." *Environment* (December 1996): 4–15.

Kauffman, Stuart. *At Home in the Universe: The Search for Laws of Self-Organization and Complexity.* New York: Oxford University Press, 1995.

Kegley, Charles W., Jr., and Eugene R. Wittkopf. *The Global Agenda: Issues and Perspectives.* Boston: McGraw-Hill, 1998.

Kiss, Alexandre, and Dinah Shelton. *International Environmental Law.* New York: Transnational Publishers, 1991.

Klitgaard, Robert. *Controlling Corruption.* Berkeley: University of California Press, 1988.

Kluchevsky, V. O. *Course in Russian History.* Vol. I. Moscow: Politizdat, 1987.

Kondratieff, Nikolai Dmitrievich. "The Long Waves in Economic Life." *Review of Economic and Statistics* 17 (part 2) (November 1925).

———. *The Works of Nikolai D. Kondratiev.* Trans. Stephen S. Wilson, eds. Natalia Makasheva, Warren J. Samuels, and Vincent Barnett. London: Pickering and Chatto, 1998.

Koppen, I. J. "The Role of the European Court of Justice." In *European Integration and Environmental Policy,* edited by J. D. Lifferink, P. D. Lowe, and A. P. J. Mol. London: Belhaven Press, 1993.

Kravis, Irving B., Alan Heston, and Robert Summers. *International Comparisons of Real Product and Purchasing Power.* Baltimore: Johns Hopkins University Press, 1978.

Kropotkin, Peter. *Evolution and Environment.* Montreal: Black Rose Books, 1995.

Krugman, Paul R., and Maurice Obstfeld. *International Economics.* Reading, MA: Addison-Wesley, 1997.

Kuik, Onno, Paul Peters, and Nico Schrijver, eds. *Joint Implementation to Curb Climate Change: Legal and Economic Aspects.* Dordrecht, Netherlands: Kluwer, 1994.

Lang, Winfried, ed. *Sustainable Development and International Law.* London: Graham and Trotman, 1995.

Larkin, P. "An Epitaph for the Concept of Maximum Sustained Yield." *Transactions of the American Fisheries Society* 106, 1977.

Lasswell, Harold D. *A Pre-View of Policy Sciences.* New York: Elsevier, 1971.

Latouch, S. *Post-Development.* London: Zed Books, 1993.

Law Reform Commission of Canada. *Protection of Life: Crimes against the Environment.* Ottawa: Law Reform Commission, 1985.

Lee, N. K. "Options for Environmental Policy." *Science* 182 (November 30, 1973): 911–12.

Leibenstein, Harvey. *A Theory of Economic-Demographic Development.* Princeton, NJ: Princeton University Press, 1954.

Leonard, H. Jeffrey. *Are Environmental Regulations Driving United States Industry Overseas? An Issue Report.* Washington, DC: Conservation Foundation, 1984.

Leontief, Wassily, et al. *A Study on the Impact of Prospective Economic Issues and Policies on the International Development Strategy.* New York: United Nations, 1977.

Lesser, William. *Sustainable Use of Genetic Resources under the Convention of Biodiversity.* New York: CAB International, 1998.

Lévi-Strauss, Claude. *The Savage Mind.* Chicago: University of Chicago Press, 1966.

Lewis, Martin, and Karen Wigen. *The Myth of Continents. A Critique of Metageography.* Berkeley: University of California Press, 1997.

Lewis, Sanford. *The Good Neighbor Handbook.* Acton, MA: The Good Neighbor Project, 1992.

Liefferink, D., P. Lowe, and T. Mol. *European Integration and Environmental Policy*. London: Belhaven, 1993.

Lindberg, Kreg, and Donald Hawkins. *Ecotourism: A Guide for Planners and Managers*. North Bennington, VT: Ecotourism Society, 1993.

Linnerooth-Bayer, Joanne, and Benjamin Davy. *Hazardous Waste Cleanup and Facility Siting in Central Europe: The Austrian Case, Report to the Bundesministerium für Wissenchaft und Forschung*. Laxenburg. Austria: International Institute for Applied Systems Analysis, 1994.

Litfin, Karen. *Ozone Discourses: Science and Politics in Global Environmental Protection*. New York: Columbia University Press, 1994.

Lopez-Ocana, Ramon. "Effectiveness of International Regimes Dealing with Desertification from the Perspective of the South." In *Global Governance: Drawing Insights from the Environmental Experience*, edited by Oran Young, 125–35. Cambridge, MA: MIT Press, 1997.

Lovelock, James. *Gaia: A New Look at Life on Earth*. New York: Oxford University Press, 1979.

Lovins, Amory. *Soft Energy Paths*. Cambridge, MA: Ballinger, 1977.

Low, Nicholas, and Brendan Gleeson. *Justice, Society and Nature: An Exploration of Political Ecology*. London: Routledge, 1998.

Ludwig, D., R. Hilborn, and C. Walters, "Uncertainty, Resource Exploitation, and Conservation: Lessons from History." *Science* 260 (2) (April 1993).

Lyster, Simon. *International Wildlife Law*. Cambridge: Grotius, 1985.

———. "Effectiveness of International Regimes Dealing with Biological Diversity from the Perspective of the North." In *Global Environmental Change and International Governance*, edited by Oran Young, George Demko, and Kilaparti Ramakrishna. Hanover, NH: University Press of New England, 1996.

MacNeill, Jim, Pieter Winsemius, and Taizo Yakushiji. *Beyond Interdependence: The Meshing of World Economy and the Earth's Ecology*. Oxford: Oxford University Press, 1991.

Mander, Jerry, and Edward Goldsmith, eds. *The Case Against the Global Economy*. San Francisco: Sierra Club, 1996.

Martell, Luke. *Ecology and Society*. Cambridge: Polity, 1994.

Maslow, Abraham H. *Motivation and Personality*. New York: Harper, 1954.

Max-Neef, Manfred, Antonio Marin Elizalde, and Martin Hopenhayn. *Human Scale Development: An Option for the Future*. Uppsala: Dag Hammarskjold Foundation, 1990.

May, Peter H., and Ronaldo Serôa da Motta, eds. *Pricing the Planet: Economic Analysis for Sustainable Development*. New York: Columbia University Press, 1996.

McConnell, Grant. *Private Power and American Democracy*. New York: Knopf, 1966.

McCormick, John. *Acid Earth: The Global Threat of Acid Pollution*. London: Earthscan, 1989.

———. *The Global Environmental Movement*. London: Belhaven, 1989.

———, ed. *Acid Earth: The Politics of Acidification*. Washington, DC: Island Press, 1997.

McDougal, Myres, Harold D. Lasswell, and Lung-chou Chen. "The Basic Policies of a Comprehensive Public Order of Human Dignity." In *Human Rights and Public Order: The Basic Policies of an International Law of Human Dignity*, edited by Myres McDougal, Harold D. Lasswell, and Lung-chou Chen. New Haven: Yale University Press, 1980.

McMichael, Philip. *Development and Social Change, a Global Perspective*. Thousand Oaks, CA: Pine Forge Press, 1996.

McNeely, J. *Economics and Biological Diversity: Developing and Using Economic Incentives to Conserve Biological Resources*. Gland, Switzerland: IUCN, 1988.

Meadows, Donella H., Dennis L. Meadows, and Jorgen Randers. *Beyond the Limits: Global Collapse or a Sustainable Society: Sequel to the Limits to Growth*. London: Earthscan, 1992.

———, and William W. Behrens III. *The Limits to Growth: A Report for the Club of Rome's Project on the Predicament of Mankind*. New York: Universe, 1972.

Mechnikov, Lev Ilich. *La civilisation et les grands fleuves historiques*. Paris: Hachette, 1889.

————.*Civilization and Great Historical Rivers. Geographical Theory of Development of Modern Societies.* Moscow, 1898 (in Russian).

Meffe, G. K., et al.1997. *Principles of Conservation Biology.* Sunderland, MA: Sinauer Associates, 1997.

Mega, Voula. *European Cities in Search of Sustainability: A Panorama of Urban Innovations in the European Union.* Dublin: European Foundation for the Improvement of Living and Working Conditions, 1997.

Mies, Maria, and Vandana Shiva, eds. *Ecofeminism.* New Delhi: Kali for Women, 1993.

Milbraith, Lester W. "Culture and the Environment in the United States." *Environmental Management* 9 (1985): 161–72.

Miller, George Tyler. *Living in the Environment.* Belmont, CA: Wadsworth, 1992.

Ministry of Environment of the Russian Federation. *State of the Environment.* Moscow: Gidrometizdat, 1997 (in Russian).

Mishan, Ezra. *The Costs of Economic Growth.* London: Staples Press, 1967.

Mitchell, Bruce. *Resource and Environmental Management.* Harlow: Longman, 1997.

Mitchell, Ronald. *Intentional Oil Pollution of the Sea: Environmental Policy and Treaty Compliance.* Cambridge, MA: MIT Press, 1994.

Molina, Mario J., and F. Sherwood Rowland. "Stratospheric Sink for Chlorofluoromethanes: Chlorine Atom-Catalysed Destruction of Ozone." *Nature* 249 (June 1974): 810–12.

Moltke, Konrad von. *The Maastricht Treaty and the Winnipeg Principles on Trade and Sustainable Development.* Winnipeg: International Institute for Sustainable Development, 1995.

————. *International Environmental Management, Trade Regimes and Sustainability.* Winnipeg: International Institute for Sustainable Development, 1996.

Monbiot, George. "The Tourist Trap." *The Guardian Supplement.* December 12, 1998, 7.

Moore, David B., and Gerald J. Schmitz, eds. *Debating Development Discourse: Institutional and Popular Perspectives.* New York: St. Martin's Press, 1995.

Munn, R. E., J. W. M. la Rivière, and N. van Lookeren Campagne, eds. *Policy Making in an Era of Global Environmental Change.* Dordrecht, Netherlands: Kluwer, 1996.

Munro, David A., and Martin W. Holdgate, eds. *Caring for the Earth: A Strategy for Sustainable Living.* Gland, Switzerland: IUCN, UNEP, WWF, 1991.

Myrdal, Gunnar. *Economic Theory and Underdeveloped Regions.* London: Duckworth, 1957.

Naisbett, John. *Global Paradox: The Bigger the World Economy, the More Powerful Its Smallest Players.* New York: William Morrow, 1994.

Nelson, James Gordon, and Rafal Serafin, eds. *National Parks and Protected Areas: Keystones to Conservation and Sustainable Development.* New York: Springer, 1997.

Noman, Omar. *Economic Development and Environmental Policy.* London: Kegan Paul International, 1996.

Norgaard, Richard. "Coevolutionary Development Potential." *Land Economics* 60(2) (1984): 160–73.

————. *Development Betrayed: The End of the Progress and a Co-Evolutionary Revisioning of the Future.* London: Routledge, 1994.

North, Klaus. *Environmental Business Management: An Introduction.* Geneva: International Labour Office, 1997.

Noss, Reed E., and Allen Y. Cooperrider. *Saving the Nature's Legacy: Protecting and Restoring Biodiversity.* Washington, DC: Island Press, 1996.

O'Connor, James. "The Second Contradiction of Capitalism." In *The Greening of Marxism,* edited by Ted Benton, 197–221. New York: Guilford Press, 1996.

Olson, Mancur. *The Logic of Collective Action.* Cambridge, MA: Harvard University Press, 1965.

Ophuls, William. *Ecology and the Politics of Scarcity: Prologue to the Political Theory of the Steady State.* San Francisco: W. H. Freeman, 1977.

Opschoor. Johannes Baptist. *Environment, Economy, and Sustainable Development.* Groningen: Wolters-Noordhoff, 1992.

Organization for Economic Cooperation and Development (OECD). *Renewable Natural Resources: Economic Incentives for Improved Management* Paris: OECD, 1989.

———. *Responding to Climate Change: Selected Economic Issues.* Paris: OECD, 1991.

———. *Climate Change: Designing a Tradeable Permit System.* Paris: OECD, 1992.

———. *Convention on Climate Change: Economic Aspects of Negotiations.* Paris: OECD, 1992.

———. *Workshop on Taxation and Environment in European Economies in Transition. Rappoteurs' Summary of the Discussions and Conclusions.* Paris: OECD, 1993.

———. *Managing the Environment.* Paris: OECD, 1994.

———. *Natural Resource Accounts: Taking Stock in OECD Countries.* Paris: OECD, 1994.

———. *Subsidies and Environment: Exploring the Linkages.* Paris: OECD, 1996.

———. *Evaluating Economic Instruments for Environmental Policy.* Paris: OECD, 1997.

———. *Globalization and Environment: Preliminary Perspectives.* Paris: OECD, 1997.

———. *Sustainable Consumption and Production: Clarifying the Concepts.* Paris: OECD, 1997.

O'Riordan, Timothy. "The Challenge for Environmentalism." In *New Models in Geography,* edited by Richard Peet and Nigel Thrift, 77–102. London: Unwin Hyman, 1989.

———, ed. *Ecotaxation.* New York: St. Martin's Press, 1997.

———, and James Cameron, eds. *Interpreting the Precautionary Principle.* London: Earthscan, 1994.

Ottinger, R. "Incorporating Environmental Externalities through Pollution Taxes." Geneva, Switzerland: World Clean Energy Conference, 4–7 November 1991.

Oxman, Bernard H., David O. Caron, and Charles L. O. Buderi, eds. *The Law of the Sea: U.S. Policy Dilemma.* San Francisco: ICS Press, 1983.

Pahre, Robert. *Leading Questions: How Hegemony Affects the International Political Economy.* Ann Arbor: University of Michigan Press, 1999.

Panayotou, Theodore. *Green Markets: The Economics of Sustainable Development.* Cambridge, MA: Harvard Institute for International Development, 1993.

Park, Chris C. *Acid Rain: Rhetoric and Reality.* New York: Routledge, 1987.

Parsons, Talcott. *The Evolution of Societies.* Englewood Cliffs, NJ: Prentice-Hall, 1977.

Pearce, David, ed. *Blueprint 2: Greening the World Economy.* London: Earthscan, 1991.

———, ed. *Blueprint 3: Measuring Sustainable Development.* London: Earthscan, 1994.

———, ed. *Blueprint 4: Capturing Global Environmental Value.* London: Earthscan, 1995.

———, Edward Barbier, and Anil Markandya. *Sustainable Development and Environment in the Third World.* Worchester: Edward Elgar, 1990.

———, Anil Markandya, and Edward Barbier. *Blueprint for a Green Economy.* London: Earthscan, 1989.

———, and Dominic Moran. *The Economic Value of Biodiversity.* London: Earthscan, 1994.

———, and Paul Steele. *Private Financing for Sustainable Development.* New York: United Nations Development Programme, 1996.

———, and R. Kerry Turner. *Economics of Natural Resources and the Environment. Terms of Environment.* Washington, DC: U.S. Environmental Protection Agency, 1992.

Pepper, David. *Political Roots of Ecological Environmentalism.* London: Croom Helm, 1984.

———. *Modern Environmentalism.* London and New York: Routledge, 1996.

Perman, Roger, Yue Ma, and James McGilvray. *Natural Resource and Environmental Economics.* London: Longman, 1996.

Peterson, M. J. *Managing the Frozen South: The Creation and Evolution of the Antarctic Treaty System.* Berkeley: University of California Press, 1988.

Pimbert, M. B., and J. N. Pretty. *Parks, People and Professionals: Putting "Participation" into Protected Area Management.* United Nations Research Institute for Social Development discussion paper DP 57:1–60. New York: United Nations, 1995.

Pimm, Stuart. *The Balance of Nature*. Chicago: University of Chicago Press, 1991.

Pirages, Dennis, ed. *Building Sustainable Societies: A Blueprint for a Post-industrial World*. Armonk, NY: M. E. Sharpe, 1996.

Poore, Duncan, ed. *No Timber without Trees*. London: Earthscan, 1989.

Porter, Gareth, and Edith Brown. *Global Environmental Politics*. Boulder, CO: Westview Press, 1996.

Porter, Michael. *Competitive Advantage: Creating and Sustaining Superior Performance*. New York: Free Press, 1986.

———. "America's Green Strategy." *Scientific American* (April 1991).

Post, Jan, and Carl Lundin, eds. *Guidelines for Integrated Coastal Zone Management*. Washington, DC: World Bank, 1996.

President's Council on Sustainable Development. *Eco-Efficiency Task Force Report*. Washington, DC: President's Council on Sustainable Development, 1997.

Prigogine, Ilya, and Isabelle Stenders. *Order Out of Chaos: Man's New Dialogue with Nature*. New York: Bantam, 1984.

Protasov, V. F., and A. V. Molchanov. *Ecology, Health and Nature Use in Russia*. Moscow: Finansy I Statistika, 1995 (in Russian).

Pullman, Richard. "Redefining Security." *International Security* 8 (1983): 129–53.

Purchase, Graham. *Anarchism and Ecology*. Montreal: Black Rose Books, 1997.

Purdue, Derrick. Seeds of Change: Conserving Biodiversity and Social Movements. Doctoral dissertation, University of the West of England, Bristol, England, 1998.

Putnam, Robert D. *Making Democracy Work*. Princeton, NJ: Princeton University Press, 1993.

Raberg, Per, ed. *The Life Region: The Social and Cultural Ecology of Sustainable Development*. London: Routledge, 1997.

Ravallion, Martin. *Poverty Comparisons*. Washington, DC: World Bank, 1992.

Rawls, John. *A Theory of Justice*. Cambridge, MA: Harvard University Press, 1971.

Reed, David, ed. *Structural Adjustment, the Environment, and Sustainable Development*. London: Earthscan, 1996.

Reimers, Nikolai Fedorovich. *Protection of Nature and Environment*. Moscow: Prosveschenie, 1992 (in Russian).

Reimers, Nikolai Fedorovich, and Feliks Robertovich Shtil'mark. *Specially Protected Areas*. Moscow: Mysl, 1978 (in Russian).

Rendon, Rodolfo. "Regimes Dealing with Biological Diversity from the Perspective of the South." In *Global Environmental Change and International Governance*, edited by Oran Young, George Demko, and Kilaparti Ramakrishna, 173–77. Hanover, NH: University Press of New England, 1996.

Rifkin, Jeremy, with Ted Howard. *Entropy: Into the Greenhouse World*. New York: Bantam, 1989.

Roan, Sharon. *Ozone Crisis: The 15 Year Evolution of a Sudden Global Emergency*. New York: John Wiley, 1989.

Rodenburg, Eric, Dan Tunstall, and Frederik van Bolhuis. *Environmental Indicators for Global Cooperation*. GEF Working Paper #11. Washington, DC: Global Environment Facility, 1995.

Roe, D., B. Dalal-Clayton, R. Hughes. *A Directory of Impact Assessment Guidelines*. London: International Institute for Environment and Development, 1995.

Roe, Emery. *Taking Complexity Seriously: Policy Analysis, Triangulation, and Sustainable Development*. Boston: Kluwer, 1998.

Rondinellli, Dennis A., and Michael A. Berry. "Industry's Role in Air Quality Improvement: Environmental Management Opportunities for the 21st Century." *Environmental Quality Management* 7(4) (1997): 31–44.

Rostow, Walt W. *The Stages of Economic Growth: A Non-Communist Manifesto*. Cambridge: Cambridge University Press, 1960.

Rump, Paul. *State of the Environment Reporting: Source Book of Methods and Approaches.* Ottawa: State of the Environment Reporting Directorate, 1996.

Sachs, A., ed. *Eco-justice: Linking Human Rights and the Environment.* Washington, DC: Worldwatch, 1995.

Sachs, Wolfgang, et al. *Greening the North: A Post-Industrial Blueprint for Ecology and Equity.* London: Zed Books, 1998.

Sale, Kirkpatrick. *Dwellers in the Land: The Bioregional Vision.* San Francisco: Sierra Club, 1985.

———. *The Green Revolution: The American Environmental Movement 1962–1992.* New York: Hill and Wang, 1993.

Samuelson, Paul A., and William D. Nordhaus. *Economics.* 14th ed. New York: McGraw-Hill, 1992.

Sand, Peter. "The Vienna Convention is Adopted." *Environment* 27(5) (June 1985): 19–20, 40–43.

Sands, Phillippe, ed. *Greening International Laws.* London: Earthscan, 1993.

Schiavone, Guiseppe. *International Organizations.* New York: St. Martin's Press, 1993.

Schmidheiny, Stephan, and Frederico Zorraquín. *Financing Change: The Financial Community, Eco-efficiency, and Sustainable Development.* Cambridge, MA: MIT Press, 1996.

Schmidtt-Bleek, F., ed. *Carnoules Declaration. Factor 10 Club.* Wuppertal: WIKUE, 1994.

Schneider, Stephen Henry. *Laboratory Earth: The Planetary Gamble We Can't Afford to Lose.* New York: Basic Books, 1994.

Schrecker, Ted, and Jean Dalgleish, eds. *Growth, Trade and Environmental Values.* London, Ont.: Westminster Institute for Ethics and Human Values, 1994.

Schrecker, Ted, ed. *Surviving Globalism: The Social and Environmental Challenges.* New York: St. Martin's Press, 1997.

Schumacher, Ernst Friedrich. *Small is Beautiful: Economics as If People Mattered.* New York: Harper and Row, 1973.

———. *Good Work.* London: Jonathan Cape, 1979.

———. *A Guide to the Perplexed.* London: Jonathan Cape, 1979.

Sen, Gita, and Caren Grown. *Development Crises and Alternative Visions: Third Women's Perspectives.* London: Earthscan, 1988.

Serageldin, Ismail. *Sustainability and the Wealth of Nations: First Steps in an Ongoing Journey.* Washington, DC: World Bank, 1996.

Serageldin, Ismail, and Joan Martin-Brown, eds. *Servicing Innovative Financing of Environmentally Sustainable Development.* Washington, DC: World Bank, 1996.

Serageldin, Ismail, and Alfredo Sfeir-Younis, eds. *Effective Financing of Environmentally Sustainable Development.* Washington, DC: World Bank, 1996.

Serageldin, Ismail, and Andrew Steer, eds. *Making Development Sustainable: From Concepts to Action.* Washington, DC: World Bank, 1994.

Sharma, A. K. *Gandhian Perspectives on Population and Development.* New Delhi: Concept Publishing Company, 1996.

Sheth, D. L. "Alternative Development as Political Practice." *Alternatives* 12 (1987): 155–171.

Shiva, Vandana. *Staying Alive: Women, Ecology, and Development.* London: Zed Books, 1989.

———. "Piracy by Patent: The Case of the Neem Tree." In *The Case Against the Global Economy,* edited by Jerry Mander and Edward Goldsmith. San Francisco: Sierra Club Books, 1996.

Shuldiner, A., and T. Raymond. *Who's in the Lobby?* Washington, DC: Center for Responsive Politics, 1998.

Shuman, James, and David Rosenau. *The Kondratieff Wave.* New York: World, 1972.

Simon, Herbert A. *Models of Man.* New York: John Wiley, 1957.

Simon, Julian, and Herman Kahn. *The Resourceful Earth.* Oxford: Basil Blackwell, 1984.

Sjostedt, Gunnar, ed. *International Environmental Negotiations: Process, Issues and Contexts.* Stockholm: FRN, 1991.

Soloviov, S. M. *History of Russia.* Vol. I. Moscow: Golos (in Russian), 1993.

Solow, Robert M. *Growth Theory: An Exposition.* New York: Oxford University Press, 1970.

Soroos, Marvin S. *The Endangered Atmosphere: Preserving a Global Commons.* Columbia: University of South Carolina Press, 1997.

———. "The Tragedy of the Commons in Global Perspective." In *The Global Agenda: Issues and Perspectives,* edited by Charles W. Kegley, Jr. and Eugene R. Wittkop, 473–86. Boston: McGraw-Hill, 1998.

Sprout, Harold, and Margaret Sprout. *The Ecological Perspective on Human Affairs.* Princeton: Princeton University Press, 1965.

Srivastava, Jitendra, Nigel J. H. Smith, and Douglas Forno. *Biodiversity and Agriculture: Implications for Conservation and Development.* Washington, DC: World Bank, 1996.

Stanchinsky, V. *Variability of Organisms and its Importance for Evolution.* Smolensk, 1927 (in Russian).

Stern, Paul C. "A Second Environmental Science: Human-Environment Interactions," *Science* 260 (25 June 1993): 1897–99.

Stigler, George J. *The Citizen and the State: Essays on Regulation.* Chicago: University of Chicago Press, 1975.

Stoett, Peter. *The International Politics of Whaling.* Vancouver: University of British Columbia Press, 1997.

Stokes, Kenneth M. *Man and the Biosphere: Toward a Coevolutionary Political Economy.* Armonk, NY: M. E. Sharpe, 1992.

Streeten, Paul (with Shahid Javed Burki). *First Things First: Meeting Basic Human Needs in the Developing Countries.* New York: Oxford University Press, 1981.

Susskind, Lawrence E. *Environmental Diplomacy: Negotiating More Effective Global Agreements.* New York: Oxford University Press, 1994.

Suzuki, Y., K. Ueta, and S. Mori, eds. *Global Environmental Security: From Protection to Prevention.* Berlin: Springer, 1996.

Taylor, Lance, and Ute Pieper. *Reconciling Economic Reform and Sustainable Human Development: Social Consequences of Neo-liberalism.* New York: Office of Development Studies, United Nations Development Programme, 1996.

Tessitore, John, and Susan Woolfson, eds. *A Global Agenda: Issues Before the 48th General Assembly of the United Nations.* Lanham, MD: University Press of America, 1993.

Therivel, Riki, et al. *Strategic Environmental Assessment.* London: Earthscan, 1992.

Theys, Jacques. "Environmental Accounting in Development Policy: The French Experience." In *Environmental Accounting for Sustainable Development,* edited by Yusuf J. Ahmad, Salah El Serafy and Ernst Lutz. Washington, DC: World Bank, 1989.

Thrupp, Lori Ann. *Cultivating Diversity, Agrobiodiversity and Food Security.* Washington, DC: World Resources Institute, 1998.

Tietenberg, Tom. *Environmental and Natural Resource Economics.* 4th ed. New York: HarperCollins, 1996.

Timmer, C. Peter. *Agriculture and the State: Growth, Employment, and Poverty in Developing Countries.* Ithaca, NY: Cornell University Press, 1991.

Timmerman, P. "Myths and Paradigms of Interactions between Development and Environment." In *Sustainable Development of the Biosphere,* edited by William C. Clark and R. Munn. Cambridge: Cambridge University Press, 1986.

Tolba, Mostafa K., and Osama A. El-Kholy, eds. *The World Environment 1972–1992: Two Decades of Challenge.* London: Chapman and Hall, 1992.

Tollison, Robert D. "Rent Seeking: A Survey," *Kyklos,* 35 (1982): 575–602.

Trainer, Ted. *The Conserver Society: Alternatives for Sustainability.* London and Atlantic Highlands, NJ: Zed Books, 1995.

Trzyna, Thaddeus, ed. *A Sustainable World: Defining and Measuring Sustainable Development.*

Sacramento: International Center for the Environment and Public Policy, California Institute of Public Affairs, 1995.

Turco, R. P., O. B. Toon, T. P. Ackerman, J. B. Pollack, and C. Sagan. "Nuclear Winter: Global Consequences of Multiple Nuclear Weapons Explosions." *Science* 222 (December 23, 1983): 1283.

United Nations Conference on Environment and Development (UNCED). *Agenda 21.* Rio de Janeiro, Brazil, June 3–June 14, 1992.

United Nations Conference on Trade and Development (UNCTAD). *Environmental Management in Transnational Corporations.* New York: United Nations Publications, 1993.

———. *Trade and Environment Report.* New York: United Nations Publications, 1994.

United Nations Development Programme. *Human Development Report 1997.* New York: Oxford University Press, 1997.

———. *Human Development Report 1998.* New York: Oxford University Press, 1998.

United Nations Economic Commission for Europe. *Relationship between Economic Development and Conservation of Flora, Fauna and their Habitats.* New York: United Nations, 10 December, 1990.

United Nations Economic and Social Council. *Draft Integrated Report on Environmental Financing.* New York: UNESCO, 1995.

———. *Guidelines on Environmental Management in CITs.* New York: United Nations, 1994.

———. *Statistics on Economic Aspects of Environmental Protection.* New York: United Nations, 15 March 1990.

———. *Guidelines on Environmental Management.* New York: United Nations, 1994.

———. *Industry and Environment.* New York: United Nations, 1995.

United Nations Educational, Social and Scientific Organization (UNESCO). *Use and Conservation of the Biosphere.* Paris: UNESCO, 1970.

United Nations Environment Programme (UNEP). *Economic Instruments for Environmental Protection.* New York: UNEP, 1993.

———. *Global Biodiversity Assessment.* New York: UNEP, 1995.

———. *The Proposed Program.* Nairobi: UNEP, 1975.

———. *Survey of Information Systems Related to Environmentally Sound Technologies.* Nairobi: UNEP, 1997.

United Nations Statistical Division. *Glossary of Environment Statistics.* New York: United Nations, 1997.

United Nations Statistical Office. *Integrated Environmental and Economic Accounting. Handbook of National Accounting, Studies in Methods.* New York: United Nations, 1993.

United States National Science and Technology Council (USNSTC). *Our Changing Planet: The 95 Global Change Research Program.* Washington, DC: USNSTC, 1994.

Uno, Kimio. *Environmental Options: Accounting for Sustainability.* Dordrecht, Netherlands: Kluwer, 1995.

———, and Peter Bartelmus, eds. *Environmental Accounting in Theory and Practice.* Dordrecht, Netherlands: Kluwer, 1998.

Urbanska, Krystyna, Nigel Webb, and Peter Edwards. *Restoration Ecology and Sustainable Development.* Cambridge: Cambridge University Press, 1997.

Van Dieren, Wouter, ed. *Taking Nature into Account.* New York: Springer-Verlag, 1996.

van Lookeren Campagne, N. "Interviews with J. M. H. van Engelshoven, N. G. Ketting, H. H. van den Kroonenberg, E. van Lennep, O. H. A. van Royen, and P. Winsemius." In *Policy Making in an Era of Global Environmental Change,* edited by R. E. Munn, J. W. M. la Rivière, and N. van Lookeren Campagne. Boston: Kluwer, 1996.

Vavilov, Nikolai Ivanovich. *Studies on the Origin of Cultivated Plants.* Leningrad: Institute of Applied Botany and Plant Breeding, 1926 (in Russian).

Vernadsky, Vladimir. *Biosfera.* Moscow, 1926 (in Russian).

Vernon, Raymond, and Louis Wells. *The Economic Environment of International Business.* Engle-wood Cliffs, NJ: Prentice-Hall, 1986.

Vig, Norman, and Michael Craft, eds. *Environmental Policy in the 1990s.* 3d ed. Washington, DC: Congressional Quarterly Press, 1997.

Viscusi, Kip W., ed. "Risk-Risk Analysis Symposium." *Journal of Risk and Uncertainty* 8 (1994): 5–122.

Visvanathan, Shiv. "Mrs. Brundtland's Disenchanted Cosmos." *Alternatives* 26(3) (1990): 377–84.

Vivekanandan, B. *International Concerns of European Social Democrats.* New York: St. Martin's Press, 1997.

Voorhees, John, and Robert Woellner. *International Environmental Risk Management: ISO 14000 and the Systems Approach.* Boca Raton, FL: Lewis Publishers, 1998.

Wackernagel, Mathias, and William Rees. *Our Ecological Footprint.* Gabriola Island, British Co-lumbia: New Society Publishers, 1996.

Walker, Sandra L. *Environmental Protection versus Trade Liberalization: Finding the Balance.* Tra-vaux et Recherches series no. 27. Brussels: Publications des Facultes Universitaires Saint-Louis, 1993.

Wapner, Paul. *Environmental Activism and World Civic Politics.* Albany: SUNY Press, 1996.

Weber, Max. *The Protestant Ethic and the Spirit of Capitalism.* New York: Charles Scribner's Sons, 1958.

Weenen, Hans van. *Design for Sustainable Development: Concepts and Ideas.* Dublin: European Foundation for the Improvement of Living and Working Conditions, 1997.

Weiner, Douglas. *Models of Nature.* Bloomington: Indiana University Press, 1988.

Welford, Richard. *Hijacking Environmentalism: Corporate Responses to Sustainable Development.* London: Earthscan, 1997.

Weller, Gunter, and Patricia A. Anderson, eds. *Assessing the Consequences of Climate Change for Alaska and the Bering Sea Region.* Fairbanks: University of Alaska Center for Global Change and Arctic System Research, 1999.

Wells, Michael, and Katrina Brandon. *People and Parks: Linking Protected Area Management with Local Communities.* Washington, DC: World Bank, World Wildlife Fund, and U.S. Agency for International Development, 1992.

Werksman, Jacob, ed. *Greening International Institutions.* London: Earthscan, 1996.

Westing, Arthur, ed. *Environmental Warfare: A Technical, Legal and Policy Appraisal.* London: Taylor and Francis, 1984.

White, Robert M. "World Climate Conference: Climate at the Millennium." *Environment* 21 (3) (April 1979): 31–33.

Williamson, John, ed. *IMF Conditionality.* Washington, DC: Institute for International Economics, 1986.

Winsemius, Pieter, and Walter Hahn. "Environmental Option Assessment." *Columbia Journal of World Business* 27 (3–4) (1992): 248–66.

Wirth, John D. "The Trail Smelter Dispute: Canadians and Americans Confront Transboundary Pollution, 1927–41." *Environmental History* 1 (2) (April 1996): 34–51.

Wolf, Amanda Marie. *Quotas in International Environmental Agreements.* London: Earthscan, 1997.

World Commission on Environment and Development. *Our Common Future.* Oxford: Oxford Uni-versity Press, 1987.

World Food Program. *World Food Aid: Experiences of Recipients and Donors.* London: James Cur-rey and Heinemann, 1993.

Yergin, Daniel. *The Prize: The Epic Quest for Oil, Money and Power.* New York: Simon and Schus-ter, 1991.

Young, Michael. *Sustainable Investment and Resource Use.* Park Ridge, NJ: Parthenon Publishing Group, 1992.

Young, Oran. *International Governance: Protecting the Environment in a Stateless Society*. Ithaca: Cornell University Press, 1994.

———, ed. *Global Governance: Drawing Insights from the Environmental Experience*. Cambridge, MA: MIT Press, 1997.

Zaelke, Durwood, Paul Orbuch, and Robert F. Housman, eds. *Trade and the Environment: Law, Economics, and Policy*. Washington, DC: Island Press, 1993.

Periodicals

Environmental Alert
Environment and Development Economics
Environment Matters
European Environmental Law Review
International Society for Ecological Economics. Newsletter
Journal of Environmental Law
Vital Signs
Worldwatch

Index

Note: Page numbers on which entry definitions appear are indicated in bold type.

Best practical environmental option (BPEO), 207

Big 10, **250**

Bilateral aid, 132, 139. *See also* Foreign assistance

Bilge dumping, **334**

Biocenosis, **34**

Biocentrism, **120**, 121

Biocolonialism, **124**

Biodiversity, 2, 3, 7, 22, 31, 36, **39**, 47, 50, 64–65, 91, 117, 139, 147, 152, 157, 169–71, 181, 185, 192, 230, 233, 235, 252, 255–56, 269, 273–83, 312, 332; indices, 50

Biodynamic agriculture, **263**, 266

Biogeocenosis, **34**, 283

Biogeochemical cycle, **40**

Biological and toxic weapons, **157**, 158

Biological and Toxic Weapons Convention. *See* Convention on the Prohibition of the Development, Production and Stockpiling of Bacteriological (Biological) and Toxic Weapons and their Destruction

Biopiracy, 117, 276

Bioprospecting, **276**

Bioregion, **34**, 67, 80

Bioregionalism, 67, **80**, 81

Biosafety, **273**, 277, 284

Biosphere, **35**, 39–40, 42–43, 66, 85, 120, 193, 238–39, 249, 277, 340, 343, 348; reserves, 238, 249, **277**

Biotechnology, **110**, 117, 146, 273, 277

Blocking (veto) state, **228**, 309, 312–13

Blueprint for Survival, **94**

Bogdanov, Alexander, 66, 183

Bonn Convention. *See* Convention on the Conservation of Migratory Species of Wild Animals

Bookchin, Murray, 82, 86

Boom-and-bust cycles, 22–23, 259

Boserup, Ester, 89, 90, 119

Boserupian thesis, **91**, 119

Bottom-up development, **10**, 16–17

Boulding, Kenneth E., 88

Bounded rationality, **54**, 61

Braidotti, Rosi, 18, 83, 251

Brazil, 7, 9, 24, 94, 146, 152, 197, 219, 230, 256, 300, 340, 347

Bretton Woods system, **133–34**. *See also* International Monetary Fund; World Bank Group

Brewer, Garry, 55

British Environmental Protection Act of 1990, 33

Brundtland Commission. *See* World Commission on Environment and Development

Brunei, 8, 153

Brunner, Ronald, 55, 186, 308

Bubble policy, 115, 199, **207**

Buddhism, 67, 91

Buddhist model of development, **91**; economics, 97

Buffer zone, 193, **277–78**

Bulgaria, 5, 27

Bundle of entitlements, **113**

Bureau of Land Management, U.S., 49, 210

Burkina Faso, 28, 285, 302

Burma, 14, 91

Business Council of Sustainable Development (BCSD), 225

Business organizations, 215–16, 225–27

Cairo Conference on Population and Development. *See* United Nations Conference on Population and Development

Calvert-Henderson Quality of Life Indicators, 27

Canada, 7, 27, 29, 135, 150, 153, 155, 160, 164, 177, 200–2, 220–21, 231, 243, 256, 271–72, 297, 299, 302, 308, 318, 322, 327–31, 338, 341, 344, 346

Canada/USA International Joint Commission (IJC), **308**

Capacity, 21, **177**; building, **10**, 73, 140

Capital, 1–6, 14–17, 19, 20, 21, 23–24, 27, 32, 48, 53, 74–76, 79, 80, 82, 84, 92, 96, 98, 120, 126–28, 130–34, 136, 137, 138, 140, 144, 151, 162–64, 168, 174, 184, 185, 233, 241, 261, 262, 271, 295, 328, 331; accumulation, 2, 83; formation, 2

Capitalism, 2–4, 83–84, 217

Capital market liberalization, **129**

Capital-output ratio, 19

Capital-Resource Substitution, **2**

Capture theory, **210**

Carbon cycle, 2

Carbon dioxide, 6, 31, 41, 152, 196, 201, 209, 278, 288–90, 297, 351; equivalent, **288**

Carbon tax, 96, **195–96**, 201

Cardoso, Fernando Henrique, 81

Caribbean Basin Initiative, 143

Caribbean Development Bank, 137

Caribbean region, 135, 137, 143, 232, 302, 322, 326, 347

Caring for the Earth, **95**

Car-pooling incentives, **350**

Carrying capacity, **32–33**, 65, 84, 87, 91, 108, 264, 285; human, 32, 34, 36, **105**

Catadromous species, **326**

Catalytic converters, 192, **350–51**

Catastrophe theory, **54**, 94, 121

Central Africa, 7, 322

Daly, Herman E., 13, 14, 30, 66, 73, 74, 125, 161, 183

Danish bottle case, **141**

Data activity index (DAI), **138**

Debt, 20, 129, **130–31**, 145, 299; conversion, **131**; debt-for-nature swaps (DFNS), **131–32**; relief, 131–32, 151

Decision on Trade and Environment, 152

Decision on Trade in Services and Environment, 152

Declaration of the Development of Forests, **298**

Declaration of Environmental Policies and Procedures Relating to Economic Development, **134**

Declaration on the Human Environment, **95**, 231, 248

Declining relative prices, **260**

Decreasing returns to scale, 20

Deep ecology, **80**, 87, 89

Deforestation, 36, 108, 134, 147, 158, 240, 250, 262, 272, **296–97**, 301–4, 312, 316–17

Delinking, **126**

Demand-side management, 218, 221, 305

Democracy, 4, 12,14, 49, 57, 76, 84, 122, 206, 253

Demographic transition, 104, 106

Denmark, 141, 153–54, 231, 338, 346

Dependency theories, 11, 78, **81**

Deposit refund systems, **197**

Desertification, 8, 158, 240, 259, 261, **302**, 303–4, 312, 316

Design for Environment, **222**

de Souza, Anthony, 11

Developed countries, **5**

Developmentalism, 19, 77

Development Alternatives with Women for a New Era (DAWN), **251**

Development decades, **95**

Development economics, 2

Development models, **12**, 13, 18, 90–91

Development planning, 18, 29

Development strategies, 26, 59, 90–91, 177, 178

Development theory, 9, 20

Diesel fuel, 25, 110, 350–51

Diminishing returns, 20

Direct regulations, **189**

Dirty dozen, **205**, **260**

Discount rate, 108, **168**

Discounted utilitarian interpretation, **76**

Disease, 34, 40, 214, 234, 265, 277, 287

Disincentive, **194**

DIVERSITAS program, 235

Dose-response technique, **182**

Draft Covenant on Environmental Conser-

vation and Sustainable Use of Natural Resources (1991), 50

Draft International Covenant on Environment and Development, 243

Driftnets, **326**

Dumping: of products, 143; of wastes, 126, 231, 254, 313, 322, 324, 334–36, 338

Duties, 126, 128, **142**, 143–47, 155, 247

Dynamic equilibrium, 41–42, 66, 273

Earmarking of revenues, **197**

Earth Charter, **96**

Earth Council, 96, **251**

Earth First! (E1!), 88, **252**

Earth Summit. *See* United Nations Conference on Environment and Development

Earthwatch Institute (also Earthwatch), 175, **252**

East Africa, 137, 322

East African Development Bank, 137, 322

East Asia, 9, 26, 143, 153, 265

East vs. West, 6

Eastern Europe, 25, 101, 134, 153, 160, 177–78, 190–91, 197, 232, 255, 316, 347

Eco-anarchism, 80, **81–82**, 86, 88

Eco-catastrophists, 69, 120

Ecocentrism, **122**

Ecocide, **158**

Eco-Cooperation Agreements, **147**

Ecodevelopment, 74, **82**

Eco-domestic product (EDP), **161**

Eco-dumping, **126**

Eco-efficiency, 214, **216**

Eco-fascism, **82**, 87–83

Ecofeminism, **82**, 88

Ecological address, **80**

Ecological capital, 2, 126–27, **161**

Ecological disruption, 95

Ecological economics, **65**

Ecological energetics paradigm, 66, 92, 306

Ecological entity, **40**

Ecological footprint, **33**, 149

Ecological imperialism, **40**, 126

Ecological integrity, **33**

Ecological rucksacks, **45**, 46

Ecological sabotage, **81**

Ecological shadow, **126**

Ecological triad, **58**

Ecology, **67**

Eco-Marxism, **83**

Economic development, 2, 4, 6, 11–12, 14–16, 20–21, 26, 28, 30, 58, 65–66, 73, 82, 85, 89, 90–91, 94–98, 100, 104, 122, 127, 133–34, 137, 140, 143–44, 153, 177, 194, 211, 239, 251, 255–56, 323, 352. *See also* Economic growth

Economic distortions, **210**

International Monetary Fund (IMF), 8, 130–34, **136–37**, 142, 151, 233, 329

International Network of Green Planners (INGP), **177**

International Oceanographic Commission (IOC), 235, 236

International Organization for Standardization (ISO), 129, **148**, 149, 223, 226

International Organization of Consumer Unions (IOCU), **227**

International organizations, 7, 77, 131, 137, 140, 151, 238, 246, 250, 251, 257, 300, 302, 314, 317, 329, 339

International Referral System for Sources of Environmental Information (INFOTERRA), **176**, 233

International Register of Potentially Toxic Chemicals (IRPTC), 176, 233, **313**

International Seabed Area, **339**

International Seabed Authority, **339**

International Telecommunications Satellite Organization (Intelsat), **346–47**

International Telecommunications Union (ITU), **347**

International trade, 101, 125–26, 135, 142–47, 150, 180, 241, 263, 273, 275, 281, 283, 286, 301–2, 309

International tradable emission permits, **201**

International Tribunal for the Law of the Sea, **321**

International Tropical Timber Agreement (ITTA), **301**

International Tropical Timber Organization (ITTO), **301**

International Undertaking on Plant Genetic Resources (IUPGR), **287**

International Union for the Protection of Nature (IUPN). *See* World Conservation Union

International watercourses. *See* Transboundary water resources

International Whaling Commission. *See* International Commission for the Regulation of Whaling

Intrinsic value, **181**

Inuits, 231, 291

Investment, 1, 5, 10, 12–16, 19–26, 29–30, 48, 52, 105, 117–18, 127–28, 137, 292, 296, 329–30, 352; allocation, 16; capital, 5; gap, **130**, 131

Iron triangle, **191**

ISO-14000, **149**, 223, 226

Italy, 7, 27, 154–55, 160, 204, 267, 346

Japan: agriculture in, 265–66; business sector environmental initiatives, 219, 226, 253; economic development in, 7, 26–27, 224; ecotourism in, 279; environmental conceptions, 265–66; environmental policies, 177, 223; greenhouse gas emissions, 292; industrial structure, 128–29; international economic initiatives, 129, 253; international environmental initiatives, 301, 322; international fisheries policies, 331; nongovernmental organizations in, 266; participation in international organizations, 132, 134, 152, 154; research on environmentally friendly products, 215; space initiatives, 346; subsidies in, 261; trade policies, 144, 223, 301, 322, 331, 346

Joint Implementation (JI), **246–47**, 292–93

Kahn, Herman, 32, 92, 121

Keep options open principle, **208**

Key area approach, **171**, **279**

Keystone species, 277

Klamath Mountains, 34

Kleptocracy, **211**

Kondratieff, Nikolai, 21

Kondratieff cycles, **21**

Kropotkin, Piotr Alekseyevich, 81, **86**, 87–88

Kyoto Protocol, 196, 244, 254, 292, **293**, 294

Labor, 1, 10, 15, 19, 20, 23, 26, 28, 47, 56, 81, 83, 106, 109, 109, 125, 127, 128, 143, 144, 145, 154, 155, 174, 179, 182, 183, 203, 216, 233, 260, 327

Land: accounts, 162; biodiversity of, 35; clearing, 289; conservation, 87, 131, 171, 180, 324; degradation, 91, 99, 149, 158, 188, 290, 297, 302–3; as ecosystem component, 34, 36, 170, 323, 325; ethic, **82**; international agreements on conservation of, 288–90, 297–99, 301–2, 304, 311, 318–19, 322–25, 335, 336, 346; management, 49, 170, 193, 210, 240; marginal, 7; ownership, 18, 183, 261, 298–99, 324; pollution, 191, 214; pressures on, 259; prices, 37; public, 49, 171, 193, 298; purchases, 258; quality, 188; reclamation, 70, 193, **324**; reform, 262, 302; regulation, 188, 216, 280–82, 304, 318; as resource, 1–2, 4, 19, 22–23, 33, 87, 149, 168, 181, 210, 261, 282, 311, 322; rotation, 261–62; scarcity, 119, 267; stewardship, 87; use, 58, 132, 171, 259–66, 280; use planning, 49, 169–70, **171**, 238, 268, 301; use-values, 171

Land use planning, **171**

Lash, Scott, 64–65

Latin America: biodiversity in, 147; community participation in, 17; debt-for-nature swaps in, 131–32; development, 9–11; devel-

Resource and pollutant flow accounts (RPFA), **164**

Resource curse, 9; hypothesis, **17**

Resource cycles, 23. *See also* Boom-and-bust cycles

Resource economics, 68, **69**, 257

Resource intensity, 16, **47**

Resource rent, **118**

Resource scarcity, 13, 21, 92, 117, **119**, 180, 250, 308, 310, 330

Resource surplus, 5

Responsible care, **221**

Restoration ecology, 41, **70**, 165

Resurrection of nature, 56

Returns to scale, 20, 23

Revealed comparative advantage (RCA), 125, **129**

Right to development, **249**

Rio Conference. *See* United Nations Conference on Environment and Development

Riparian habitat, **44**; rights, 308, **310**

Risk, environmental, 2, 144–45, 224, 226, 247

Rostow's stages of economic growth, **17**

Russia, 5, 29; conservation in, 282; cultural resources, 3; deforestation in, 254, 297; development in, 17, 200; disaster preparedness, 316; energy policies, 24; environmental conceptions and models, 34–35, 58, 65–66, 80, 86–87, 92–93, 183, 276, 279, 282–83; environmental policies, 282; international economic initiatives, 7; international environmental initiatives, 231, 253, 293–94; international fishing, 329, 331, 338; international scientific cooperation, 99, 338; international security, 160–61; radioactivity in, 315; space initiatives, 341, 346; urban planning, 14

Russian Far East, 201, 254

Sagan, Carl, 160

Sale, Kirkpatrick, 34–35, 59

Sanitation, 8, 25, 220, 234, 263, 318

Satellite accounts, 162, **164**, 165

Satellites, space, 327, 340–42, 345–47; artificial, **341**

Saudi Arabia, 6, 8, 307–8

Savings, 12, 18–19, 28, 73, 75, 125; gap, 2

Schumacher, Ernst F., 91, 97

Scientific Committee on Problems of the Environment (SCOPE), **241**

Sea-level rise, **290–91**, 292

Seas, 32, 154, 281, 290, 310, 336, 338; enclosed, **326**; freedom of the, 319, 326; high, 113, 123, 318, **319**, 320, 326, 330–33, 335; presential, **320**; semi-enclosed, **318**, 322, 324;

territorial, 318–19, **320**, 326, 337. *See also* Transboundary water resources

Second law of thermodynamics, 40, 66

Second world, **9**

Sector loan, **132**

Sectoral management, **325**

Security: environmental, 96, 160–61, 237; food, 39, 261, **264–65**, 266–69; national, 86, 101, 156, 160–61, 245–46, 254, 293, 318; navigational, 320; personal, 30, 37, 106, 265; social, 84

Selective logging, 296, **298**

Self-reliant economic growth, 14, 126

Seminatural environment, **36**, 37

Sensible sustainability, **73**

Service sector, 10, 15

Sewage charge, **202**

Shadow price, 23, **168**

Shadow projects, **168**

Shapley value, **291**

Shared natural resources, **117**, 245

Shelter, 25–26, 37, 107, 133, 231–31, 297

Shifting cultivation. *See* Swidden agriculture

Side payments, **206**

Sierra Club, 250, **256**

Simon, Julian L., 13, 32, 92, 121–22

Singapore, 9, 153

Sink capacity, 31, 126, 290

Size distribution of income, **107**. *See also* Income distribution

Small is Beautiful (Schumacher), **98**

Social accounting, 95

Social capital, 1, 3, **4**

Social cost, 6, 39, 74, **168**

Social ecology, 81–82, **88**

Social institutions, **103**

Social justice, 73, 77, 91 96, 103, 241

Social safety nets, **132**

Socialist bloc, 6, 9, 27, 230

Society for Conservation Biology, 65

Soft energy path, 66, **308**

Soft law, **249–50**, 273, 287

Soft loans, 130, **132**

Soil, 13, 33–35, 38–40, 43 46, 58, 65, 87, 105, 110, 115, 162, 259, **261**, 279, 283, 287, 297, 309, 312; erosion, 49, 67, 72, 166, 182, 192, 195, 207, 235, **260**, 263, 265–66, 269

South Africa, 24, 143

South Centre, 33–34

South Commission, **239–40**

South Korea, 9, 26, 144

South Pacific Commission (SPC), **156**

South Pacific Forum, **156**

South Pacific Regional Environment Program (SPREP), **156**

Southeast Asia, 9, 265

Soviet Union, 25, 59, 94, 157–58, 160–61, 177, 230, 239, 254, 266, 283, 315–16, 339, 341–42, 344; concept of development in, 19. *See also* Russia

Space: commercialization, **342**; debris, **342**; international station, **346**; law, **344**; pollution, **342**; race, **342**

Spaceship economy, **88–89**

Spasms of extinction, 273, **276**

Special area management, **325**

Special drawing rights (SDRS), **132**

Special Programme for Food Security (SPFS), **267–68**

Species, 23, 39–40, **44**, 66–67, 275, 293; anadromous, **325**, 338; catadromous, **326**; conservation of, 64, 157, 186, 192, 274, 277, 280, 282, 287; distribution of, 34–36; diversity, 36, 44, 47, 50, 180, 272; endangered, 170–71, 186, 192, 217, 252, 257, 273, **274**, 275, 281, 283–86, 297–99; evolution of, 42; existence value of, 38; exotic, 72, 78, 326–27; extinction of, 42–43, 46, 272–76, 281, 291, 297; fish, 200, 325–34; flagship, 260; harmful, 72, 283; highly-migratory, 209, **328**, 336; interdependence of, 72, 79, 248, 283; keystone, **276**; marine, 291, 326; migratory, 209, 285, 328, 336; non-migratory, 200; protected, 43, 282, 285; rare, 3, 281; resilience of, 36, 105; restoration of, 70; surviving, 276; tree, 46, 298–99; valued, 49

Spirituality, 2, 26, 45, 79, 89, 97, 120

Stability: chemical, 348; economic, 111, 155–56; of ecosystems, 33, **36**, 39; of energy markets, 307; environmental, 291; international, 154, 229; monetary, 136; sociopolitical, 86, 156, 302

Stakeholder, 16–17, **18**, 49, 76, 170, 177, 179, 204–6, 224, 277; negotiation, 16, 18

Standard of living, 25, 29, 155, 304; measures of, 26, 29, 53

Standard-setting, **192–93**

State enterprises, 5, 131, 212

Statement of Forest Principles, 231–32, **299**

State-of-the-environment reports (SERS), **176**

Stationary state, 73

Steady-state economy, 73

Stewardship, 87, **89**, 97, 184, 300, 327; forest, 300; product, 110, 149, **215**, 221

Stewart, Francis, 25

Stockholm Conference (1972). *See* United Nations Conference on the Human Environment

Straddling stocks, 320, **330–31**, 332

Strategic lawsuits against public participation (SLAPPS), 79

Stress-Response Environmental Statistical System, **164**, 311–12

Strong sustainability, 2, 4, **74**

Structural adjustment, 131, 136–37, 221; loans (SALS), **132**, 137

Subnatural environment, **36–37**

Subsidiarity, 150, **209**

Subsidies, **23**, 24–25, 68, 75, 96, 112–13, 125, 142, 188, 211–12, 266; agricultural, 261; energy, 25, 305–6; export, 142, 147, 149; fishing, 329–30; food, forest exploitation, 134; import, 259; technology, 197; transportation, 350; water, 24

Subsistence economy, 60, 79–80, 114, 291, 333

Substitutability: of capital, 2, 73; of materials, 214

Superfund, 98, **315**

Supporting state, **229**

Surrogate market, **184**

Sustainability, **74**; agricultural, **266**; community, **73**; indicators, **53–54**; sensible, **73**; societal, **103**; strong, **74**; weak, **75**

Sustainable agriculture, 257, **266**, 267–69

Sustainable development: agendas for, 94–98, 178, 214, 216, 232, 298–300; as an organizational objective, 152–55, 156, 177–78, 226, 232–35; 243–44, 253–55, 306; conceptions of, 2, 34, **74**, 75–78, 85, 140, 234, 241, 247, 251–52, 276–77, 282, 303; measurement of 34, 150–51, 163–65, 176–77; models of, 90–94; of agriculture, 257, **266**, 267–69; of fishing, 268–69, of forests, 299–300; of marine resources, 322–23; requisites of, 17, 22, 24, 32, 59, 74, 77, 98, 110, 129, 133, 205, 252–53, 257

Sustainable forestry, 298, **299**, 300–301

Sustainable national income (SNI). *See* Accounting

Sustainable natural resource management, **75**

Sustainable society, 84, 95, 98–102, **103**, 219, 257

Swidden agriculture, **261–62**

Swing state, **230**

Synecology, **67**

Synergism, **44**

System of environmental and economic accounts, 52, 166

System of national accounts (SNA), **164**

Taiwan, 9, 14, 144, 153, 329, 331

Take-off stage, 12, 18

Taoism, 67

Taxes, 18, 21, 23, 107, 125, 128, 142, 144, 194, 199, **202**, 213, 217, 246; and benefit principle of taxation, **194**; carbon, 96, **195**; effluent, **198**; emission, **198**; energy, 190; environmental, 164, 186, 196, 203; Pigovian, **194**,

United Nations Development decades. *See* Development decades

United Nations Development Programme (UNDP), 27, 134, **139–40**

United Nations Educational, Social and Cultural Organization (UNESCO), 8, 117, 193, **232**, 235, 237–40, 277, 285, 295, 338, 344

United Nations Environment Programme (UNEP), 7–8, 82, 95, 134, 139, 175–77, 231, **232**, 240, 243, 252, 273, 280, 282, 294, 314, 321

United Nations Food and Agriculture Organization (FAO), 7, 117, 134, 233, 249, 261, 263–65, 267, **268**, 269, 273, 287, 200, 303, 314, 329–30, 333

United Nations General Assembly, 97–98, 151, 157, 231–34, 241, 242–43, 268, 291, 309, 320, 326, 343,–45

United Nations Industrial Development Organization (UNIDO), 6, 7, **140**, 233, 273

United Nations International Decade on Natural Disaster Reduction, **318**

United Nations Non-Governmental Liaison Office (UN-NGLS), **233**

United Nations Population Fund (UNFPA), **227**

United Nations Programme on Space Applications, **347**

United Nations Revolving Fund for Natural Resources Exploration, **233**

United Nations World Charter for Nature, **98**

United Nations University (UNU), **239**

United States, 6–7, 19, 59, 146, 223; Agency for International Development (USAID), **140–41**; animal rights movement in, 78; Army Corps of Engineers, 60; Bureau of Land Management, 210; business groups' environmental initiatives, 218–20; deforestation in, 297; desertification in, 302; domestic environmental actions and policies, 141, 173, 176, 187, 202, 205, 215, 274, 280, 288, 305, 315, 324, 328, 350–51; economic growth attitudes in, 122; environmental conceptions in, 180, 273; environmental movements in, 80, 108, 122, 148, 218, 253–57, 297, 311; federal government, 17, 91–92, 135, 140, 250, 256, 340–41; Forest Service, 49, 210; interest groups, 79, 212; international environmental actions, 91–92, 96, 129, 134, 148, 150, 229–30, 272, 277, 286, 293, 299, 308–9, 322, 328, 338–39, 349; international fisheries initiatives, 327–28, 331, 338–39; international population initiatives, 228; international security initiatives, 158–61, 341; land-use regulations, 193; migration in, 159–60; nuclear industry, 317; plant breeder

rights in, 117; polluting industries in, 128–29; population issues, 228; relations with Canada, 155, 272, 308, 327–28; relations with Mexico, 143, 155; relations with the Soviet Union, 341–42, 346; role in establishing international bodies, 135–36, 153, 155, 157, 160–61, 229–32, 271, 204,1343,4346; space initiatives, 341–42; Superfund program, **98**; toxics exports, 260

Unregulated market, 22

Urban agriculture, **266**

Urbanization, 105–6, 126, 149–50, 170, 231, 238, 266, 323–24, 341

Uruguay Round, 145, **151**. *See also* General Agreement on Tariffs and Trade

Use value, 171, 180–**81**, 184

User cost approach, **165**

Valuation, 22, 37–38, 41, 69, 112–13, 164, 167–68, 175, 179–86

Vavilov centers, **276**

Vicarious benefit, **182**

Victim pays principle (VPP), **195**

Vienna Convention on Protection of the Ozone Layer, **349**

Voluntary groups, 4, 77. *See also* Nongovernmental organizations

Wastes, 6, 20, 31–32, 35, 38, 45–46, 48, 53, 65, 67–69, 76, 88, 94, 110, 116, 126, 149, 161, 164, 192, 231, 254, 324, 334–35; agricultural, 260, 266, 334; export of, 124, 143, 312–13; hazardous, 44, 101, 143, 188, 312, **313**, 336, 342; industrial, 110, 188, 214–16, 219–20, 224; nuclear, 254, 306, **313**, 314, 321, 336, 342; solid, 342; water, 196, 311, 334

Water: charges, 208; conservation and management, 110, 134, 240, 268, 283, 286, 308–11; as ecosystem component, 32–35, 38, 40, 43, 45, 87, 105, 266; erosion effects of, 261, quality, 148, 162, 170, 183, 188, 192, 205, 108, 238; pollution, 45, 70, 149, 162, 166, 168, 171, 188, 198, 214, 259, 278, 287, 334–37; as resource, 46, 49, 65, 69, 180, 196, 221, 260–61, 265; rights, 308–11; sanitary, 8–9, 13, 25, 29, 46, 85, 175, 180, 234; scarcity, 91, 105, 119–20, 266, 290, 293, 308–11; subsidies, 24–25; as symbol, 67; treatment, 188, 311

Water Environment Federation (WEF), **311**

Waterfowl, 286–87

Water-resource management, **311**

Weak sustainability, 52, 73, **75**

Welfare economics, **70–71**

Well-being, 12–13, 26, 30, 33, 37, 53–54, 72, 77, 106–7, 112, 172, 231

West, 6–7, 12, 56, 63, 66–69, 77–78, 88, 91–92, 95, 97, 117–19, 125–26, 143, 145, 149, 160, 255, 267, 274, 345

Western Hemisphere Convention. *See* Convention on Nature Protection and Wildlife Preservation in the Western Hemisphere

Wilderness, **45**, 79, 193–94; protection, 95, 186, 252, 256, 281–83

Wilderness Society, 87, 250

Willingness to accept, 182, **185**

Willingness to pay, 165, 175, 180–82, **185**

Winnipeg Principles on Trade and Sustainable Development, **150**

Women: approaches for improving status and involvement of, 85–90, 133, 251, 269; empowerment of, 140, 227; environmental activism of, 251, 260, 329; rights of, 83, 140, 249; status of, 27–28, 30, 56, 82, 86, 105, 133, 143, 267

Women and development approach (WAD), **89**

Women in development approach (WID), 85, **89–90**, 250

World Bank, 8, 31, 48, 53, 60, 107, 124, 131–36, **137**, 138–40, 162–63, 165, 178, 256, 265, 267, 279, 281, 299, 301, 303

World Bank Group, 135–36, **137**, 138, 142, 150–51, 254

World Business Council for Sustainable Development, 226

World Charter for Nature. *See* United Nations World Charter for Nature

World Climate Conferences, **294**

World Climate Programme, **240**

World Climate Research Programme (WCRP), **295**

World Commission on Environment and Development (WCED), 74, 234, 241, **240**, 251

World Conservation Strategy (WCS), 47, 74, 95, **281**

World Conservation Union (IUCN), 47, 95, 243, **257**, 280–82

World Court. *See* International Court of Justice

World Economic Forum (WEF), **226**

World Food Conference, 267

World Food Council (WFC), **269**

World Food Programme (WFP), 264, **269**

World Food Summit, 265, **269**

World Health Organization (WHO), **234**, 234, 263–64

World Heritage Convention, **284**

World Meteorological Organization (WMO), 175, 233, 240, 293–94, **295**

World Resources Institute (WRI), **257**

World Summit Plan of Action for World Food Security, 40

World system theory, 78

World Trade Organization (WTO), 114, 128, 142, 148, 151, **152**

World War II, 134

World Wildlife Fund for Nature, 95, **257**, 282

Worldwatch Institute, **257**

Young, Oran, 58, 100, 102–3, 118, 153, 284–86, 291, 304

Zakazniks, 282

Zapovedniks, 66, **282**

Zero-cost improvements, **209**

Zero population growth, 95, **228**

Zoning, 116, 173, **193**

Library of Congress Cataloging-in-Publication Data
Guide to sustainable development and environmental policy /
edited by Natalia Mirovitskaya and William Ascher.
Includes bibliographical references and index.
ISBN 0–8223–2735–x (cloth : alk. paper)
ISBN 0–8223–2745–7 (pbk. : alk. paper)
1. Sustainable development.   2. Environmental policy.
I. Mirovitskaia, N. S. (Nataliia Sergeevna)   II. Ascher, William.
HC79.E5 G85 2001   338.9'27—dc21   2001033688